MATHEMATICS
THE
FIRST STEP

MATHEMATICS

THE FIRST STEP

The beginner's choice for engineering exams preparation

Book for JEE Mains/Advanced, NTSE, KVPY, Olympiad, IIT Foundation + CAT

TOPICS

SET THEORY

NUMBER THEORY

MODULUS AND WAVY CURVE

QUADRATIC EQUATIONS AND EXPRESSIONS

By

RAMESH CHANDRA B.TECH IIT KANPUR

Notion Press

Old No. 38, New No. 6
McNichols Road, Chetpet
Chennai - 600 031

First Published by Notion Press 2017
Copyright © Ramesh Chandra 2017
All Rights Reserved.

ISBN 978-1-946714-70-1

TABLE OF CONTENTS

ABOUT THE BOOK

This book intended exclusively for the usage of students, teachers and persons who are related to competitive exams. The book is based on our experience over the past 8 years and design on the basis of current competitive level of Engineering like IIT JEE mains/ Advanced, MHCET, BITSAT, NTSE, KVPY, Olympiad, IIT Foundation + CAT and other state engineering exams in India, where 1194938 i.e. around 12 Lakh of students (Year 2016) write a single engineering exam. As an educator, I understand the student's need of these topics and the difficulties faces by students in transition from standard 10th to 11th class. As students enter their 11th standard, they find a substantial change in the course content and level of difficulty. They find some totally new concepts of Mathematics, widely used in Physics and Chemistry. They may be completely unfamiliar with concepts of absolute value, Interval Methods, Set Notation, inequalities etc.

The book has been prepared for them to learn the concepts of algebra from basic to advanced level of thinking. The book is prepared to serve as a bridge for 10th to 11th standards, CAT aspirants etc. Software engineers can also be in benefit in writing the code due to concepts clarity.

The book contain the following Learning Methodology.

(i)   Basic concepts and easy learning.

(ii)   Necessary examples and Experiments for beginners level to expert.

(iii)   Psychology of student's brain and their thinking.

(iv)   Pictorial view of problems and solutions.

(v)   Challenging problems (Ultimate Finish – for Top All India Rankers between 1 - 500).

(vi)   Exercises and Assignments to test the understanding and growing knowledge.

(vii)   Sample Test Paper to have experience before actual exam.

(viii) Puzzles and interactive learning to keep interest.

(ix)   How to make notes to up-to-date and add your thinking inside the book.

(x)   Archive of IIT-JEE Mains/ Advanced.

In order to get maximum benefit from this material *"Mathsarc learning Advice"* gives overleaf, has to be followed.

Wishing you, a very successful year ahead!

Ramesh Chandra, B.Tech IIT Kanpur

MATHSARC EDUCATION LEARNING ADVICE

Study and understand the concept that author want to deliver. Take pen and paper while studying the book and solve the solved problems, examples, exercises etc before seeing the solution. It will help you in concept learning, strengthening and brain development. If you find something new concept, idea, thinking then write it at "*How to make note section*".

Note down the doubts and difficulties that you face in the book and ensure that these must be removed at anyhow. You can discuss the problem among friends, your teachers, even you can write to us at "*pncbyrc@gmail.com*".

The purpose of the assignments is to give you to practice in solving various levels and varieties of problems. Each problem has some important concept which it highlights. When you do a problem from an assignment, make sure that you have completed the earlier part of the book, have committed the formulae to your memory and have solved the solved problems (most of them on your own before seeing the solution). Do not open the book to refer to formulae/theoretical concepts while doing the assignment problems unless it is absolutely essential to do so.

Do full justice to the exercises and assignment problems, Even if you do not get the answer to a problem, keep typing on your own and only approach your friends or teachers after making lot of attempts.

Do not look at the answer and try to work backwards. This would defeat the purpose of doing the problem. Remember the purpose of doing an assignment problem is not simply to get the answer (it is only evidence that you solved it correctly) but to develop your ability to think. Try to introduce twists and turns in given problem to create similar problems.

Regards
Mathsarc Team

Mathsarc Education
A learning place to fulfill your dreams of success!

SET THEORY

"In mathematics, the art of proposing a question must be held of higher value than solving it."

Georg Cantor

ABOUT SET THEORY

Set theory is the branch of mathematical logic that studies sets, which informally are collections of objects. Although any type of object that can be collected into a set, set theory is applied most often to objects that are relevant to mathematics.

The language of set theory can be used in the definitions of nearly all mathematical objects. The purpose of set theory is not practical application in the same way that, for example, Fourier analysis has practical applications.

To most mathematicians (i.e. those who are not themselves set theorists), the value of set theory is not in any particular theorem but in the language it gives us. Nowadays even computer scientists describe their basic concept - Turing machines - in the language of set theory.

Many uses of diagrams outside mathematics use a combination of naive set theory and classical Boolean logic. The technical language of many disciplines use the terms union, intersection, complement of sets, and use a correspondence between (logical) combination of conditions on elements and combination of subset creation and operations.

Without knowing set theory, speaking to a mathematician will be like speaking to a Frenchman. You don't speak French, and he refuses to speak English (I am just joking); mathematicians are nice people. They will explain in English if you don't speak set theory.

"The essence of mathematics lies in its freedom"

Definition: A set is a well defined collection of objects (of any nature) & denoted by capital Roman letters.

Let us have some examples to get the definition correctly. Consider the following collections

(1) List of 10 best cricket player in the world.

(2) 5 clever students of your class.

(3) Vowels in English alphabets.

(4) 5 electronic gadgets that you uses in daily life.

From above four examples, you can see that the answer of example (1), (2) and (5) changes person to person and hence are not well defined collections, hence they are not sets but in example (3), every one write the answer as: a, e, i, o, u. so this is well defined collection, hence it's a set.

Definition: The objects in a set are called the *elements* or *members* of the set.

Set Notation: Let a set 'A' have members 1, 2, 4, 5 & 6 then notation for this set is A = {1, 2, 4, 5, 6}.

For making a set of members we write members enclosed inside curly braces { } and each member are separated by a comma (,).

Now consider the set A = {1, 2, 4, 5, 6}. In this case we can say about the element '1'.

'1' is an element of set A

Or

'1' is a member of set A

Or

'1' belongs to set A

Here: '1' belongs to set A is equivalent to writing $1 \in A$ and '9' does not belongs to set A is equivalent to writing $9 \notin A$. So you can see the difference in saying and writing the same thing.

Ex. Let set A = {0, 2, 3, 5, 8}. Insert the appropriate symbol $\in$ or $\notin$ in the blank spaces:

(i) 2 .... A   (ii) 7 .... A   (iii) 5 ....A

Solution: (i) $\in$ (ii) $\notin$ (iii) $\in$

SOME IMPORTANT SET NOTATIONS, REPRESENTATIONS AND THEIR MEANINGS

Notation	Meaning or Definition	Notation	Meaning or Definition
$\in$	Belongs to	$\cong$	Congruent to
$\notin$	Not an element of; does not belongs to	$\cup$	Union, OR
$\approx$	Approximately equal to	$\cap$	Intersection, AND
$\neq$	Not equal to	$\exists$	There exist
$\equiv$	Identical or equivalent to	$\nexists$	There is no
$\sim$	Similar to	$\rightarrow$	Implication
$\forall$	For all	$\leftrightarrow$	If and only if, equivalent, Iff
$\therefore$	Therefore	ϕ	Empty set, Null Set, Void Set
$\because$	Because	$:=$	Defined equal
$:$	Such That	$\subset$	Proper sub set
$\square$	End of proof	$\supset$	Proper Super subset
$\geq$	Greater than or equal to	$\subseteq$	Subset
$\leq$	Less than or equal to	$\not\subset$	Not a proper subset
[a, b]	Closed interval $x \in R, a \leq x \leq b$	(a, b)	Open interval $x \in R, a < x < b$
$\perp$	Perpendicular to	A^c or A'	Not A, Complement of set A
		Separator, Such That		

Some more notations, in real & Complex number system

Set of Natural numbers (N):	N = {1, 2, 3, 4, 5, 6,..........}
Set of all integers (Z Or I):	I = Z = {0, ± 1, ±2, ±3, ... }
Set of all rational numbers:	Q
Set of Irrational Numbers:	Q' or Q^c
Set of positive integers (I⁺):	N = I⁺ = {1, 2, 3, 4, 5, 6,..........}
Set of real numbers (R):	R
Set of Complex Numbers (C):	C
Set of Whole Numbers (W):	W = {0, 1, 2, 3, 4, 5, 6,..........}
Set of positive real numbers (R⁺):	R⁺

These are the standard notations that we need to remember in the case of mathematical language. There are two ways to represent a set

(A)   TABULAR / ROASTER FORM

(B)   SET BUILDER FORM

Consider a set A in two different representations

A = {0, 1, 2, 3, 4, 5, 6, 7, 8, 9}   : Tabular / Roaster Form

A = {x: x ∈ W & x < 10}   : Set Builder Form

You can observe that the elements of the sets are enclosed inside curly braces in the case of Tabular / Roaster form, while in set builder form, there is some coding language, when you decode it then you will get the same set as shown in tabular form.

A common format, to write a tabular form in Set builder form

Ex.   A = {x: x has property P};

Meaning: The set A is the collection of elements 'x' which follow the some property P.

Let us have an NCERT Problem to describe it in details.

Ex1.   Write the following set builder form of sets in roster/tabular form:

(i)   A = {x: x is an integer and –3 < x < 7}

(ii)   B = {x: x is a natural number less than 6}

(iii)   C = {x: x is a two-digit natural number such that the sum of its digits is 8}

(iv)   D = {x: x is prime numbers, which are divisor of 60}

(v)   E = the set of all letters in the word TRIGONOMETRY

(vi)   F = the set of all letters in the word BETTER

(vii)   G = {x: x is an odd natural number}

(viii) H = {x: x is an integer, x² ≤ 4}

(ix)   I = {x: x is an integer, $-\dfrac{1}{2} < x < \dfrac{9}{2}$}

(x)   J = {x: x is a letter in the word "LOYAL"}

(xi)   K = {x: x is a month of a year not having 31 days}

(xii)   L = {x: x is a consonant in the English alphabet which precedes k}.

Solutions:

(i)   Set A is a collection of integer 'x' in between – 3 and 7 so set A = {- 2, - 1, 0, 1, 2, 3, 4, 5, 6}.

(ii)   Set B is a collection of natural number 'x' which are less than 6 hence set B = {1, 3, 2, 5, 4}. Here you can observe that in roaster form, order of elements inside curly braces is immaterial.

(iii)   C = {17, 26, 35, 44, 53, 62, 71}.

(iv)   D = {2, 3, 5}.

(v)   E = {T, R, I, G, O, N, M, E, Y}. Here you can observe that repetition of elements is immaterial. You can get the answer (of question why?) in equality of two sets.

(vi)   F = {B, E, T, R}.

(vii)   G = {1, 3, 5, 7, 9, ...}. A set containing infinite number of elements called infinite set, Hence this set is an infinite set.

(viii) Set H is a collection of integers, following the property $x^2 \leq 4$, Hence H = {-2, -1, 0, 1, 2}.

(ix)   I = {0, 1, 2, 3, 4}.

(x)   J = {L, O, Y, A}.

(xi)   K = {Feb, Apr, Jun, Sep, Nov}.

(xii)   L = {b, c, d, f, g, h, j}.

Ex2.   Describe the following sets in roaster form (tabular form):

(i)   A = {x: x ∈ Z and |x| <5}.

(ii) B = {x: x^3 = - 1, x ∈ R }.

(iii) C = {x: x = 2n – 1 which is less than 20 and n ∈ N}.

(iv) D = {x: x = n/ (n+1), n∈ N}.

Solutions:

(i)   Set A is the collection of integers 'x' whose absolute value is less than 5.
Hence set A = {- 4, - 3, - 2, - 1, 0, 1, 2, 3, 4}.

(ii)   Set B is a collection of real roots of the equation x^3 = - 1.

$x^3 + 1 = 0 \rightarrow (x + 1)(x^2 - x + 1) = 0$, so equation have roots - 1, $\dfrac{1 \pm \sqrt{-3}}{2} = \dfrac{1 \pm i\sqrt{3}}{2}$, You can see, we have three roots in total but have only one real root. Hence set B = {- 1}.

(iii)   C = {1, 3, 5, 7, 9, 11, 13, 15, 17, 19}.

(iv)   D = {1/2, 2/3, 3/4, 4/5, 5/6,....................}

Ex3.   Write the following sets in roaster form:

(i) A = {$3^n - 2^n$: n∈N and 1 ≤ n ≤ 5 }   (ii) B = {a_n : n∈ N, a_{n+1} = 2 a_n and a_1 = 3}

(iii) C = {a_n : n∈N, a_{n+2} = a_{n+1} + a_n , a_1 = 1, a_2 = 2}

Solutions:

(i)   In this case, Set A is collection of $3^n - 2^n$, where n = 1, 2, 3, 4 & 5. Hence A = {1, 5, 19, 65, 211}.

(ii)   Set B is collection of $a_n \rightarrow$ B = {$a_1, a_2, a_3, a_4, \ldots\ldots$}. So here we needs to find the values of $a_1, a_2, a_3,$ a_4, etc. a_1 = 3 (given), hence a_2 = 2 a_1 = 6, a_3 =12, a_4 = 24,…….. so B = {3, 6, 12, 24, 48,…….}.

(iii)   Set C is collection of a_n ie. C = {$a_1, a_2, a_3, a_4, \ldots\ldots$}. So here we needs to find the values of a_3, a_4, etc. a_3 = 1 + 2 = 3, a_4 = a_2 + a_3 = 2 + 3 = 5, ……. So C = {1, 2, 3, 5, 8, 13, 21,……..}

I know, you have learned that how to convert set builder form into roaster form and also observed the way of writing into reverse way. So let us move on to reverse conversion. First solve your own and then see the solution.

Ex4. Write the following sets in set- builder form:

(i)   A = $\left\{ \dfrac{1}{3}, \dfrac{2}{4}, \dfrac{3}{5}, \dfrac{4}{6}, \dfrac{5}{7}, \dfrac{6}{8}, \dfrac{7}{9}, \dfrac{8}{10} \right\}$

(ii)   B = {1, 4, 9, 16,……..}.

(iii)   C = {1, 2, 5, 10, 17, 26,……..}

(iv)   D = {5, 7, 11, 13, 17, 19, 23, 29}

(v)   E = {5, 9, 13, 17, 21,……….}

(vi)   F = {14, 21, 28, 35, 42,……, 98}

(vii)   G = {0, 5, 10, 15,…………}

Solutions:

(i)   A = {x : x = n/(n + 2), n∈N & n < 9}. You can also write as A = {x : x = n/(n + 2), n∈N & n ≤ 8}.

(ii)   B = {y: y = n^2, n∈N}. *I know you have written like* B = {x: x = n^2, n∈N}. *Both are correct.*

(iii)   C = {x: x = n^2 + 1, n ∈ I} or C = {x: x = n^2 + 1, n ∈ W} both are correct.

(iv)   D = {x: x is a prime number, 5 ≤ x < 30} or D = {x: x is a prime number, 5 ≤ x ≤ 29}.

(v)   E = {x : x = 4n + 1, n ∈ N}.

(vi)   F = {x : x = 7n, n ∈ I, 2 ≤ n ≤ 14} or F = {x : x = 7n, n ∈ N, 2 ≤ n ≤ 14}.

(vii)   G = {x : x = 5k, k∈ W}.

From here you can see we can write a single roaster form set in multiple set builder form. This is same as you are describing the particular thing in different way but all are having same meaning.

I know you have learned the concepts, let's refresh your mind by solving a puzzle. Just solve it and enjoy learning

PUZZLE # 1

Consider the following statements and answer the question.

(i)   You are a king/Queen having a KINGDOM.

(ii)   You have 5 soldiers in it.

(iii)   Each soldier has five cages.

(iv)   Each cage has 5 lions.

(v)   Each lion have 4 lionesses.

How many foots the kingdom have?

At the moment, we need to check our knowledge that we learn or understand from above set theory so we have a exercise 1 to solving the purpose. Please complete it and proceed.

EXERCISE - 1

1. Write the following sets into set builder form.

   (i) $\{1, \pi\}$

   (ii) $\{0, \pi, 2\pi, 3\pi, 4\pi, \ldots\ldots\ldots\}$

   (iii) $\{0, 1, 2, 2^2, 2^3, 2^4, \ldots\ldots\ldots 2^{100}\}$

   (iv) $\{1\times2, 3\times4, 5\times8, 7\times16, 9\times32, \ldots\ldots\ldots\}$

   (v) $\left\{ \begin{bmatrix} 1 & 0 \\ 0 & 1 \end{bmatrix}, \begin{bmatrix} 1 & 2^1 \\ 0 & 3 \end{bmatrix}, \begin{bmatrix} 1 & 2^3 \\ 0 & 4 \end{bmatrix}, \begin{bmatrix} 1 & 2^4 \\ 0 & 7 \end{bmatrix}, \begin{bmatrix} 1 & 2^7 \\ 0 & 11 \end{bmatrix}, \ldots\ldots\ldots \right\}$

   (vi) $\{1, -1, i, -i\}$ where $i = \sqrt{(-1)}$

   (vii) $\{1, 4, 10, 16, 64, 22, 256, 28, \ldots\ldots\ldots\}$

2. Write the following sets into tabular form.

   (i) $\left\{ b : b = A_n = \begin{bmatrix} a_{11} \\ a_{21} \end{bmatrix}_{2\times 1}, a_{ij} = i^n + nj, \ i = 1,2 \ \& \ j = 1, \ n \in N \right\}$

   (ii) $\{(a,b) \mid a^2 + b^2 = 25, a \in I, b \in I^+\}$.

   (iii) $\{(x, y): 2x - 3y = 1 \ \& \ x + 4y = 6, x, y \in R\}$.

   (iv) $\{(x, y, z) \mid x^2 + |y^2 - y| + |2z - z^2| = 0, x, y, z \in R\}$. Where $|.|$ = absolute value

   (v) $\{(x, y): x^2 = y, 2x - y + 1 = 0, x, y \in R\}$.

   (vi) $\{x: x = \min\{a, b\}$ where $a = x^2, b = x^3, a, b \in I, |x| < 4, x \in R\}$.

EQUALITY OF TWO SETS

Definition: The two sets A, B are said to be equal only if each and every element of set A belong to set B and vice versa. Equality of sets can be express as A = B.

Let us have an example.

Consider the following sets, A = {1, 3, 4, 5}, B = {1, 3, 3, 4, 5} & C = {1, 4, 5, 3}.

By definition you can justify A = B = C. Here you can have following observations

   1) Repetition of elements in a set is immaterial.

   2) Ordering of elements inside curly braces (set) is immaterial.

   3) n(A) = n(B) = n(C). Where n (A) = Number of elements in set A or Cardinality of set A.

Note: (i) If n(A) = 0, Then set A is Empty Set or Null Set or Void set ($\emptyset$).

   (ii) If n(A) = 1, then set A is called singleton set.

   (iii) If n(A) = 2, then set A is called doublet set.

   (iv) If n(A) = 3, then set A is called triplet set.

   (v) If n(A) = ∞, then set A is called infinite set.

   (vi) If n(A) = finite natural number, then set A is called finite set.

Ex 1. Check whether the given sets A and B are equal or not

    (i)   A = {a, e, i, o, u} & B = {x | x is a letter of word 'euo o eia'}. ☺

    (ii)   A = {G, Y, P, S} & B = {x: x is a letter of word 'GYPSY'}.

    (iii)   A = {x | $x^2 - 3x + 2 = 0$, x ∈ I } & B = {1, 2, 3}.

Solutions

    (i)   Equal Sets, A = B   (ii)   Equal Sets, A = B

    (iii)   Unequal Sets, A ≠ B as element 3 of set B is not present in set A.

Ex2. In the following, state whether A = B or not:

    (i)   A = {a, b, c, d}   B = {d, c, b, a}

    (ii)   A = {4, 8, 12, 16}   B = {8, 4, 16, 18}

    (iii)   A = {2, 4, 6, 8, 10}   B = {x: x is positive even integer and x ≤ 10}

    (iv)   A = {x: x is a multiple of 10}   B = {10, 15, 20, 25, 30, . . .}

Solutions

    (i)   A = B, as order of elements in set is immaterial.

    (ii)   A ≠ B, as 12 ∉ B and 18 ∉A.

    (iii)   A = B, as Set B = {2, 4, 6, 8, 10} = A.

    (iv)   A ≠ B, as A = {10, 20, 30, 40, 50, ………} and B = {10, 15, 20, 25, 30, . . .} where 15, 25 ∉ A.

Ex3. Is the following pair of sets equal? Give reasons.

    (i)   A = {2, 3}, B = {x: x is solution of $x^2 + 5x + 6 = 0$}

    (ii)   A = { x: x is a letter in the word FOLLOW} & B = { y : y is a letter in the word WOLF}

Solutions

    (i)   A = B   (ii)   A = B

SUBSETS

Consider two sets A = {0, 1, 3, 5} and B = {0, 1, 2, 3, 4, 5, 6}, here you can observe that every elements of set A are present in set B. Hence we can say "A is a subset of set B" and written as "A ⊆ B" and you can also observe that 6∈ B and 6 ∉ A, 2∈B and 2 ∉ A, 4∉ A, Hence "B is not a subset of set A" and written as "B ⊄ A ".

Definition: The set A is said to be subset of set B, if each and every elements of set A are present in set B. And denoted as A ⊆ B.

We can also write the definition as 'A ⊆ B if a ∈A → a∈B'.

Observations:

    (i)   A ⊆ A, i.e. any set is a subset of itself.

    (ii)   φ is an empty set; hence it's a subset of every set.

    (iii)   If A ⊆ B then we can also say that B ⊇ A, i.e. Set B is a **Superset** of set A.

    (iv)   A = B if and only if A ⊆ B and B ⊆ A.

Ex1. Consider the sets φ, A = { 1, 3, 4 }, B = {1, 4, 5, 9}, C = {1, 3, 4, 5, 7, 9}. Insert the symbol ⊆ or ⊄ between each of the following pair of sets:

(i) φ . . . B (ii) A . . . B (iii) A . . . C (iv) B . . . C

Solutions: (i) ⊆ (ii) ⊄ (iii) ⊆ (iv) ⊆

Ex2. Make correct statements by filling the symbols ⊆ or ⊄ in the blank spaces. (NCERT)

(i) {x: x is a student of Class XI of your school}. . . . .{x: x student of your school}

(ii) {x: x is a circle in the plane} . . . . {x: x is a circle in the same plane with radius 1 unit}

(iii) {x: x is a triangle in a plane}. . . . . {x: x is a rectangle in the plane}

(iv) {x: x is an equilateral triangle in a plane}. . . . . . {x: x is a triangle in the same plane}

(v) {x: x is an even natural number}. . . . . {x: x is an integer}

Solutions

(i) ⊆ (ii) ⊄ (iii) ⊄ (iv) ⊆

(v) ⊆

Ex3. In each of the following, determine whether the statement is true or false. If it is true, prove it and if it is false, give an example.

(i) If A⊆ B and x ∉ B, then x ∉ A. (ii) If x∈ A & A ⊄ B, then x ∈ B.

(ii) If x ∈ A & A ⊆ B, then x ∈ B.

Solutions

(i) **TRUE;** Let a∈A and A ⊆ B → a ∈ B and x is any element that is not in B then it's a must that it will also not be in set A, i.e. if x ∉B then x ∉ A.

(ii) **FALSE;** Consider the sets A = {1, 4, 6, 7, 8} and B = {0, 4, 6, 8, 9}, 7∈A & A ⊄ B and 7 ∉ B.

(iii) **TRUE;** Since, A is a subset of set B hence every element of set A must lies in set B. Therefore if x ∈ A & A ⊆ B, then x ∈ B

PROPER SUBSETS

Definition: Set A is called a proper subset of set B if each and every elements of set A lies in set B but converse is not true.

In other words you can say "A is a subset of set B and there is at least one element of set B which is not in set A".

Notes:

(i) 'A is a proper subset of set B' written as 'A ⊂ B'. Here you can observe that equality sign is missing.

(ii) If set A ⊂ B and B ⊂ C then A ⊂ C.

Ex1. Let A = {0, 1, 4}, B = {0, 1, 2, 3, 4, 5} & C = {0, 4, 4, 1} then A ⊆ B, A ⊂ B, A = C, A⊆ C, C ⊆ A But the statement 'A ⊂ C' Wrong.

Ex2. Let A = {1, 2, {3, 4}, 5}. Identify the correct statements and give the reason?

(i) {3, 4} ⊆ A (ii) {3, 4} ∈ A (iii) {{3, 4}} ⊂ A (iv) 1 ∈ A

(v) 1 ⊂ A (vi) {1, 2, 5} ⊆ A (vii) {1, 2, 5} ∈ A (viii) {1, 2, 3} ⊆ A

(ix) φ ∈ A (x) φ ⊂ A (xi) {φ} ⊆ A

Solutions:

Here you can observe 1, 2, {3, 4} & 5 are elements of set A. You may be surprised that set {3, 4} is behaving like as an element. We will work on it, when we were studying power set. So let's discuss the current problem.

    (i)   **INCORRECT:** $\{3, 4\}$ is an element of set A hence $\{3, 4\} \in A$. also $3 \notin A$, $4 \notin A$, so $\{3, 4\} \not\subset A$.

    (ii)   **CORRECT:** $\{3, 4\}$ is an element of set A hence $\{3, 4\} \in A$.

    (iii)   **CORRECT:** $\{3, 4\}$ is an element of set A hence $\{3, 4\} \in A$. Hence $\{\{3, 4\}\} \subset A$

    (iv)   **CORRECT:** 1 is an element of set A, hence $1 \in A$.

    (v)   **INCORRECT:** as the proper subset $(\subset)$ applies only between two sets and 1 is not a set.

    (vi)   **CORRECT:** as 1, 2 & 5 are elements of set A, hence $\{1, 2, 5\} \subset A$.

    (vii)   **INCORRECT:** as $\in$ apply between an element and a set and the element $\{1, 2, 5\} \notin A$

    (viii)   **INCORRECT:** as $3 \notin A$, hence $\{1, 2, 3\} \not\subset A$.

    (ix)   **INCORRECT:** as ϕ is behaving like as an element and $\phi \notin A$.

    (x)   **CORRECT:** as ϕ (empty set) is a subset of every set.

    (xi)   **INCORRECT:** as $\phi \notin A$, hence $\{\phi\} \not\subset A$.

I hope you get the concept correctly. If there is any doubt, you can write to us facebook/Mathsarc. Let's move on next concept, Number of subsets.

Ex3.   Consider a set A = $\{1, 3, 5\}$. List all the subsets of set A.

Sol: (i)   Number of set having zero elements i.e. Empty set = 1.

    (ii)   Number of singleton sets i.e. Sets having only one element ($\{1\}$, $\{3\}$, $\{5\}$) = 3.

    (iii)   Number of doublets sets i.e. sets having only 2 elements ($\{1, 3\}$, $\{1, 5\}$, $\{3, 5\}$) = 3.

    (iv)   Number of subsets having 3 elements ($\{1, 3, 5\}$) = 1.

Hence, total number of subsets of set A = 1 + 3 + 3 + 1 = 8. Here you can observe that 8 include the counting of empty set and the set itself.

No. of proper subset of set A = 8 − 1 = 7.

THEOREM 1: If $n(A) = m$, then number of subsets of set A = 2^m. It includes the empty set and the set A itself in counting.

PROOF: Consider a set A = $\{x_1, x_2, x_3, \ldots\ldots\ldots, x_m\}$. Here $n(A) = m$.

Let us have an empty set B at initial and we are making a subset B of set A, so we will take

elements from set A and placed inside the set B. Now x_1 has two choices (i) Either it will be a

member of set B or (ii) It may not be a member, similarly $x_2, x_3, x_4, \ldots\ldots x_m$ have 2 choices. Hence

by multiplication principle of counting total number of subsets

= $2 \times 2 \times 2 \times 2 \ldots\ldots \times 2$ (m times) = 2m.

ALTERNATE THINKING:

We need some ^nC_r notation to explain the concept. Its meaning is selection of 'r' object from n given distinct objects. So let's move on the proof part.

Number of subset containing zero elements = $^nC_0$.

Number of subset containing 1 elements = $^nC_1$.

Number of subset containing 2 elements = $^nC_2$.

....

Number of subset containing n elements = ^nC_n.

So total number of subset = $^nC_0 + {}^nC_1 + {}^nC_2 + \ldots\ldots\ldots + {}^nC_n = 2^n$.

Ex 4. Consider a set A = {a, e, i, o, u}. Find the solutions of following questions.

(i) Number of sunsets of set A.

(ii) Number of proper subsets of set A.

(iii) Number of subsets of set A, excluding empty set ϕ and set A.

(iv) Number of subsets of set A having element 'a'.

(v) Number of subsets of set A having element 'a' but not 'e'.

(vi) Number of subsets of set A having at-most 4 elements and 'a' remains its element.

Solutions

(i) By statement, number of subsets of set A = 2^5 = 32.

(ii) Number of proper subsets of sets A = 2^5 – 1 = 31. We are removing the counting of set A.

(iii) Number of subsets of set A, excluding empty set ϕ and set A = 2^5 – 2 = 30.

(iv) The element 'a' has only one choice to be included in the subset and rests have 2 choices either be in or out. Hence, Number of subsets of set A having element 'a' = 1×2^4 = 16.

(v) The elements 'a' and 'e' have only 1 choice and rests have 2 choices to be a member of subset. Hence, answer = $1 \times 1 \times 2 \times 2 \times 2 = 8$.

(vi) Consider B is a required subset having element 'a', then

(a) Number of subsets such that n(B) = 2 are 4 i.e. {a, e}, { a, i}, { a, o},{ a, u} . $^4C_1$ = 4.

(b) Number of subsets such that n(B) = 3 are 6 i.e. {a, e, i}, {a, e, o}, {a, e, u}, {a, i, o}, {a, i, u}, {a, o, u}. $^4C_2$ = 6.

(c) Similarly, Number of subsets such that n(B) = 4 are 4. i.e $^4C_3$ = 4.

Hence answer to the question = 4 + 6 + 4 = 14.

Ex5. If A = {8^n – 7n – 1: n$\in$N} and B = {49(n - 1): n $\in$ N}. Then show that A $\subseteq$ B.

Solution:

As per binomial theorem $(1+ 7)^n = {}^nC_0 + {}^nC_1 7 + {}^nC_2 (7)^2 + \ldots\ldots + {}^nC_n (7)^n$.

i.e. $8^n = 1 + 7n + {}^nC_2 (7)^2 + \ldots\ldots + {}^nC_n (7)^n \rightarrow 8^n - 7n - 1 = (7)^2 \times$(Some Integer). Which is divisible by 49 but it will not form a table of 49.

B ={0, 49, 98, 147, 196,........ } . Hence every elements of set A will be inside the set B, So A $\subseteq$ B.

Ex6. Find the cardinality of set A (i.e. n(A)), where A = {x: x =$[a_{ij}]_{2 \times 2}$, $a_{ij} \in$ {0, 1, 2} & det(x) = 0}. Given, If $A = \begin{vmatrix} a & b \\ c & d \end{vmatrix}$, then det(A) = ad – bc.

Solution:

Let $A = \begin{vmatrix} a & b \\ c & d \end{vmatrix}$, where a, b, c, d$\in$ {0, 1, 2}. det(A) = ad – bc = 0 $\rightarrow$ ad = bc. If a, b, c, d$\in$ {0, 1, 2}, then product of two such numbers can be 0, 1, 2 or 4. So let's make cases in which ad = bc.

S. No.	a d = b c	REPRESENTATION	NO. OF SUCH MATRICES
		a d = b c	
		0 0 0 0	
		0 1 0 1	
1	0 = 0	0 2 0 2	$5 \times 5 = 25$
		1 0 1 0	
		2 0 2 0	
2	1 = 1	1×1 1×1	1
3	2 = 2	1×2 1×2	$2! \times 2! = 4$
4	4 = 4	2×2 2×2	1

Hence, total number of elements in set A = 25 + 1 + 4 + 1 = 31.

Ex7. Two finite sets have m and n elements. The number of subsets of the first set is 112 more than that of the second set. The value of m and n are, respectively

(A) 4, 7   (B) 7, 4   (C) 4, 4   (D) none of these

Solution:

Let first set be A and second set is B then n(A) = m & n(B) = n as per question.

According to question we can have the equation, $2^m = 2^n + 112$.

Now to get the values of m and n, we need to apply hit and trial method but it may take time to solve the problem, so be logical, and now think about the number m such that 2^m is just greater than 112. I think we have $2^7 = 128 = 112 + 16 = 112 + 2^4$.

I hope you got the required relation, $2^7 = 112 + 2^4$. So m = 7 & n = 4. Hence Answer = (B).

Note: (i) There is an important relation that we needs to remember are: $\mathbb{N} \subset \mathbb{W} \subset \mathbb{Z}$ or $\mathbb{I} \subset \mathbb{Q} \subset \mathbb{R} \subset \mathbb{C}$.

INTERVAL AS A SUBSET OF R

Consider a, b $\in$ R, as two finite real numbers such that a < b then a set {y: a < y < b, y$\in$ R} is called open interval a to b and denoted as "(a, b)". In this set the number 'a' and 'b' are excluded. There are some other cases also; I am explaining them, one by one.

(i) **x $\in$ (a, b)** $\equiv$ x$\in$ {y: a < y < b, y$\in$ R} and called as Open Interval. It excluded x = a, x = b.

(ii) **x $\in$ [a, b]** $\equiv$ x$\in$ {y: a $\leq$ y $\leq$ b, y$\in$ R} and called closed interval. It include both, x = a, x = b.

(iii) **x $\in$ (a, b]** $\equiv$ x$\in$ {y: a < y $\leq$ b, y$\in$ R}, x = a excluded and x = b included in the set.

(iv) **x $\in$ [a, b)** $\equiv$ x$\in$ {y: a $\leq$ y < b, y$\in$ R}, x = a included and x = b excluded in the set.

(v) **x $\in$ [b, ∞)** $\equiv$ x$\in$ {y: y $\geq$ b, y$\in$ R}.

(vi) **x$\in$(- ∞, ∞)** $\equiv$ x $\in$ {y: y$\in$ R}

Here we have a question in mind that can we write as 'x ∈ [b, ∞], i.e ∞ **has a closed bracket?'**

Answer is No. Since '∞' is not a finite number so we never – ever have closed bracket at ∞ or - ∞. Even ∞ is not a real number and also not a complex number. It's not following the rules of real numbers as 2 + 3 = 5, Here ∞ + ∞ = ∞, ∞∞ = ∞. ∞ is just, a thinking.

PICTORIAL UNDERSTANDING OF DIFFERENT SUB SET OF REAL NUMBERS ON NUMBER LINES

REPRESENTATION	MEANING

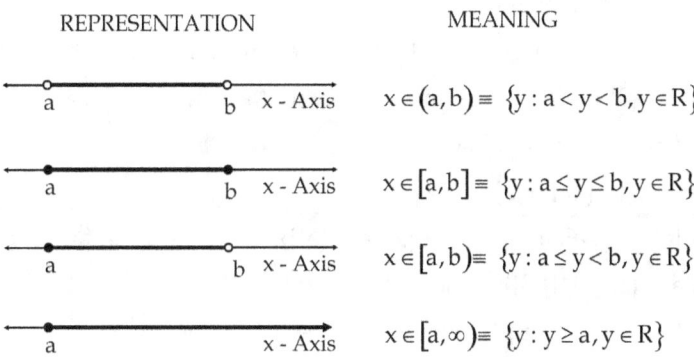

$x \in (a,b) \equiv \{y : a < y < b, y \in R\}$

$x \in [a,b] \equiv \{y : a \le y \le b, y \in R\}$

$x \in [a,b) \equiv \{y : a \le y < b, y \in R\}$

$x \in [a,\infty) \equiv \{y : y \ge a, y \in R\}$

There are other cases also, you can derived them using above concept and your brain.

Note: b – a = Length of any interval (a, b) or [a, b] or (a, b] or [a, b).

POWER SET

Let us have an example to understand it better. Consider a set A = {0, 1, 2}. There are 2^3 subsets of set A, lets write them at one place.

φ = {}, {0}, {1}, {2}, {0, 1}, {0, 2}, {1, 2}, {0, 1, 2}. Now if we make a set of all these subsets then the resultant set is called power set of set A and denoted as P(A).

So, P(A) = {{}, {0}, {1}, {2}, {0, 1}, {0, 2}, {1, 2}, {0, 1, 2}}. Here you can observe that a set is behaving like as an element. Even every element in set P (A) is a set.

Definition: The collection of all the subset of set A called power set of set A and denoted by P(A).

Note: Number of elements in P(A) = Number of subsets of set A. If n(A) = m, then number of elements in P(A) = 2^m. i.e. n(P(A)) = 2^m.

UNIVERSAL SET

Universal set is a set on demand basis. It changes sets to set. Lets us think of a concept of system and surrounding, then UNIVERSE = SYSTEM + SURROUNDING. Let us have some sets.

A = {0, 1, 2}, B = {1, 2, 3, 5}, C = {2, 4, 6, 8, 0}. The minimum requirements for a universal set of set A, B and C is {0, 1, 2, 3, 4, 5, 6, 8}. You can assume this set to be Universal set for set A, B and C. Universal set is usually denoted by U or X.

Let us have a set if integers, then we can have set of rational numbers as a universal set or even we can have set of real numbers as a universal set. I hope! You got the understanding of universal set.

Ex1. Find the number of elements in following sets?
A = {{0}, φ, {φ}, {φ, {φ}}}.   B = {x: x is even integers & x < 29}.
C = {x: - 1 ≤ x ≤ 2, x∈ Q}.

Solution: n (A) = 4, n (B) = ∞, n(C) = ∞.

Ex2. Examine whether following statements are true or false. Explain them with reason.

(i) $\phi \in \phi$

(ii) $0 \in \phi$

(iii) $\phi = \{0\}$

(iv) $\phi \in \{\phi\}$

(v) $\{\{1\}, 2, 3\} = \{1, \{2\}, 3\}$

(vi) $\{a\} \in \{a, \{a\}, \{\{a\}\}\}$

(vii) $\{a\} \subseteq \{a, \{a\}, \{\{a\}\}\}$

(viii) $2 \subset \{1, 2, 3\}$

(ix) $\{\{1\}\} \subset \{\{1, 2, 3\}\}$

Solutions:

(i)　FALSE:　as ϕ is not an element of empty set ϕ.

(ii)　FALSE:　as ϕ is an empty set so it do not have any element, hence $0 \notin \phi$.

(iii)　FALSE:　as $0 \in \{0\}$ but $0 \notin \phi$. Even $n(\phi) = 0$, $n(\{0\}) = 1$, so they can't be equal.

(iv)　TRUE:　as ϕ is an element of set $\{\phi\}$ hence $\phi \in \{\phi\}$.

(v)　FALSE:　as $2, \{1\} \notin \{1, \{2\}, 3\}$.

(vi)　TRUE:　as $\{a\}$ is an element of set $\{a, \{a\}, \{\{a\}\}\}$, hence $\{a\} \in \{a, \{a\}, \{\{a\}\}\}$.

(vii)　TRUE:　as $a \in \{a, \{a\}, \{\{a\}\}\}$, hence $\{a\} \subseteq \{a, \{a\}, \{\{a\}\}\}$.

(viii) FALSE:　as the sign '$\subset$' apply between two sets but 2 is not a set.

(ix)　FALSE:　as $\{1\} \notin \{\{1, 2, 3\}\}$, hence $\{\{1\}\} \not\subset \{\{1, 2, 3\}\}$.

Lets us have an exercise to get the concepts. Solve them before moving to next topic.

EXERCISE - 2

1. Which of the following statements are true and which of them are false for given set A = {4, 11, 17, 21}.

   (i)　$\{\phi\} \subset A$

   (ii)　$\{11\} \in A$

   (iii)　$21 \not\subset A$

   (iv)　$11 \subset A$

   (v)　$\{11\} \subset A$

2. Which of the following statements are correct?

   (i)　$\{\phi\} = \{0\}$

   (ii)　$\{a, b, c\} = \{b, a, c\}$

   (iii)　$\{1, 2, \{3\}\} = \{\{1\}, 2, 3\}$

   (iv)　$\{\{1\}, 2, \{3\}\} = \{\{1\}, \{3\}, 2\}$

   (v)　$\{\phi\} \in \{\{\phi\}\}$

   (vi)　$\{\{1, 2\}, \{2\}, \{2, 3\}\} = \{\{1\}, \{1, 2\}, \{3\}\}$

   (vii)　$\{3\} \in \{2, 3, 4\}$

   (viii)　$\phi \subset \{\{\phi\}\}$

   (ix)　$a \in \{\{a\}, \{\{a\}\}\}$

   (x)　$4 \in \{4\}$

3. Which of the following statements are true for the set A = $\{\phi, \{\phi\}, 1, \{1, \phi\}, 7\}$?

   (i)　$7 \subset A$

   (ii)　$\{\{\phi\}\} \subset A$

   (iii)　$\{\{7\}, \{1\}\} \not\subset A$

   (iv)　$\{\phi, \{\phi\}, \{1, \phi\}\} \subset A$

   (v)　$\{1\} \in A$

4. Find the power sets of A = $\{\{\phi\}\}$, B = $\{\phi, \{\phi\}\}$, C = $\{1, 2, \{1, 3\}\}$.

VENN DIAGRAMS

Venn diagrams have vast roll in the field of sets theory, logic, etc. Venn diagram is named after English logician, "*JOHN VENN*". The diagram consists of closed curves. The universal sets, usually denoted by rectangle and its subsets are denoted by circles.

Let us have an example as shown in figure.

Ex1. Universal set U = {1, 2, 3, 4, 5, 9}.

Its subsets A = {1, 2, 3, 4} and B = {2}.

You can observe here as B ⊂ A ⊂ U.

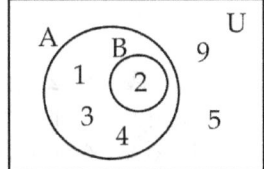

OPERATIONS ON SETS

An operation on two or more sets is a set. This is similar to addition, subtraction etc. Here we have some rules of operation based on different definitions.

(A) **UNION** (Symbol "∪")

Definition:

The union of two set A and B is a set C which contains all the elements of set A as well as B only.

Look at the Venn's diagram, shaded region represents A∪B.

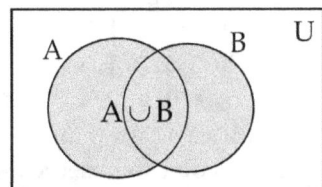

Ex1. Let A = {1, 2, 4, 5, 6} and B = {3, 4, 5, 7, 8} then C = A∪B = {1, 2, 3, 4, 5, 6, 7, 8}.
A∪B = {x: x∈A or x∈B} and usually read as "A union B".

Ex2. The student's s_1, s_2, s_4 & s_6 played football and student's s_1, s_3 & s_8 played cricket then the students played either football or cricket are, s_1, s_2, s_4 , s_6, s_3, s_8. So let A = {s_1, s_2, s_4 , s_6} & B = {s_1, s_3, s_8} then the union of sets A & B is A∪B = { s_1, s_2, s_4 , s_6, s_3, s_8}.

Ex3. Let A = {1, 2, 3, 4}, B = {2, 3, 7} and U = {1, 2, 3, 4, 5, 7}, then Venn's diagram representing the A∪B is as shown in figure by shaded region. Where A∪B = {1, 2, 3, 4, 7}.

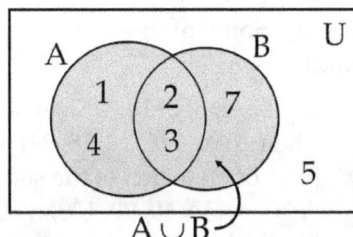

Similarly we can have the definition for union of multiple sets as well.

Note: The set of all the objects which belongs to atleast one of the set A, B, C,......... is called union of these sets and denoted by A∪B∪C....... Thus, A∪B∪C....... ={x: x∈A or x∈B or x∈ C or ....}.

PROPERTIES OF UNION

(i) A ∪ A = A ....(Idempotent Law)

(ii) A∪φ = A ....(φ is an identity of ∪, Law of identity element)

(iii) A ∪ B = B ∪ A ....(Commutative Law)

(iv) A∪ (B∪C) = (A∪B) ∪C ....(Associative Law)

(v) A ∪ U = U ....(Universal Law)

(B) **INTERSECTION** (Symbol "∩")

Definition:

The intersection of two sets A and B is defined as the set of common elements of both the sets A and B and denoted by A∩B.

i.e. A∩B = {x: x∈A and x∈B}.

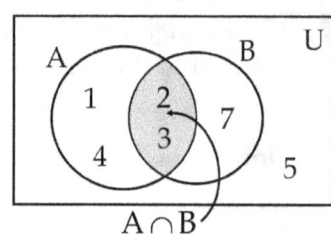

Ex1. Let A = {1, 2, 3, 4}, B = {2, 3, 7} and U = {1, 2, 3, 4, 5, 7}, then Venn's diagram representing the A∩B is as shown in figure by shaded region. Where A∩B = {2, 3}.

Ex2. Consider the following sets:
A = {1, 2, 3, 4, 5}.
B = {2, 4, 5, 9}.
C = {1, 4, 6}.
D = {0, 1}.
U = {0, 1, 2, 3, 4, 5, 6, 7, 8, 9}
Then, A∩B = {2, 4, 5}, A∩B∩C = {4} and
A∩B∩C∩D = ϕ (empty set).
Here you can observe its Venn's diagram as shown.

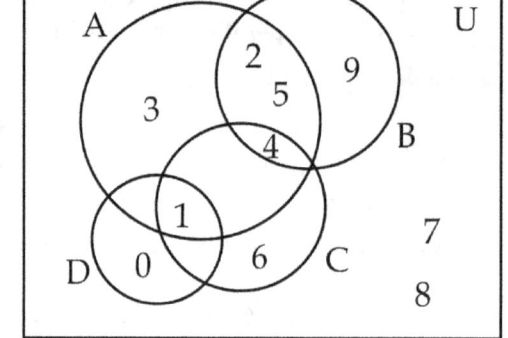

Note: The set of all the objects which are common in the sets A, B, C, …… is called the intersection of these sets and denoted by A∩B∩C……… Thus, A∩B∩C……. = {x: x∈A and x∈B and x∈ C and ……}.

Ex3. Consider the set A = {x : x = 2n, n∈N, x ≤ 400}, B = {x: x = 3n, n∈ W, x ≤ 300} and C = {y | y = 30n, n ∈ N, y ≤ 600}. Then find the solution of following questions.
(i) n(A∩B)   (ii) n(A∩C)   (iii) n(A∩B∩C)   (iv) A∩B
(v) Some of the elements of set A∩B.

Solution:

A = {2, 4, 6, 8, 10,……., 400}, B = {0, 3, 6, 9, 12,……, 300} and C = {30, 60, 90, ……, 600}
Then A∩B = {6, 12, 18, 24, 30, ……. ,300} i.e. multiple of LCM (2, 3) = 6. Hence n(A∩B) = 50.
Sum of elements of the set A∩B = 6 + 12 + 18 + 24 + 30+ ……. + 300 = 50(6 + 300)/2 = 7650.
A∩C = {30, 60, 90, 120, …..,390} i.e. multiple of LCM(2, 30) = 30. Hence n(A∩C) = 13.
Similarly, A∩B∩C = {30, 60, 90, 120, ……, 300} i.e. multiple of LCM(2, 3, 30) = 30.
Hence, n(A∩B∩C) = 10.

Ex4. Let A = {1, 2, 3, 4, 5} & B = {2, 4, 5, 7} then find the number of set C such that C ⊂ B & C ⊄ A.

Solution:

Consider the figure as shown. Since set C follow the conditions C ⊂ B & C ⊄ A. Hence element '7' must lie inside the set C. Now the remaining elements of set B, i.e. 2, 4 & 5 have two choices to include in the set C. Hence total number of such set C = 2^3 = 8. *(Is it correct answer?) No!!*

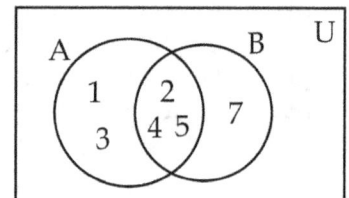

Since, Set C is a proper subset of set B, but 8 count the set B also. Hence correct answer to the question = 8 - 1 = 7.

PROPERTIES OF INTERSECTION

(i) $A \cap A = A$ ....(Idempotent Law)

(ii) $A \cap \phi = \phi$ and $A \cap U = A$ ....(Law of ϕ and U)

(iii) $A \cap B = B \cap A$ ....(Commutative law)

(iv) $(A \cap B) \cap C = A \cap (B \cap C)$ ....(Associative law)

(v) $A \cap (B \cup C) = (A \cap B) \cup (A \cap C)$ .... (Distributive law) i.e. distribution of $\cap$ over $\cup$

You can easily understand above law using Venn's diagram. Just explore your mind. Let us have a logical problem.

LoEx1. Consider a society, in which Doctors, Engineers, IAS officers, Teachers and Chartered Accountants are living. I am showing the details of society using Venn's diagram as shown. Where sets symbols are

E = Engineers

T = Teachers

C = Chartered Accountants

D = Doctors

I = IAS officers

Then select the correct statements

(A) Some doctors are teachers

(B) Engineers, Doctors, Chartered accountants are Teachers

(C) Teachers, Engineers, Doctors and Chartered accountants are IAS officers

(D) Some person in the society who are doctor, teacher are also IAS officers.

Ans. A, D

LoEx.2 Which one of the following diagrams given below illustrates the relationship among Officers, Women and Birds?

Ans. D

Note: (a) $A_1 \cap A_2 \cap A_3 \cap A_4 \cap A_5 \cap \ldots \ldots \cap A_n = \bigcap_{i=1}^{n} A_i$.

(b) $A_1 \cup A_2 \cup A_3 \cup A_4 \cup A_5 \cup \ldots \ldots \cup A_n = \bigcup_{i=1}^{n} A_i$.

(c) If $A \cap B = \phi$ then sets A and B are called **Disjoint Sets**.

Ex5. Suppose $A_1, A_2 \ldots, A_{30}$ are thirty sets, each with five elements and $B_1, B_2, \ldots, B_n$ are n sets each with three elements. Let $\bigcup_{i=1}^{30} A_i = \bigcup_{j=1}^{n} B_j = S$, Assume that each element of S belongs to exactly ten of the A_i's and exactly to nine of the B_j's. Find the value of n.

Solution:

Number of elements in $S = A_1 \cup A_2 \cup A_3 \cup A_4 \cup A_5 \cup \ldots \ldots \cup A_{30} = 5 \times 30 = 150$. Are you agreeing? No.

I hope you got the problem. Here, we have count a single member of set A_i multiple times.

Since each element of set S belongs to exactly ten A_i's so $n(S) = 150/10 = 15 \ldots \ldots \ldots$ (i). Think!

Let $a \in S$ then the set has nature $S = \left\{ \underbrace{a, a, a, a, a, \ldots .a}_{10 \text{ TIMES}}, \ldots \ldots \ldots \right\}$. The same thing is happening

for every elements of set S. So, $10 \times n(S) = 5 \times 30 \rightarrow n(S) = 150/10 = 15$.

Similarly, Number of elements in $S = B_1 \cup B_2 \cup B_3 \cup B_4 \cup B_5 \cup \ldots \ldots \cup B_n = 3 \times n = 3n$. Here we have cont the same member of set B, multiple times.

Since each element of set S belongs to exactly nine B_i's so $n(S) = 3n/9 = n/3 \ldots \ldots$ (ii)

From (i) and (ii) $n(S) = 15 = n/3 \rightarrow n = 45$.

Ex6. Let $S = \{1, 2, 3, 4\}$. The total number of unordered pairs of disjoint subset of S is equal to.

(A) 25   (B) 34   (C) 42   (D) 41

Solution

The sets A and B partitioned the universal set in three parts, so let start filling elements of S into these partitions. Hence every element of S has three choices to fill, i.e. either it goes to set A or in Set B or neither in A nor in B. Hence total number of ways to do so $= 3 \times 3 \times 3 \times 3 = 81$.

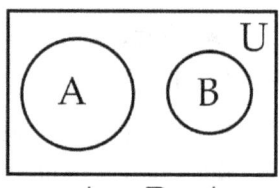

$A \cap B = \phi$

But there is a problem of unordered pair. Let take two cases
(I) $A = \{1, 3\}$ and $B = \{2\}$
(II) $A = \{2\}$ and $B = \{1, 3\}$
In case of ordered pair, these are different cases but in the case of unordered pair, these are same cases, hence every case has counted twice except the case $A = \phi$, $B = \phi$ (Counted only once).
So Correct Answer $= 1 + (80/2) = 41$. (D)

(C)   DIFFERENCE (Symbol "-")

Definition:

The difference A – B in order, is defined as the set of elements of set A which are not present in set B. i.e. $A - B = \{x : x \in A, x \notin B\}$.

Look at the Venn's Diagram for A – B.

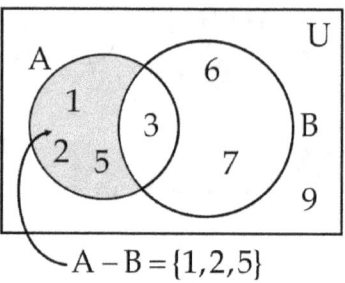

$A - B = \{1, 2, 5\}$

Hare $A = \{1, 2, 3, 5\}$, $B = \{3, 6, 7\}$ and $U = \{1, 2, 3, 5, 6, 7, 9\}$. Hence $A - B = \{1, 2, 5\}$. (Shaded region in figure)

Ex1. Let $A = \{1, 2, 3\}$, $B = \{2, 3, 4, 5\}$ then $A - B = \{1\}$ and $B - A = \{4, 5\}$.

Note: Here you can observe that A – B, A∩B and B – A are mutually disjoints sets.

Ex2. Let $P = \{\theta : \sin\theta - \cos\theta = \sqrt{2}\cos\theta\}$ and $Q = \{\theta : \sin\theta + \cos\theta = \sqrt{2}\sin\theta\}$ be two sets. Then,
(A) $P \subset Q$ and $Q - P \neq \emptyset$.   (B) $Q \not\subset P$
(C) $P \not\subset Q$ (D) $P = Q$

Solution

The question is based on set theory and trigonometric equation. Beginners can leave this example, but later on you have to do such kind of problem to be successful.

Consider the set P. $\sin \theta - \cos\theta = \sqrt{2} \cos\theta \rightarrow \sin\theta = (1+\sqrt{2}) \cos\theta$

$\rightarrow \sin\theta = (1+\sqrt{2})(1 - \sqrt{2}) \cos\theta / (1 - \sqrt{2})$.

$\rightarrow (1 - \sqrt{2}) \sin\theta = - \cos\theta \rightarrow \sin\theta + \cos\theta = \sqrt{2} \sin\theta$ which is the equation in set Q. So, we can say P = Q. Hence (D) is correct option. (A) Option may be the correct one but in this case Q – P = ∅.

COMPLEMENT OF A SET

Consider a Universal set U = {0, 1, 2, 3, 5, 7, 8} and it's a subset A = {0, 2, 5}, then a set A′ or A^c called complement of set A, where A′ =A^c = {1, 3, 7, 8}. You can observe that A′ = {x: x∈U & x∉ A}.

Definition:

If set A is a subset of universal set U, then the difference set U – A is called complement of set A and it's denoted by A′ or A^c or $\overline{A}$.

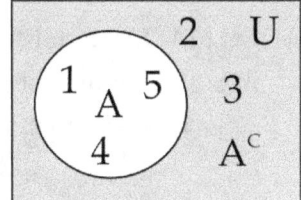

Thus, A^c = U – A = {x: x∈U and x∉ A}.

Look at the Venn's diagram. Shaded region represents A^C.

Ex1. Set U = {1, 2, 3, 4, 5}, A = {1, 4, 5} then A^C = {2, 3}.

PROPERTIES OF COMPLEMENT OF SET A

(i)   A∩A′ = φ and A∪A′ = U. ....(Complement Laws)

(ii)   (A∪B)′ = A′∩B′ and (A∩B)′ = A′∪B′. ....(De Morgan's Law)

(iii)   (A′)′ = A. ....(Law of double complement)

(iv)   U′ = φ and φ′ = U. ....(Law of empty set and universal set)

You can easily verify above properties using Venn's diagrams.

VENN'S DIAGRAM FOR 3 SETS A, B & C (Region Identification)

This is the topic, where most of the students get confused so I am explaining in depth by assuming some case.

Let us have three sets S, A and C in a college and making communities among them. Where

S = Set of students who have interest in Science stream.

A = Set of students who have interest in Art's stream.

C = Set of students who have interest in Commerce stream.

Then, there will be 8 types of communities in this college as shown in Venn's diagram.

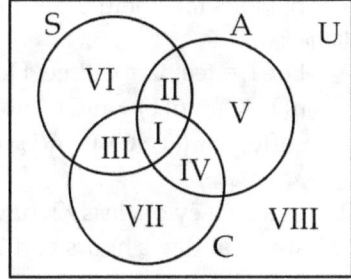

I: Community of students who have interest in the entire three streams. (S∩A∩C)

II: Community of students who have interest in Arts and Science stream only. (S∩A∩C′)

III: Community of students who have interest in Commerce and Science stream only. (S∩A′∩C)

IV: Community of students who have interest in Arts and Commerce stream only. (S′∩A∩C)

V: Community of students who have interest in Arts stream only. (S′∩A∩C′)

VI: Community of students who have interest in Science stream only. (S∩A′∩C′)

VII: Community of students who have interest in Commerce stream only. (S′∩A′∩C)

VIII: Community of students who do not have interest in the given streams. (S∪A∪C)′

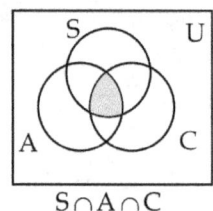 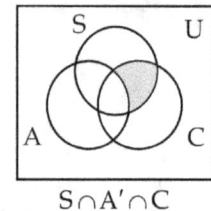 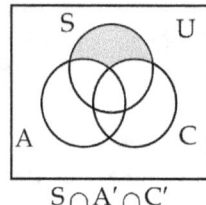 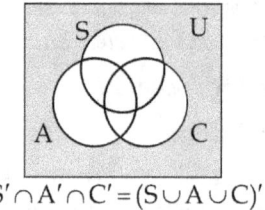

S∩A∩C   S∩A′∩C   S∩A′∩C′   S′∩A′∩C′=(S∪A∪C)′

INCLUSION – EXCLUSION PRINCIPLE

THEOREM:

$$n(A \cup B) = n(A) + n(B) - n(A \cap B)$$

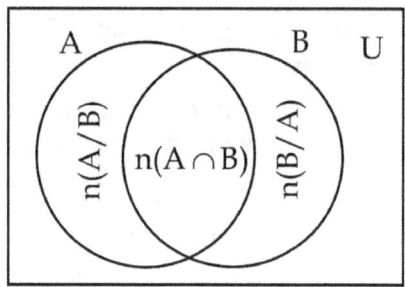

Proof: Let n(A/B) = number of element in set A only.

n(A∩B) = number of element in A∩B.

→ n(A∪B) = n(A/B) + n(B/A) + n(A∩B)

= n(A) - n(A∩B) + n(B) - n(A∩B) + n(A∩B).

= n(A∪B) = n(A) + n(B) - n(A∩B).

Note: n(A/B) = n(A - B) = n(A) - n(A∩B).

Ex1. If A and B are two sets such that n(A) = 150, n(B) = 250 and n(A∪B) = 300. Find the value of n(A - B) and n(B - A).

Solution:

n(A∩B) = n(A) + n(B) - n(A∪B).

n(A∩B) = 150 + 250 - 300 = 100.

So n(A - B) = n(A) - n(A∩B) = 150 - 100 = 50.

Similarly, n(B - A) = n(B) - n(A∩B) = 250 - 100 = 150.

Ex2. In a group of 100 persons, 80 takes tea, 30 take coffee and 20 take both of tea and coffee. How many persons take neither tea nor coffee?

Solution:

Let T = tea, C = Coffee. Hence, as per question, n(T) = 80, n(C) = 30 and n(T∩C) = 20.

n(T∪C) = n(T) + n(C) - n(T∩C) = 80 + 30 - 20 = 90. Hence, Number of persons take neither tea nor coffee = n((T∪C)′) = n(U) - n(T∪C) = 100 - 90 = 10.

Ex3. In a survey of class XI, having 65 strength found that every one love to study. 50 students love to study in day's hours and 20 loves to study at night. How many students love to study in both hour's?

Solution:

Let N = Night and D = Day. So n(D∪N) = 65, n(D) = 50, n(N) = 20.

Hence, n(A∩B) = n(A) + n(B) - n(A∪B). So n(A∩B) = 50 + 20 - 65 = 5.

Ex4. In a survey of 400 students in a school, 100 were listed as taking apple juice, 150 as taking orange juice and 75 were listed as taking both apple as well as orange juice. Find how many students were taking neither apple juice nor orange juice. (NCERT)

Solution:

Let U denote the set of surveyed students and A denote the set of students taking apple juice and B denote the set of students taking orange juice. Then $n(U) = 400$, $n(A) = 100$, $n(B) = 150$

and $n(A \cap B) = 75$. Now $n(A' \cap B') = n(A \cup B)' = n(U) - n(A \cup B)$

$= n(U) - n(A) - n(B) + n(A \cap B)$

$= 400 - 100 - 150 + 75 = 225$

Hence 225 students were taking neither apple juice nor orange juice.

Ex5. There are 200 individuals with a skin disorder, 120 had been exposed to the chemical C_1, 50 to chemical C_2, and 30 to both the chemicals C_1 and C_2. Find the number of individuals exposed to

(i)   Chemical C_1 but not chemical C_2

(ii)   Chemical C_2 but not chemical C_1

(iii) Chemical C_1 or chemical C_2 (NCERT)

Solution:

Let U denote the universal set consisting of individuals suffering from the skin disorder, A denote the set of individuals exposed to the chemical C_1 and B denote the set of individuals exposed to the chemical C_2. Here $n(U) = 200$, $n(A) = 120$, $n(B) = 50$ and $n(A \cap B) = 30$.

(i)   From the Venn diagram given in Figure, we have

$A = (A - B) \cup (A \cap B)$.

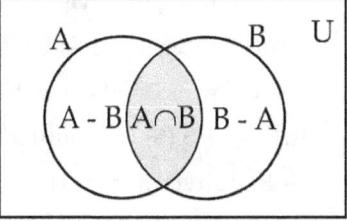

$n(A) = n(A - B) + n(A \cap B)$ (Since A - B) and A ∩ B are disjoint.)

or $n(A - B) = n(A) - n(A \cap B) = 120 - 30 = 90$

Hence, the number of individuals exposed to chemical C_1 but not to chemical C_2 is 90.

(ii)   From the Figure, we have $B = (B - A) \cup (A \cap B)$.

So, $n(B) = n(B - A) + n(A \cap B)$ (Since B - A and A ∩B are disjoint.)

or $n(B - A) = n(B) - n(A \cap B) = 50 - 30 = 20$

Thus, the number of individuals exposed to chemical C_2 and not to chemical C_1 is 20.

(iii) The number of individuals exposed either to chemical C_1 or to chemical C_2, i.e.

$n(A \cup B) = n(A) + n(B) - n(A \cap B) = 120 + 50 - 30 = 140$.

Ex6. Consider a set $A = \{1, 2, 3, 4, 5, \ldots, 1000\}$. Number of element in the set A such that it is divisible by either 2 or 5?

Solution:

let $n(2)$ = number of element in the set A that are divisible by 2

So $n(2) = 500$. Similarly $n(5) = 200$.

$n(2 \cap 5)$ = number of element in the set A that are divisible by both 2 and 5. i.e. 10, So $n(2 \cap 5) = 100$.

→ $n(2 \cup 5) = n(2) + n(5) - n(2 \cap 5)$

→ $n(2 \cup 5) = 500 + 200 - 100 = 600$.

Let us represent the same problem in different manner.

Ex7. How many integers between 1 and 1000, both inclusive, do not share a common factor with 1000?

Solution:

Observe that $1000 = 2^3 5^3$, and thus from the 1000 integers we must weed out those that have a factor of 2 or of 5 in their prime factorization.

If A_2 denotes the set of those integers divisible by 2 in the interval [1, 1000] then clearly $n(A_2) = 500$.

Similarly, if A_5 denotes the set of those integers divisible by 5 i.e. $n(A_5) = 200$. Also $n(A_2 \cap A_5) = 100$.

This means that there are $n(A_2 \cup A_5) = 500 + 200 - 100 = 600$ integers in the interval [1, 1000] sharing at least a factor with 1000, thus there are $1000 - 600 = 400$ integers in [1, 1000] that do not share a a a common factor with 1000.

Ex8. Of 40 people, 28 swim and 16 run. It is also known that 10 both swim and run. How many among the 40 neither swims nor run?

Solution:

Let A denote the set of swimmers and B the set of runners. Then $n(A \cup B) = n(A) + n(B) - n(A \cap B)$

$= 28 + 16 - 10 = 34$, meaning that there are 34 people that either swim or run.

Therefore, the number of people that neither swim nor run is $40 - 34 = 6$.

Ex9. Find the sum of all the integers from 1 to 100 such that integer is divisible by either 4 or 5.

Solution:

Sum of integers divisible by 4 is $S(4) = 4 + 8 + 12 + \ldots..100 = 4(1+2+3+4\ldots\ldots25) = 4 \times 25 \times 26/2 = 1300$.

Sum of integers divisible by 5 is $S(5) = 5+10+15+\ldots\ldots+100 = 5(1+2+\ldots+20) = 5 \times 20 \times 21/2 = 1050$.

Now if we say $S(4) + S(5)$ is the required sum then we must be wrong. Why?

In $S(4) + S(5)$ we counted 20, 40, 60,....100 twice so we have to remove this sum.

$\rightarrow$ Required Sum = S (4) + S (5) – S(L.C.M. of (4, 5)).

Where, S(L.C.M. of (4, 5)) = 20 (1 + 2 + 3 + 4 + 5) = 300. So required sum = 1300 + 1050 - 300 = 2050.

Ex10. If A and B are two sets containing 3 and 6 elements respectively, then find the minimum and maximum number of elements in the set $A \cup B$.

Solution:

$n(A \cup B) = n(A) + n(B) - n(A \cap B)$.

For $n(A \cup B)|_{MIN}$, $n(A \cap B)$ should be maximum, hence $n(A \cap B)|_{MAX} = 3$ (If $A \subseteq B$).

Hence, $n(A \cup B)|_{MIN} = 3 + 6 - 3 = 6$.

For $n(A \cup B)|_{MAX}$, $n(A \cap B)$ should be minimum, hence $n(A \cap B)|_{MIN} = 0$ (If $A \cap B = \phi$).

Hence, $n(A \cup B)|_{MAX} = 3 + 6 - 0 = 9$.

Here we have applied and learned the inclusion – exclusion principle on two sets. It can be applied on multiple sets also. Let's apply it for three sets. You need to apply your brain to get the proof.

THEOREM:

$n (A \cup B \cup C) = n(A) + n(B) + n(C) - n(A \cap B) - n(B \cap C) - n(A \cap C) + n(A \cap B \cap C)$.

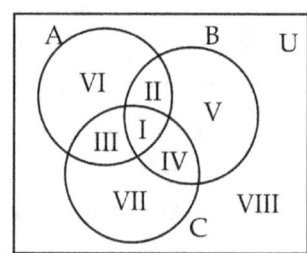

Proof: Here I have divided the regions and named as I, II, III,...., VIII.

So, $n(A \cup B \cup C) = n(A) + n(V) + n(IV) + n(VII)$.

$= n(A) + [n(V) + n(IV) + n(I) + n(II)] + n(VII) - n(I) - n(II)$.

$= n(A) + n(B) + n(VII) - n(I) - n(II)$.

$= n(A) + n(B) + [n(C) - n(I) - n(III) - n(IV)] - n(I) - n(II)$.

$= n(A) + n(B) + n(C) - [(n(I) + n(III)) + (n(IV) + n(I)) + n(II)]$.

$= n(A) + n(B) + n(C) - [n(A \cap C) + n(B \cap C) + (n(II) + n(I)) - n(I)]$.

$= n(A) + n(B) + n(C) - [n(A \cap C) + n(B \cap C) + n(A \cap B) - n(I)]$.

$= n(A) + n(B) + n(C) - [n(A \cap C) + n(B \cap C) + n(A \cap B) - n(A \cap B \cap C)]$.

$= n(A) + n(B) + n(C) - n(A \cap B) - n(B \cap C) - n(A \cap C) + n(A \cap B \cap C)$.

Note: This can be extends to more than 3 sets as well. Here I am writing the formula for the same.

(i) $\displaystyle\bigcup_{i=1}^{n} A_i = \sum_{i=1}^{n} A_i - \sum_{1 \le i < j \le n} A_i \cap A_j + \sum_{1 \le i < j < k \le n} A_i \cap A_j \cap A_k - \ldots\ldots + (-1)^{n-1}\left(\bigcap_{i=1}^{n} A_i\right)$.

(ii)   De Morgan Laws for multiple sets.

(a)   If $A_1, A_2, A_3, \ldots, A_n$ are multiple sets then, $A_1^C \cap A_2^C \cap A_3^C \cap \ldots \cap A_n^C = \left(A_1 \cup A_2 \cup \ldots \cup A_n\right)^C$.

In another words, $\displaystyle\bigcap_{i=1}^{n} A_i^C = \left(\bigcup_{i=1}^{n} A_i\right)^C$.

(b)   If $A_1, A_2, A_3, \ldots, A_n$ are multiple sets then, $\left(A_1 \cap A_2 \cap A_3 \cap \ldots \cap A_n\right)^C = \left(A_1^C \cup A_2^C \cup \ldots \cup A_n^C\right)$.

In another words, $\displaystyle\left(\bigcap_{i=1}^{n} A_i\right)^C = \bigcup_{i=1}^{n} A_i^C$.

(iii)   Maximum region generated by intersections of four sets = 15 as shown in figure.

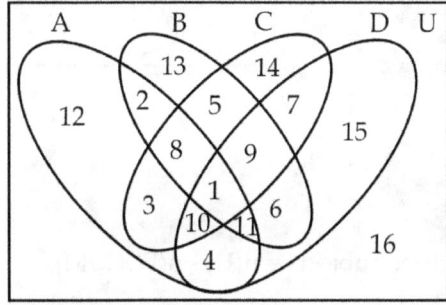

(iv)   Distributive law of multiple sets

(a)   If $A_1, A_2, A_3, \ldots, A_n$ are multiple sets then,

$A \cap (B_1 \cup B_2 \cup B_3 \cup \ldots \cup B_n) = (A \cap B_1) \cup (A \cap B_2) \cup (A \cap B_3) \cup \ldots\ldots\ldots \cup (A \cap B_n)$. i.e. $A \cap \left(\displaystyle\bigcup_{i=1}^{n} B_i\right) = \bigcup_{i=1}^{n} (A \cap B_i)$.

(b)   If $A_1, A_2, A_3, \ldots, A_n$ are multiple sets then,

$A \cup (B_1 \cap B_2 \cap B_3 \cap \ldots \cap B_n) = (A \cup B_1) \cap (A \cup B_2) \cap (A \cup B_3) \cap \ldots \ldots \cap (A \cup B_n)$. i.e. $A \cup \left(\displaystyle\bigcap_{i=1}^{n} B_i\right) = \bigcap_{i=1}^{n} (A \cup B_i)$

I knew your mind get tired! So let's have a mind refreshing problem. Just enjoy and feel learning.

PUZZLE # 2

If ABC × DEED = ABCABC; where A, B, C, D and E are different digits, what are the values of D and E?

PUZZLE # 3

Which of the following is/ are corrects?

(A) Only two options are correct

(B) C is correct

(C) B & D are correct

(D) At - least one is correct

Ex11. A Class has 175 students. Following is the description showing the number of students studying one or more of the following subjects in this class.

Mathematics 100, Physics 70, Chemistry 46, Mathematics and Physics 30, Mathematics and Chemistry 28, Physics and Chemistry 23, mathematics, Physics and Chemistry 18.

How many students are enrolled in Mathematics alone, Physics alone and Chemistry alone? Are there students who have not offered any of these three subjects?

Solution:

Consider P = Physics, C = Chemistry, M = Mathematics.

In such kind of problems, always start from P∩C∩M part. Doing so makes, problem easier.

n(P∩C∩M) = n(I) = 18, n(II) = n(P∩M) - n(P∩C∩M) = 12.

Similarly, n(III) = 23 – 18 = 5, n(IV) = 28 – 18 = 10.

Number of students enrolled in mathematics alone

= n(V) = n(M) – n(I) – n(II) – n(IV)

= 100 – 18 – 12 – 10 = 60.

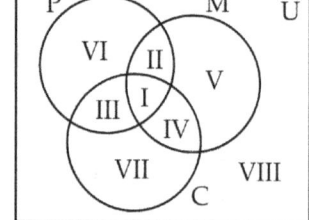

Similarly, Number of students enrolled in Physics alone = 70 – 18 – 12 – 5 = 35.

Number of students enrolled in chemistry alone = 46 – 18 – 5 – 10 = 13.

n(P∪C∪M) = 100 + 70 + 46 – 30 – 28 – 23 + 18 = 153.

Number of students who have not offered any of these three subjects = n(U) - n(P∪C∪M)

= 175 – 153 = 22.

Ex12. A college awarded 38 medals in football, 15 in basketball and 20 in cricket. If these medals went to a total of 58 men and only three men got medals in all the three sports, how many received medals in exactly two of the three sports?

Solution:

Consider the figure as shown in which F = Football, B = Basketball and C = Cricket.

So n(F) = 38, n(B) = 15, n(C) = 20, n(F∪B∪C) = 58 and n(F∩B∩C) = 3.

The number of medals in exactly two of the three sports = a + b + c.

a + b + c = a+ d + b + d + c + d – 3d

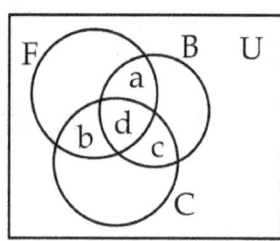

$= [n(F \cap B) + n(F \cap C) + n(B \cap C)] - 3\, n(F \cap B \cap C).$

$= [n(F) + n(B) + n(C) + n(F \cap B \cap C) - n(F \cup B \cup C)] - 3\, n(F \cap B \cap C).$

$= 38 + 15 + 20 + 3 - 58 - 3 \times 3 = 9.$ So, the number of medals in exactly two of the three sports = 9.

Ex13. In a survey of 25 students, it was found that 15 had taken mathematics, 12 had taken physics and 11 had taken chemistry, 5 had taken mathematics and chemistry, 9 had taken mathematics and physics, 4 had taken physics and chemistry and 3 had taken all three subjects. Find the number of students that had taken

(i) Only chemistry

(ii) only mathematics

(iii) Only physics

(iv) physics and chemistry but not mathematics

(v) Mathematics & physics but not chemistry

(vi) only one of the subjects

(vii) At least one of three subjects

(viii) none of three subjects.

Solution:

Consider the figure as shown. In such type of question always start from central region i.e. $P \cap C \cap M$ and then intersection of two sets and then single set.

Given: $n(M) = 15$, $n(P) = 12$, $n(C) = 11$, $n(M \cap C) = 5$, $n(M \cap P) = 9$, $n(P \cap C) = 4$, $n(P \cap C \cap M) = 3$.

Assume, Set P, C and M are initially vacant, then just Fill 3 in place of $P \cap C \cap M$. Now $n(M \cap C) = 5$ and $n(P \cap C \cap M) = 3$ so $n(P' \cap C \cap M) = 2$.

$n(M \cap P) = 9$ and $n(P \cap C \cap M) = 3$ so $n(P \cap C' \cap M) = 6$.

Similarly, fill the other parts and then you will get the figure as shown above. Now answer your questions.

The number of students that had taken

(i) Only chemistry = 5.   (ii) Only mathematics = 4.   (iii) Only physics = 2.

(iv) Physics and chemistry but not mathematics = 1.

(v) Mathematics and physics but not chemistry = 6.

(vi) Only one of the subjects = 2 + 5 + 4 = 11.

(vii) At least one of three subjects = $n(P \cup C \cup M) = n(M) + 5 + 1 + 2 = 23$ (Observe from Figure.)

(viii) None of three subjects = $n(U) - n(P \cup C \cup M) = 25 - 23 = 2.$

Ex14. The Expansion of $(1+x)^n = \sum_{r=0}^{n} {}^nC_r x^r = {}^nC_0 + {}^nC_1 x + {}^nC_2 x^2 + {}^nC_3 x^3 + \ldots\ldots + {}^nC_n x^n$. Where $n \in W$ and the number of terms in $(1+x)^n = n + 1$. Hence find the number of different terms in the expansion of $(1+x)^{150} + (1+x^2)^{100} + (1+x^3)^{90}$.

Solution:

Total number of term is equal to the total number of different power of x. so for $(1+x)^{150}$ contain x^0, x^1, x^2, ..., x^{150}. Let $A = \{x^0, x^1, x^2, ..., x^{150}\}$ or $A = \{0, 1, 2, 3, ..., 150\}$

Similarly for $(1+x^2)^{100}$, Let $B = \{0, 2, 4, 6, 8, ..., 200\}$ and for $(1+x^3)^{90}$

Let C = {0, 3, 6, 9, 12, ..., 270}. So Total number of term = n(A∪B∪C).Think! Why?

n(A∪B∪C) = n(A) + n(B) + n(C) – n(A∩B) – n(A∩C) – n(B∩C) + n(A∩B∩C).

n(A) = 151, n(B) = 101, n(C) = 91. A∩B = {0, 2, 4,.......150} so n(A∩B) = 76

A∩C = {0, 3, 6, 9, 12, ..., 150} so n(A∩C) = 51. B∩C = {0, 6, 12, 18,....., 198} so n(B∩C) = 34.

A∩B∩C = {0, 6, 12, 18,.......150} so n(A∩B∩C) = 26. Hence

n(A∪B∪C) = (151 + 101 + 91) – (76 + 34 + 51) + 26 = 208.

Ex15. Consider the following sets A = {(x, y): xy < 1, x, y ∈ R}, B = {(x, y): $x^2 + y^2 \le 5/2$, x, y ∈ R} &

C = {(x, y): y = 2x, x, y ∈ R}, then

(A)   Length of interval of set A∩B∩C = $\sqrt{10}$.

(B)   If P (a, b) is a ordered pair, where a, b ∈ I, then number of points P (a, b) ∈ A∩B∩C', are 6.

(C)   Let Q(a, b) is a member of set B, then maximum value of expression $a^2 + b^2 – 6a – 4b + 14$, is $(33/2) + \sqrt{130}$.

(D)   n (A∩B'∩C) = 2.

Solution:

Please take some time to solve the problem, only then you can get this solution. Observe and analyze the figure correctly.

Boundary Excluded

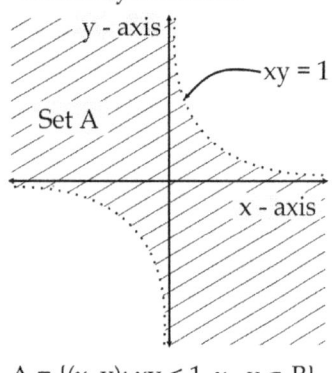

A = {(x, y): xy < 1, x , y ∈ R}

Boundary Included

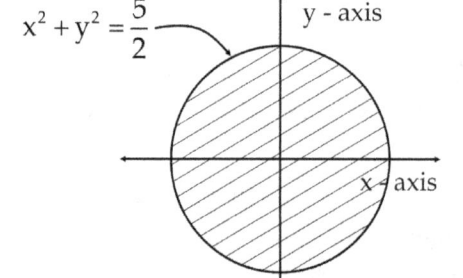

$B = \left\{ (x, y) : x^2 + y^2 \le \dfrac{5}{2}, x, y \in R \right\}$

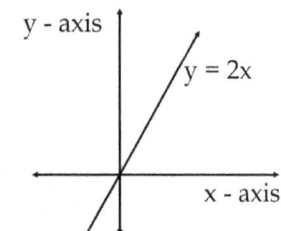

$C = \left\{ (x, y) : y = 2x, x, y \in R \right\}$

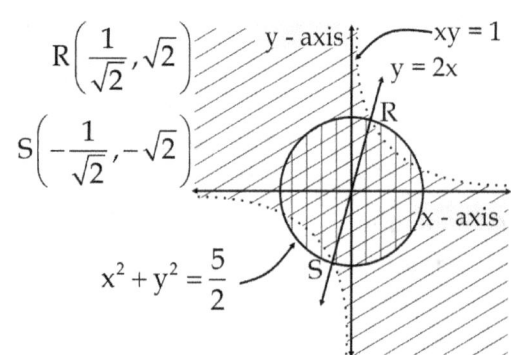

FIGURE – (A)

(A)   A∩B∩C is the line segment RS. Hence RS = $\sqrt{10}$. Where, R and S are points of intersection of curves as shown in above figure (A).

(B)   The region A∩B∩C' is as shown in figure right side.

So total points P(a, b), a, b ∈ I, satisfy A∩B∩C' are 6, as shown in figure.

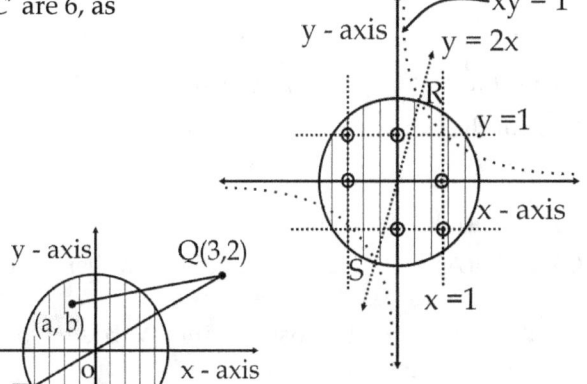

(C)   Consider the figure as shown.

$a^2 + b^2 - 6a - 4b + 14$

$= (a - 3)^2 + (b - 2)^2 + 1$

Max $(a^2 + b^2 - 6a - 4b + 14)$

$= $ Max $((a - 3)^2 + (b - 2)^2 + 1)$

$= QT^2 + 1$

$= (OQ + OT)^2 + 1$

$= (33/2) + \sqrt{130}$.

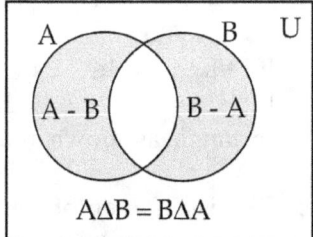

(D)   n (A∩B'∩C) = 0.

Hence correct answers are A, B and C.

SYMMETRIC DIFFERENCE OF TWO SETS (Symbol Δ)

The topic is not in IIT JEE syllabus but for knowledge, I am explaining it.

If A and B are any two sets then symmetric difference AΔB = BΔA = (A - B) ∪ (B - A).

AΔB is represented by shaded region as shown in figure.

You can observe AΔB = (A∪B) – (A∩B).

Ex1.   If A = {0, 1, 3, 5, 6, 7} and B = {2, 4, 5, 7, 8, 9}. Find AΔB.

Ans.   AΔB = {0, 2, 2, 3, 4, 6, 8, 9}.

Ex2.   If A = {x: 0 < x ≤ 5, x∈R} and B = {0, 1, 2, 3, 4, 6}, then find the symmetric difference of set's A and B.

Solution:

A = (0, 5] and B = {0, 1, 2, 3, 4, 6}.

A∪B = (0, 5] ∪ {6} and A∩B = {1, 2, 3, 4}.

AΔB = (A∪B) – (A∩B). So

AΔB = (0, 5] ∪ {6} – {1, 2, 3, 4}.

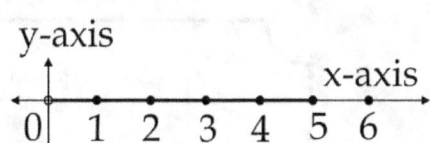

CARTESIAN CROSS PRODUCT (Symbol ×)

Let A and B are two non empty sets then, the Cartesian product of A and B is denoted by A×B and is defined as set of all the ordered pair (a, b) such that a∈A and b∈B.

Thus, A×B = {(a, b): ∀ a∈A and b∈B}. If A =∅ and B = and B = ∅ then A×B = ∅.

Note: In general, (a, b) ≠ (b, a) if (a, b) is a ordered pair and in the case of unordered pair (a, b) = (b, a). If ordered pair (a, b) = (b, a) → a = b.

Ex1. Let A = {1, 2, 3} and B = {a, e}. Find A×B and B×A.
Solution:

A×B = {(1, a), (1, e), (2, a), (2, e), (3, a), (3, e)}.

B×A = {(a, 1), (a, 2), (a, 3), (e, 1), (e, 2), (e, 3)}.

Look at the figure shown for A×B.

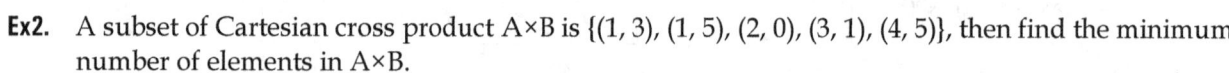

Note: (1) If A×B = B×A → A = B.

(2) If n(A) = n and n(B) = m, then n(A×B) = n(B×A) = nm.

(3) If n(A) = n and n(B) = m, then number of subsets of A×B = 2^{mn}.

Ex2. A subset of Cartesian cross product A×B is {(1, 3), (1, 5), (2, 0), (3, 1), (4, 5)}, then find the minimum number of elements in A×B.
Solution:

For minimum number of elements in A×B, the set A = {1, 2, 3, 4} and B = {0, 1, 3, 5}. Hence, minimum number of elements in A×B = n(A) × n(B) = 4×4 = 16.

Ex3. Find the value of x and y if ordered pair (sin x, √2) = (1, cosec y).
Solution:

Ordered pairs (a, b) = (c, d) → a = c and b = d. So

Sin x = 1 → x = nπ + (- 1)^n π/2 where n∈ I and √2 = cosec y → sin y = 1/√2 = sin (π/4) so → y = mπ + (- 1)^m π/4.

Ex4. If I ={x: x∈R, 1≤ x ≤ 4}.
J = {x: x∈R, 0 ≤ x ≤ 1}, then I × J consist of all the points P(a, b) of rectangle as shown in the Cartesian co-ordinate plane.

Note: If A is a non-empty set such that n(A) = n, then n(A×A×A) = n^3. Where A×A×A = {(a, b, c): ∀ a, b, c ∈ A}. It will represent points in rectangular Cartesian co-ordinate system.

Let's refresh your mind. Have a puzzle.

PUZZLE # 4

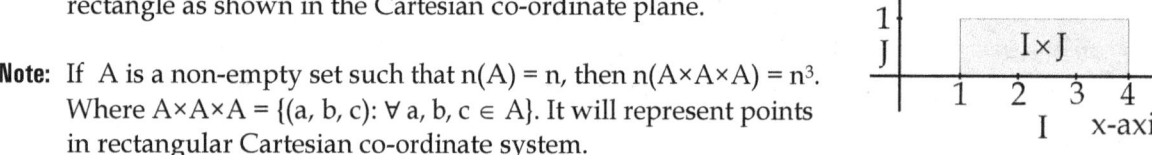

The letters A to I represent the numbers 1 to 9 in some order. Each letter represents a different number such that

• A + B + C > D + E + F > G + H + I

• E is a prime factor of G

• F > A & B + G = H , I ≠ 1

Number ABC DEFGHI?

Author: Ramesh Chandra   **Amazon / Flipkart Book :** Permutation and Combinations

Here we are having an exercise, so solve and test your learning.

EXERCISE - 3

1. In a class of 120 students numbered 1 to 120, all even numbered students opt for Physics, whose numbers are divisible by 5 opt for Chemistry and those whose numbers are divisible by 7 opt for Math. How many opt for none of the three subjects?
   (A) 19   (B) 41   (C) 21   (D) 57

2. If set $A = \{(a, b): b = e^a, a \in R\}$ and $B = \{(a, a): a \in R\}$, then
   (A) $B \subset A$ (B) $A \subset B$ (C) $A \cup B = A$ (D) $A \cap B = \phi$

3. Let $A = \{x : x$ is a multiple of 3, $x \in N\}$ and $B = \{x : x$ is a multiple of 5, $x \in N\}$. Then $A \cap B$ is given by
   (A) $\{3, 6, 9, \ldots\ldots\}$ (B) $\{5, 10, 15, 20, \ldots\}$ (C) $\{15, 30, 45, \ldots\ldots\}$ (D) ϕ

4. In a class 40% of the students enrolled for Math and 70% enrolled for Economics. If 15% of the students enrolled for both Math and Economics, what % of the students of the class did not enroll for either of the two subjects?
   (A) 5%   (B) 15%   (C) 0%   (D) 25%

5. In a class of 40 students, 12 enrolled for both English and German. 22 enrolled for German. If the students of the class enrolled for at least one of the two subjects, then how many students enrolled for only English and not German?
   (A) 30   (B) 10   (C) 18   (D) 28

6. Find the number of triples (A, B, C) where A, B, C are subset of $\{1, 2, 3, \ldots\ldots, n\}$ such that $A \cap B \cap C = \emptyset$, $A \cap B \neq \emptyset$, $B \cap C \neq \emptyset$.
   (A) $7^n + 2 \cdot 6^n - 5^n$ (B) $7^n - 2 \cdot 6^n + 5^n$ (C) $7^n - 6^n - 5^n$ (D) $7^n + 2 \cdot 6^n + 5^n$

7. If S is a set with 10 elements and $A = \{(x, y): x, y \in S, x \neq y\}$, then the number of elements in A is?
   (A) 100   (B) 90   (C) 50   (D) 45

8. $\left\{ x : \dfrac{2x-1}{x^3 + 4x^2 + 3x} \in R \right\}$ Equals
   (A) $R - \{0\}$ (B) $R - \{0, 1, 3\}$ (C) $R - \{0, -1, -3\}$ (D) $R - \{0, -1, -3, 1/2\}$

9. If set A and B are defined as $A = \{(x, y): y = 1/x, 0 \neq x \in R\}$, $B = \{(x, y): y = -x, x \in R\}$, then
   (A) $A \cap B = A$ (B) $A \cap B = B$ (C) $A \cap B = \phi$ (D) None of these.

10. In a school, 224 played cricket, 240 played hockey and 336 played basketball. Of the total, 64 played basketball and hockey; 80 played cricket and basketball and 40 played cricket and hockey; 24 played all the three games. The number of boys who play cricket but not hockey is
    (A) 128   (B) 184   (C) 160   (D) 216

11. Let $X = \{1, 2, 3, 4, 5\}$. The number of different ordered pairs (Y, Z) that can formed such that $Y \subseteq X$, $Z \subseteq X$ and $Y \cap Z$ is empty, is

    (A) 5^2 (B) 3^5 (C) 2^5 (D) 5^3

12. If P, Q and R are subsets of a set A, then $R \times (P^C \cup Q^C)^C$ equals.

    (A) $(R \times P) \cap (R \times Q)$ (B) $(R \times P) \cup (R \times Q)$ (C) $(P \times R) \cap (Q \times R)$ (D) $(P \times R) \cup (R \times Q)$

13. Let $A = \{\theta : \sin\theta = \tan\theta, \theta \in R\}$ and $B = \{\theta : \cos\theta = 1, \theta \in R\}$ be two sets, then

    (A) $A = B$ (B) $A \not\subseteq B$ (C) $B \not\subseteq A$ (D) $A \subset B$ and $B - A \neq \phi$

14. Of the 200 candidates who were interviewed for a position at a call center, 100 had a two-wheeler, 70 had a credit card and 140 had a mobile phone. 40 of them had both, a two-wheeler and a credit card, 30 had both, a credit card and a mobile phone and 60 had both, a two wheeler and mobile phone and 10 had all three. How many candidates had none of the three?

    (A) 0   (B) 20   (C) 10   (D) 18

15. Let S is the set of points inside the square, T is the set of points inside the triangle and C is the set of points inside the circle. If the triangle and circle intersect each other and are contained in the square, then incorrect option is

    (A) $S \cap T \cap C \neq \phi$ (B) $S \cup T \cup C = C$ (C) $S \cup T \cup C = S$ (D) $S \cup T = S \cup C$

16. Let $S = \{1, 2, 3, ..., n\}$ and $A = \{(a, b) \mid 1 \leq a, b \leq n\} = S \times S$. A subset B of A is said to be a good subset if $(x, x) \in B$ for every $x \in S$. Then the number of good subsets of A is:

    (A) 1   (B) 2^n (C) $2^{n(n-1)}$ (D) 2^{n^2}

Note: If you have new thinking, write here

❖   .

❖   .

❖   .

❖   .

SOME SOLVED EXAMPLES

Ex1. Consider set A = {x: x∈R, -1 < x ≤ 2 or 3 ≤ x < 6} and B = {x: x∈R, - 3 ≤ x < 1 or 2 < x ≤ 4 or x = 7}. Find the following sets

(i) A∩B   (ii) A∪B   (iii) A – B   (iv) B - A

Solution:

Look at the sets A and B as shown in figure. You can find the following.

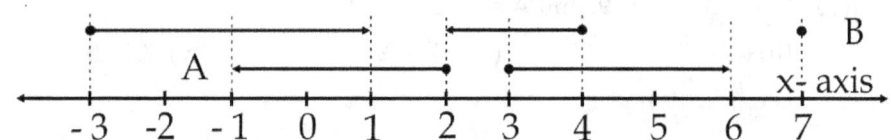

(i) A∩B = (- 1, 1) ∪ [3, 4].   (ii) A∪B = [- 3, 6] ∪ {7}

(iii) A – B = [1, 2] ∪ (4, 6)   (iv) B – A = [- 3, - 1] ∪ (2, 3) ∪ {7}.

Ex2. Consider A = {0, 1, 2, 3,...,9} and B = {a, e, i, o, u}, then find the number of subset of A×B.

Solution: n(A) = 10 and n(B) = 5 so n(A×B) = 10 × 5 = 50. So number of subsets = $2^{10 \times 5} = 2^{50}$.

Ex3. In a certain city, only two newspapers A and B are published. It is known that 25% of the city population reads A and 20% reads B, while 8% read both A and B. It is also known that 30% of those who read A but not B, look into advertisements and 40% of those who read B but not A, look into advertisements, while 50% of those who read both A and B, look into advertisements. What % of the population read in advertisement?

Solution: Let's the population of city be P.

n(A) = 0.25 P, n(B) = 0.2 P, n(A∩B) = 0.08 P.

No. of people look into advertisement

= 30% n(I) + 40% n(III) + 50 % n(II)

= 0.3 × (0.25 – 0.08)P + 0.4 (0.2 – 0.08) P + 0.5 (0.08)P = 0.139 P

So % = 0.139P ×100/ P = 13.9 %.

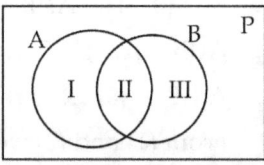

Ex4. There are 20 students in a chemistry class and 30 students in a physics class. Find the number of students who are either in physics class or in chemistry class.

(i) When the two classes meet at different hours and 10 students are enrolled in both the courses.

(ii) Two classes meet at the same hour.

Solution:

n(C) = 20 and n(P) = 30, n(C∪P) = ?

(i) n(C∩P) = 10 so n(C∪P) = 20 + 30 – 10 = 40.

(ii) n(C∩P) = φ so n(C∪P) = 20 + 30 – 0 = 50.

Ex5. **Matrix-Match**

    (A) $A = \{x: x \in R, x^2 + 9 = 9\}$ (p) subset of integers

    (B) $B = \{x: x \in R, x^2 - 4x + 4 = 0\}$ (q) Singleton set

    (C) $C = \{y: y = 2x + 3, x \in N\}$ (r) Finite set

    (D) $D = \{\theta: \sin\theta + \cos\theta = \sqrt{2}\}$ (s) Infinite set

Solution: $A \to p, q, r$; $B \to p, q, r$; $C \to p, s$ $D \to s$

Ex6. If universal set $S = \{0, 1, 2, 3, 4, 5, 6, 7, 8, 9\}$ and $A = \{1, 2, 3, 4\}$, $B = \{2, 3, 5, 6\}$, $C = \{2, 3, 7\}$ then find

    (i) A' (ii) $(A - B)'$ (iii) $B' - A'$ (iv) $A' \cap B$

    (v) $A \cup B'$ (vi) $(A \cap C)'$

Solution:

    (i) $A' = \{0, 5, 6, 7, 8, 9\}$.   (ii) $A - B = \{1, 4\}$ so $(A - B)' = \{0, 2, 3, 5, 6, 7, 8, 9\}$.

    (iii) $B' = \{0, 1, 4, 7, 8, 9\}$, so $B' - A' = \{1, 4\}$.   (iv) $A' \cap B = \{5, 6\}$.

    (v) $A \cup B' = \{0, 1, 2, 3, 4, 7, 8, 9\}$.   (vi) $(A \cap C)' = \{0, 1, 4, 5, 6, 7, 8, 9\}$.

Ex7. If $aN = \{ax: x \in N\}$, describe the set $3N \cap 7N$.

Solution:

    $3N = \{3, 6, 9, 12, 15, \ldots\ldots\}$ and $7N = \{7, 14, 21, 28, \ldots\ldots..\}$ so $3N \cap 7N = \{21, 42, 63, 84, \ldots..\} = 21N$.

Ex8. If universal set $U = \{0, 2, 3, 5, 7, 9\}$, $A = \{3, 7\}$ & $B = \{2, 5, 7, 9\}$, then find $A \cap B'$, $B \cap A'$ and verify the De Morgan's Law $(A \cup B)' = A' \cap B'$.

Solution:

    $B' = \{0, 3\}$ so $A \cap B' = \{3\}$. $A' = \{0, 2, 5, 9\}$ so $B \cap A' = \{2, 5, 9\}$.

    $A \cup B = \{2, 3, 5, 7, 9\}$ so $(A \cup B)' = \{0\}$ ……………….(1)

    $A' \cap B' = \{0\}$………………….. (2)

    From (1) and (2) we can say $(A \cup B)' = A' \cap B'$.

Ex9. In each of the following, determine whether the statement is true or false. If it is true, prove it and if it's false, give an example.

    (i) If $A \subseteq B$ and $x \notin B$, then $x \notin A$.   (ii) If $x \in A$ and $A \not\subset B$, then $x \in B$.

Solution:

    (i) Consider the figure as shown. If $A \subseteq B \to \forall a \in A \to a \in B$.

        But, if $x \notin B \to x \notin A \because A \subseteq B$, so statement is TRUE.

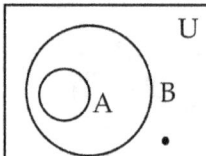

    (ii) Let $A = \{1, 2, 3, 4\}$ and $B = \{3, 4, 5\}$ you can see $1 \in A$, $A \not\subset B$ and $1 \notin B$, so statement is FALSE.

Ex10. Let $A = \{1, 2, 3, 4\}$, $B = \{1, 2, 3\}$ and $C = \{2, 4\}$. Find all sets X satisfying each pair of conditions:

    (i) $X \subseteq A$, $X \subseteq B$ and $X \subseteq C$ (ii) $X \subseteq A$ and $X \not\subset C$

Solution:

(i) let $x \in X$, Since $X \subseteq A$, $X \subseteq B$, $X \subseteq C$ so $x \in A$, $x \in B$, $x \in C$. So exhaustive set $X = \{2\}$.

Hence X can be $\phi = \{\}$ or $\{2\}$

(ii) $X \subseteq A$ and $X \not\subset C$, for this exhaustive set $X = \{1, 2, 3, 4\}$ but $X \neq \{2, 4\}$, $X \neq \{2\}$, $X \neq \{4\}$, $X \neq \phi = \{\}$.

So, possibilities of X are

    (a) $\{1, 2, 3, 4\}$ (b) $\{1, 2, 3\}$, $\{1, 2, 4\}$, $\{1, 3, 4\}$, $\{2, 3, 4\}$

    (c) $\{1, 2\}$, $\{1, 3\}$, $\{1, 4\}$, $\{2, 3\}$, $\{3, 4\}$ (d) $\{1\}$, $\{3\}$

Ex11. Let $A = \{1, 2\}$ and $B = \{1, 2, 3, 4\}$, $C = \{5, 6\}$ and $D = \{5, 6, 7, 8\}$. Verify that

    (i) $A \times (B \cap C) = (A \times B) \cap (A \times C)$ (ii) $A \times C$ is a subset of $B \times D$.

Solution: (i) $B \cap C = \{3\}$.

    $A \times (B \cap C) = \{(1, 3), (2, 3)\}$ ....(1)

    $A \times B = \{(1, 1), (1, 2), (1, 3), (1, 4), (2, 1), (2, 2), (2, 3), (2, 4)\}$.

    $A \times C = \{(1, 3), (1, 5), (1, 6), (2, 3), (2, 5), (2, 6)\}$.

    $(A \times B) \cap (A \times C) = \{(1, 3), (2, 3)\}$...................(2)

    From (1) and (2) $A \times (B \cap C) = (A \times B) \cap (A \times C)$.

    (ii) $A \subseteq C$ & $C \subseteq D$ Hence $A \times C \subseteq B \times D$

Ex 12: Find the number of 6 letter word form by exactly 3 letter of set $\{A, B, C, D, E, F\}$

Solution:

Select 3 letters in $^6C_3$ ways. Suppose you have been selected A, B, C from these 6 letters and making 6 letter word. Let

X = A do not appear in 6 letter word.

Y = B do not appear in 6 letter word.

Z = C do not appear in 6 letter word.

$n(X) = 2^6$, $n(Y) = 2^6$, $n(Z) = 2^6$.

$n(X \cap Y) = 1^6$, $n(Y \cap Z) = 1^6$, $n(X \cap Z) = 1^6$, $n(X \cap Y \cap Z) = 0^6$.

$n(A \cup B \cup C) = n(A) + n(B) + n(C) - n(A \cap B) - n(B \cap C) - n(C \cap A) + n(A \cap B \cap C)$.

$n(A \cup B \cup C) = 3 \times 2^6 - 3 \times 1^6 + 0^6 = 3 \times 63 = 189$.

So Total number of required ways = $^6C_3 \times (3^6 - 189) = 10800$.

Ex13. The set $S = \{1, 2, 3, ..., 12\}$ is to be partitioned into three sets A, B, C of equal size.

Thus $A \cup B \cup C = S$, $A \cap B = B \cap C = A \cap C = \emptyset$. The number of ways to partition S is

    (A) $\dfrac{12!}{(4!)^3}$ (B) $\dfrac{12!}{(3!)^4}$ (C) $\dfrac{12!}{3!(4!)^3}$ (D) $\dfrac{12!}{(3!)^5}$

Solution: Number of ways is $^{12}C_4 \times {}^8C_4 \times {}^4C_4$. So A, is answer.

Ex14. Assume that $M = \{(x, y) \mid x^2 + 2y^2 = 3\}$, and $N = \{(x, y) \mid y = mx + b\}$. If $M \cap N \neq \emptyset \; \forall \; m \in R$, then b takes values

(A) $\left[-\dfrac{\sqrt{6}}{2}, \dfrac{\sqrt{6}}{2}\right]$ (B) $\left(-\dfrac{\sqrt{6}}{2}, \dfrac{\sqrt{6}}{2}\right)$ (C) $\left(-\dfrac{\sqrt{6}}{2}, \dfrac{\sqrt{6}}{2}\right]$ (D) $\left[-\dfrac{2}{\sqrt{3}}, \dfrac{2}{\sqrt{3}}\right]$

Solution:

For any $m \in R$, we have $M \cap N \neq \emptyset$, which means point $(0, b)$ is on or inside the ellipse $x^2 + 2y^2 = 3$.

Therefore, $2b^2 \leq 3 \rightarrow b \in \left[-\dfrac{\sqrt{6}}{2}, \dfrac{\sqrt{6}}{2}\right]$ so option (A) is correct.

Ex15. The number of ordered pairs (m, n), where $m, n \in \{1, 2, 3, \ldots, 50\}$, such that $6^m + 9^n$ is a multiple of 5, is

(A) 1250   (B) 2500   (C) 625   (D) 500

Solution:

For being multiple of 5, unit place must be either 0 or 1.

We need to observe some pattern followed by unit digits in 6^m and 9^n.

6^1......6(Unit place), 6^2...... 6, 6^3.........6.

Here you can observe that 6^m unit digit will be 6 $\forall \; m \in \{1, 2, 3, \ldots, 50\}$.

Similarly, 9^1.....9, 9^2.......1, 9^3.........9, 9^4.......1. unit digit is following periodicity of 2. So,

$6^m + 9^n$ is a multiple of 5 only if $n \in \{1, 3, 5, 7, \ldots, 49\}$.

Hence, Number of ordered pairs $(m, n) = 50 \times 25 = 1250$, so option (A) is correct

Ex16. Let $A = \{\theta \in R \mid \cos^2(\sin\theta) + \sin^2(\cos\theta) = 1\}$ and $B = \{\theta \in R \mid \cos(\sin\theta)\sin(\cos\theta) = 0\}$. Then $A \cap B$

(A) is the empty set   (B) has exactly one element

(C) has more than one finitely many elements   (D) has infinite many elements

Solution:

$\cos^2(\sin\theta) + \sin^2(\cos\theta) = 1$, possible only if $\sin\theta = \cos\theta \rightarrow \tan\theta = 1 \rightarrow \theta = n\pi + \pi/4$, but for these θ $\cos(\sin\theta)\sin(\cos\theta) \neq 0$. Hence, $A \cap B = \emptyset$. So, option (A) is correct.

Ex17. At a certain conference of 100 people, there are 29 Indian women and 23 Indian men. Of these Indian people 4 are doctors and 24 are either men or doctors. There are no foreign doctors. How many foreigners or women doctors are attending the conference?

Solution:

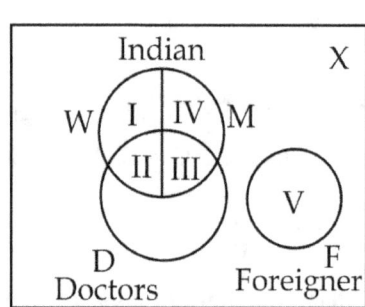

Consider the question figure as shown

$n(\text{Indian}) = 29 + 23 = 52$. $n(X) = 100$ so $n(F) = n(X) - n(\text{Indian})$ i.e. $n(F) = 100 - 52 = 48$. $n(I_W) = n(I) + n(II) = 29$.

$n(I_M) = n(IV) + n(III) = 23$. As per question $n(II) + n(III) = 4$ and $n(IV) + n(II) + n(III) = 24$ so $n(IV) = 20 \rightarrow n(III) = 23 - 20 = 3$.

$\rightarrow n(II) = 1, n(I) = 29 - 1 = 28$

So, Number of foreigners or women doctors are attending the seminar $= n(F) + n(II) = 48 + 1 = 49$.

EXERCISE - HINT AND SOLUTIONS

PUZZLE # 1 : 2 (your foots)+ 10 (soldiers foots) + 500 (lions foots) + 2000 (lionesses foots) = 2512 foots.

PUZZLE# 2 : D = 1 and E = 0.   PUZZLE# 3 : B, C, D.   PUZZLE# 4 : 719628453

ANSWER KEY EXERCISE - 1

1. (i) $\{x: (x-1)(x-\pi) = 0, x \in R\}$ (ii) $\{x: x = n\pi, n \in W\}$

   (iii) $\{a : a = a_i$, where $i = 0, 1, 2, 3, \ldots 100, 101$ & $a_k = 2^k$ for $k \neq 101$ & $a_{101} = 0\}$.

   (iv) $\{x: x = (2n-1) \times (2^n), n \in N\}$

   (v) $\left\{ x : x = \begin{vmatrix} 1 & b_i \\ 0 & a_i \end{vmatrix}, \text{ where } i \in N, a_1 = 1, a_2 = 3, a_{n+1} = a_n + a_{n-1} \ \forall n \geq 2 \ \& \ b_1 = 0, b_j = 2^{a_{j-1}} \ \forall j \geq 2 \right\}$

   (vi) $\{z: z^4 - 1 = 0, z \in C\}$ (vii) $\left\{ x : x = f(n) = \begin{vmatrix} 2^n & \text{if } n = \text{even} \\ 3n+1 & \text{if } n = \text{odd} \end{vmatrix}, \ n \in W \right\}$

2. (i) $\left\{ \begin{vmatrix} 2 \\ 3 \end{vmatrix}, \begin{vmatrix} 3 \\ 6 \end{vmatrix}, \begin{vmatrix} 4 \\ 11 \end{vmatrix}, \begin{vmatrix} 5 \\ 20 \end{vmatrix}, \ldots \right\}$ (ii) $\{(4, 3), (3, 4), (0, 5), (-4, 3), (-3, 4)\}$

   (iii) $\{(2, 1)\}$ (iv) $\{(0, 0, 0), (0, 1, 0), (0, 0, 2), (0, 1, 2)\}$

   (v) $\{1+\sqrt{2}, 1 - \sqrt{2}\}$ (vi) $\{-27, -8, -1, 4, 9\}$

ANSWER KEY EXERCISE - 2

1. (i) FALSE   (ii) FALSE   (iii) TRUE

   (iv) FALSE   (v) TRUE

2. (i) FALSE   (ii) TRUE   (iii) FALSE

   (iv) TRUE   (v) TRUE   (vi) FALSE

   (vii) FALSE   (viii) TRUE   (ix) FALSE

   (x) TRUE

3. (i) FALSE   (ii) TRUE   (iii) TRUE

   (iv) TRUE   (v) FALSE

4. $P(A) = \{\{\}, \{\{\phi\}\}\}$ $P(B) = \{\{\}, \{\phi\}, \{\{\phi\}\}, \{\phi, \{\phi\}\}\}$

   $P(C) = \{\{1\}, \{2\}, \{\{1, 3\}\}, \{1, 2\}, \{1, \{1, 3\}\}, \{2, \{1, 3\}\}, \{1, 2, \{1, 3\}\}\}$.

ANSWER KEY – EXERCISE 3

Q. NO.	ANSWER	Q. NO.	ANSWER	Q. NO.	ANSWER
1	B	6	B	11	B
2	D	7	B	12	A
3	C	8	C	13	B
4	A	9	C	14	C
5	C	10	B	15	B

HINTS / SOLUTIONS – EXERCISE · 1

1. (i) 1 and π are any two random chosen number so we can have any equation having 1 and π as only roots. So one form I have as, $\{x: (x-1)(x-\pi) = 0, x \in R\}$.

(iii) You can see, the numbers are in pattern like 2^n, but there is one problem of zero so we have defined it in another way as shown, $\{a : a = a_i$, where $i = 0, 1, 2, 3, \ldots 100, 101$ & $a_k = 2^k$ for $k \neq 101$ & $a_{101} = 0\}$.

(iv) Here you can see $1, 3, 5, 7, 9, \ldots$; are following the pattern $(2n-1)$ and $2, 4, 8, 16, \ldots$; are following the pattern 2^n hence we have $\{x: x = (2n-1) \times (2^n), n \in N\}$ as answer.

(v) The set is collection of 2×2 matrices of type $\begin{vmatrix} 1 & b \\ 0 & a \end{vmatrix}$, in which 'a' and 'b' are changing.

You can observe the series, $1, 3, 4, 7, 11, 18, \ldots$ we needs to assign as $a_1 = 1$ and $a_2 = 3$ and rests are following the pattern $a_{n+1} = a_n + a_{n-1}$. Similarly you can observe the behaviour of b as shown in answer.

$$\left\{ x : x = \begin{bmatrix} 1 & b_i \\ 0 & a_i \end{bmatrix}, \text{ where } i \in N, a_1 = 1, a_2 = 3, a_{n+1} = a_n + a_{n-1} \forall n \geq 2 \,\& \, b_1 = 0, b_j = 2^{a_{j-1}} \forall j \geq 2 \right\}$$

(vi) $z^4 - 1 = (z^2 - 1)(z^2 + 1) = (z-1)(z+1)(z^2+1) = 0 \rightarrow z = 1, -1, \pm\sqrt{(-1)}$ ie $\pm$ i.

(vii) The set is combination of two series, one following the rule $3n + 1$ if $n = $ odd and 2^n if n is even.

2. (i) $a_{ij} = i^n + nj \rightarrow a_{11} = 1 + n$ and $a_{21} = 2^n + n$ for the matrix A_n.

Hence, the answer is $\left\{ \begin{bmatrix} 2 \\ 3 \end{bmatrix}, \begin{bmatrix} 3 \\ 6 \end{bmatrix}, \begin{bmatrix} 4 \\ 11 \end{bmatrix}, \begin{bmatrix} 5 \\ 20 \end{bmatrix}, \ldots \right\}$.

(iv) $x^2 + |y^2 - y| + |2z - z^2| = 0$, Here you can observe the sum of three non – negative number is zero, hence its possible only if individual parts $x^2 = 0, |y^2 - y| = 0, |2z - z^2| = 0$.

(vi) min$\{a, b\}$ = a only if a < b for particular x, as 'a' and 'b' both are related to x.

min$\{a, b\}$ = b, $\forall x \in (-4, 0)$ and min$\{a, b\}$ = a, $\forall x \in (1, 4)$. Min $\{0, 0\}$ = $\emptyset$. Ans:$\{-27, -8, -1, 4, 9\}$

HINTS / SOLUTIONS – EXERCISE · 3

1. B

$S = \{1, 2, 3, 4, \ldots, 120\}$, $P = \{2, 4, 6, 8, \ldots, 120\}$, $C = \{5, 10, 15, 20, \ldots, 120\}$, $M = \{7, 14, 21, \ldots, 119\}$.

Where S = Students, P = Students opting Physics, C = Students opting Chemistry, M = Students opting Mathematics. So, n(S) = 120, n(P) = 60, n(C) = 24, n(M) = 17.

$P \cap C = \{10, 20, 30, \ldots, 120\}$, so n$(P \cap C)$ = 12, $P \cap M = \{14, 28, 42, \ldots, 112\}$, so n$(P \cap M)$ = 8.

$C \cap M = \{35, 70, 105, 140, 175\}$ so n$(C \cap M)$ = 5. $P \cap C \cap M = \{70\}$, so n$(P \cap C \cap M)$ = 1.

Hence, n$(P \cup M \cup C)$ = 60 + 24 + 17 – 12 – 8 – 5 + 1 = 78. Hence, No. of students opt for none of the three subjects = n$(P \cup M \cup C)'$ = n(s) - n$(P \cup M \cup C)$ = 120 – 78 = 41.

2. D

    The elements of set A lies on the curve $y = e^x$ and that of the set B, lies on the line $y = x$.

    You can observe that there is no any intersection of these two curves; hence there is no any common point.

    → $A \cap B = \phi$ is correct option.

    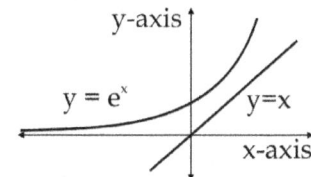

3. C

4. A

    Let the Number of students in the class = x. So, $n(M) = 0.4x$, $n(E) = 0.7x$, $n(M \cap E) = 0.15x$.

    Hence, $n(M \cup E) = 0.4x + 0.7x – 0.15x = 0.95x$.

    $n((M \cup E)') = 0.05x$ = No of students of the class did not enroll for either of the two subjects.

    % $n(M \cup E)' = (0.05x \times 100)/x = 5\%$.

5. C

    $n(U) = n(E \cup G) = 40$ (Since, the students of the class enrolled for at-least one of the two subjects)

    $n(E - G)$ = No. of students enrolled for only English and not German.

    $= n(E \cup G) – n(G) = 40 – 22 = 18$.

6. B

    I know you may be thinking of different curve as shown here. Can you guess why I made this type of figure?

    Consider the figure as shown, here $A \cap B \cap C = \phi$.

    Region I and II must have at-least one element ($A \cap B \neq \emptyset$, $B \cap C \neq \emptyset$).

    As per conditions, there are 7 partitions of universal set U, in total. If we want to fill the elements in these regions then there are 7 choices to fill, for each element.

    Hence, total number of ways to fill = 7^n, but this contains the following situations as well. $n(I) = n(A \cap B) = 0$, $n(II) = 0$. So, number of triplets (A, B, C) in which $n(I) = 0$, are = 6^n.

    Number of triplets (A, B, C) in which $n(II) = 0$, are = 6^n.

    Number of triplets (A, B, C) in which $n(I) = 0$ & $n(II) = 0$, are = 5^n.

    Number of triplets in which at least one of (I or II) is vacant = $2 \times 6^n - 5^n$.

    Hence, Number of required triplets = $7^n – (2 \times 6^n - 5^n) = 7^n – 2 \cdot 6^n + 5^n$.

7. B

    Let $S = \{x_1, x_2, x_3, \ldots., x_{10}\}$ and $x, y \in S$.

    x has 10 choices then y has 9 choices ($\because x \neq y$). So, number of elements in A = $10 \times 9 = 90$.

8. C

    $$\frac{2x-1}{x^3 + 4x^2 + 3x} = \frac{2x-1}{x(x+3)(x+1)}$$ is real, if $x \in R - \{0, -1, -3\}$.

9. C

    There no any intersections and hence no any common point. So $A \cap B = \phi$.

    Look at the figure shown below.

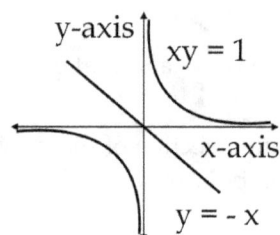

10. **B**

    Consider the figure as shown. H = Hockey, C = Cricket and B = Basketball.

    $n(C \cap H') = n(C) - n(C \cap H)$

    $n(C \cap H') = 224 - 40 = 184$

    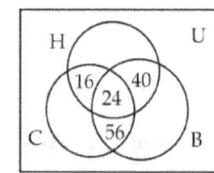

11. **B**

    Consider the figure as shown as per situation $Y \subseteq X$, $Z \subseteq X$ & $Y \cap Z = \phi$.

    $n(X) = 5$. Each of element of set X has three choices to be filled, either in Y or in Z or in $(Y' \cap Z') \cap X$.

    So, number of different ordered pairs (Y, Z) = $3 \times 3 \times 3 \times 3 \times 3 = 3^5$.

    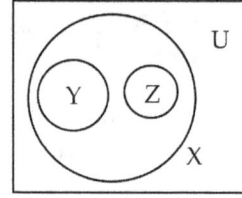

12. **A**

    $(P^C \cup Q^C)^C = P \cap Q$. Hence, $R \times (P \cap Q) = (R \times P) \cap (R \times Q)$.

13. **B**

    $Sin\theta = tan\theta \rightarrow sin\theta (1 - 1/cos\theta) = 0$. $Cos\theta \neq 0$, $cos\theta = 1$, $sin\theta = 0$.

    From here, you can observe that $B \subset A$ and $B - A = \phi$. So $A \not\subset B$ is the correct choice.

14. **C**

    Consider and think about the figure. How will you start to get the figure?

    $n(T \cup C \cup M) = (100 + 70 + 140) - (40 + 30 + 60) + 10 = 190$

    $n(U) = 120$, so number of candidates had none of the three

    $= n(U) - n(T \cup C \cup M) = 200 - 190 = 10$.

    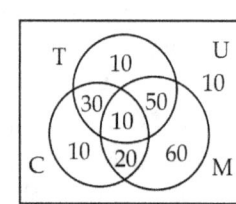

15. **B**

16. **C**

    $\forall x \in S$, $(x, x) \in A$, the remaining elements of set $A = n^2 - n = n(n - 1)$. So (x, x) must be in set B and remaining elements of set A has 2 choices of inclusion (i.e. in or out) in set B. hence number of such subsets = $2^{n(n-1)}$.

NOTES MAKING

HOW TO MAKE NOTES?

When you read the book, you will have some thinking and questions in your mind. There might be situation that you want to discuss to your teachers. Even you may have a better solution to particular question then write it for your future. You can discuss your thinking at pncbyrc@gmail.com

Que / Page No.	NOTES

POINTS TO REMEMBER:

❖ .

❖ .

❖ .

❖ .

❖ .

OTHERS: --

NUMBER THEORY

"Mathematics is the most beautiful and most powerful creation of the human spirit"

Stefan Banach *(Polish Mathematician)*

ABOUT NUMBER THEORY

Think, if we do not have numbers in our life, what may be the possible situations and what about the technology? At initial stage of development of numbers people used to exchange goods. Whatever are we having today, is the work of thousands of scientist, thinkers, Mathematics lovers. They seen beauty in Mathematics and play with numbers and develop the way of communication in mathematics.

Let's see the development of digit "1" over time.

Many older typewriters do not have a separate symbol for 1 and use the lowercase "/" instead.

After getting Zero (0) from Aryabhatta, we started Binary number system (Base 2, only 0, 1. it's revolute in computer technology). Decimal system (Base 10, digits used 0, 1, 2, 3, 4, 5, 6, 7, 8, 9). After getting Zero, We get "Whole Numbers". '0' fills the vacancies in Natural number that we initially have. It gave answers to many questions to scientist.

Now a days, we are at very far- far better situation, even we are developing numbers at its system. You may be heard of developments of mathematical constant π (Pie) i.e. an irrational number approximated as 22/7 or 3.14, Pi has one million digits. Scientist used to calculate value of it to know computers calculation speed. I am writing here up to 55 digits.

$$\pi = 3.1415926535897932384626433832795028841971693993751058.........$$

Even we have another mathematical constant "e" i.e. Euler Number, a base of natural logarithm.

$$e = \sum_{n=0}^{\infty} \frac{1}{n!} = 1 + \frac{1}{1} + \frac{1}{1 \cdot 2} + \frac{1}{1 \cdot 2 \cdot 3} + ... = 2.718281828459045235360287471352662497757724709369995.......$$

"476 AD–550 AD, an Indian taught the world. How to count? Aryabhatta invented Zero"

NATURAL NUMBERS (N)

Natural numbers are those used for counting. The numbers, 1, 2, 3, 4, 5....... are called natural numbers. Its set is denoted by N, thus N = {1, 2, 3, 4, 5,..........}.

Note: Some scientist assumes '0' as natural number. Now a day's its common in set theorist & logicians.

As per piano axioms:

(1)   '0' is natural number.

(2)   Every natural number has successor.

(3)   0 is not the successor of any natural numbers.

As per our current competitive exams, do not use '0' as natural number until unless it's said in exam. We also uses I⁺, Z⁺ symbols to represents natural numbers.

WHOLE NUMBERS (W)

The numbers 0, 1, 2, 3,...... are whole numbers. There is no fractional or decimal part and no negatives. Its set represented by W, hence W = {0, 1, 2, 3, 4, 5, 6,.........}. It's also called set of non-negative integers.

INTEGERS (Z or I)

An integer (A Latin word Integer means "whole") is a number that can be written without fractional component. The concept of additive inverse, expand the whole numbers into integers. This is represented by symbol 'Z', now a days we use 'I' as well to represent the same. These are $0, \pm 1, \pm 2, \pm 3,.....$ Hence we have $Z = I = \{0, \pm 1, \pm 2, \pm 3, \pm 4, \pm 5, ...\}$.

Z is a totally ordered set without upper or lower bound. The ordering of Z is given by:

$$\ldots\ldots -3 < -2 < -1 < 0 < 1 < 2 < 3 < \ldots\ldots$$

An integer is positive if it is greater than zero and negative if it is less than zero. Zero is defined as neither negative nor positive.

The ordering of integers is compatible with the algebraic operations in the following way:

(i)   if $a < b$ and $c < d$, then $a + c < b + d$

(ii)   if $a < b$ and $c > 0$, then $ac < bc$.

(iii)   if $a < b$ and $c < 0$, then $ac > bc$.

Ex1. If $a > b$ and $c > d$ then select the correct statement which is true always

(a) $ad + bc < ac + bd$ (b) $ac > bd$

(c) $ac > bd$ if $bd > 0$ (d) $a - c > b - d$

Solution: $a - b > 0$ and $c - d > 0 \rightarrow (a - b)(c - d) > 0 \rightarrow ac + bd > bc + ad$. Hence, (A) is the only answer.

Ex2. The integer 1, 2, 3, 4,......., 40 are written on a blackboard. The following operation is then repeated 39 times. In each repetition, any two numbers, say a and b, currently on the blackboard are erased and a new number a + b - 1 is written. What will be the number left on the board at the end?

(A) 820   (B) 821   (C) 781   (D) 819

Solution: The sum $1 + 2 + 3 + 4 + ... + 40 = 20(1 + 40) = 820$.

In first operation the sum of number on board reduced to 819. Hence in its 39th operation the number on the board will remain 820 - 39 = 781. Hence (C) is correct option.

PRIME / COMPOSITE NUMBERS

A prime number (or a prime) is a natural number greater than 1 that has no positive divisors other than 1 and itself. A natural number greater than 1 that is not a prime number is called a composite number.

Ex. 2, 3, 5, 7,..... etc are prime numbers. These numbers are having 1 and the number itself as divisors.

Here is the list of prime number less than 500.

2, 3, 5, 7, 11, 13, 17, 19, 23, 29, 31, 37, 41, 43, 47, 53, 59, 61, 67, 71, 73, 79, 83, 89, 97, 101, 103, 107, 109, 113, 127, 131, 137, 139, 149, 151, 157, 163, 167, 173, 179, 181, 191, 193, 197, 199, 211, 223, 227, 229, 233, 239, 241, 251, 257, 263, 269, 271, 277, 281, 283, 293, 307, 311, 313, 317, 331, 337, 347, 349, 353, 359, 367, 373, 379, 383, 389, 397, 401, 409, 419, 421, 431, 433, 439, 443, 449, 457, 461, 463, 467, 479, 487, 491, 499.

FUNDAMENTAL THEOREM OF ARITHMETIC

Every composite number can be expressed (factorised) as a product of primes, and this factorisation is unique, apart from the order in which the prime factors occur.

Ex. $120 = 2 \times 2 \times 2 \times 3 \times 5 = 2 \times 2 \times 3 \times 2 \times 5 = 3 \times 5 \times 2^3 = 2^3 \times 5 \times 3$

If we arrange them in ascending order of prime numbers then this factorisation is unique.

❖   Let a composite number $N = p_1 \times p_2 \times p_3 \times..........\times p_n$, where prime numbers $p_1 \leq p_2 \leq p_3 \leq....\leq p_n$. If we arrange them in ascending order then $N = p_1 \times p_2 \times p_3 \times........\times p_n$ is the unique prime factorisation.

EUCLID'S DIVISION LEMMA

Euclid's Division lemma states that for given positive integers 'a' and 'b', there exist unique integers 'q' and 'r' satisfying $a = b \times q + r$, $0 \leq r < b$.

Ex.   Let a = 18 & b = 7 then $18 = 7 \times 2 + 4$. Here we get 2 = quotient, 4 = remainder on dividing 18 by 7.

We use this lemma to get HCF of natural numbers.

Euclid's division lemma and algorithm are so closely interlinked that people often call former as the division algorithm also. Although Euclid's Division Algorithm is stated for only positive integers, it can be extended for all integers except zero ($b \neq 0$). i.e.

INTEGER DIVISION: *If a & d are integers, with d non-zero, it can be proven that there exist unique integers q and r, such that a = qd + r and $0 \leq r < |d|$. The number q is called the quotient, while r is called the remainder.*

NUMBER OF PRIME NUMBERS

The number of prime, are infinite. These are endless.
Many scientists proof the statement. Let's have Euclid's proof.

Proof: Think! You have prime numbers $p_1, p_2, p_3,...., p_n$ only, in this world. i.e. prime numbers are finite. Now think of a number $N = p_1 \times p_2 \times p_3 \times....\times p_n + 1$. You can see, N is not divisible by given existing prime numbers $p_1, p_2, p_3,..., p_n$. This means N is also a prime Number. Hence, prime numbers are infinite.

Note: (1)   The prime counting function π(n) is defined as the number of primes $\leq n \approx n/\ln(n)$ (approx).

(2)   Euler noted that the function $n^2 + n + 41$ yields prime numbers for $0 \leq n < 40$, a fact leading into deep algebraic number theory.

(3)   If N > 3, is a prime number then it would be 6n +1 or 6n − 1 form but converse is not true, where n∈ Natural numbers. As (6n + 2, 6n + 3, 6n + 4, 6n are composite numbers, so can't be prime). The statement is verified for first 10 lakh prime numbers.

(4)   Some scientist assumes '1' as a prime number, as it satisfies the definition but it fails fundamental theorem. Let a composite number be $6 = 2 \times 3 = 1 \times 2 \times 3$, which violet the uniqueness of prime factorisation. So, it's a dispute in theory.

(5)   '2' is the only even prime number.

(6)   (2, 3) is the only pair of prime number having difference '1'.

(7)   The property of being prime (or not) is called primality. To check a given number is prime or not, we have very simple and slow method called trial division. It consists of testing whether n is a multiple of any integer between 2 and $\sqrt{n}$.

(8)   Fermat primes are of the form $F_k = 2^{(2^k)} + 1$. However F_5 is composite but rest are primes.

PRIME NUMBER TESTING

Ex1.   Check the primality of the number 97.

Solution:

$\sqrt{97} \approx 9.84$, so prime numbers between 2 and 9 are 3, 5, 7 and none of them divide 97 hence the number 97 is prime.

Ex2.   Find the HCF of 30 and 75 using Euclid division lemma.

Solution:

Euclid division lemma is an algorithm that we needs to apply again and again till getting remainder '0'.

Here 75 > 30 so its algorithm starts as shown in steps.

Step 1: $75 = 30 \times 2 + 15$.

Step 2: $30 = 15 \times 2 + 0$.

Hence HCF (75, 30) = 15.

Ex3.   Find the HCF of 5 and 73 using Euclid division lemma.

Sol.   *Observe the algorithm for Euclid division. Develop relation from figure as shown.*

Step 1: $73 = 5 \times 14 + 3$.

Step 2: $5 = 3 \times 1 + 2$.

Step 3: $3 = 2 \times 1 + 1$.

Step 4: $1 = 1 \times 1 + 0$.

So HCF (5, 73) is 1.

You may have seen the method in lower classes.

step 1: $73 = 5 \times 14 + 3$.

step 2: $5 = 3 \times 1 + 2$.

step 3: $3 = 2 \times 1 + 1$.

step 4: $2 = 1 \times 2 + 0$.

Ex4.   Find positive integers x and y that satisfy both xy = 40 and 31 = 2x + 3y.

Solution:

Since 40 have the unique factorization $40 = 2^3 \times 5$, there are only 8 possibilities for the pair (x, y). These are (1, 40), (2, 20), (4, 10), (8, 5), (5, 8), (10, 4), (20, 2), (40, 1).

Only (x, y) = (8, 5) additionally satisfies 31 = 2x + 3y.

Ex5.   Suppose that 1998 is written as a product of two positive integers whose difference is as small as possible. What is this difference?

(A) 18   (B) 15   (C) 17   (D) 47

Solution:

The prime decomposition of 1998 is $1998 = 2 \times 3^3 \times 37$. The factorization of 1998 into two positive integers whose difference is as small as possible is $1998 = 37 \times 54$ and the difference is $54 - 37 = 17$.

Hence (C) is correct option

Ex6. Find all positive integers n such that $2^8 + 2^{11} + 2^n$ is a perfect square.

Solution:

Suppose that k is an integer such that $k^2 = 2^8 + 2^{11} + 2^n = 2304 + 2^n = 48^2 + 2^n$. Then

$k^2 - 48^2 = (k - 48)(k + 48) = 2^n$. For being this be true, we must have $k - 48 = 2^s$ & $k + 48 = 2^t$, where $s + t = n$. But then $2^t - 2^s = 48 - (-48) = 96 = 3 \times 2^5$, so $2^t - 2^s = 3 \times 2^5$

$\rightarrow 2^s (2^{t-s} - 1) = 3 \times 2^5$. For this to be true, the only possible case is $s = 5$ and $t - s = 2$.

So $s + t = n = 12$. Thus, the only natural number n such that $2^8 + 2^{11} + 2^n$ is a perfect square is $n = 12$.

Ex7. If $a^2 + 2b = 7$, $b^2 + 4c = -7$ and $c^2 + 6a = -14$, then the value of $a^2 + b^2 + c^2$ is

(a) 14   (b) 25   (c) 36   (d) 47

Solution: Add all the three we get

$a^2 + 2b + b^2 + 4c + c^2 + 6a = -14 \rightarrow (a + 3)^2 + (b + 1)^2 + (c + 2)^2 = 0$

$\rightarrow a = -3, b = -1, c = -2 \rightarrow a^2 + b^2 + c^2 = 14$.

EVEN / ODD INTEGERS

If we divide any integer by 2 then remainder may be either 0 or 1. If it's '0' then it's even integer else its odd integer.

❖ Even integers are of form 2k, where $k \in I$.

❖ Odd integers are of form 2k + 1, where $k \in I$.

Look at the figure of number line.

Even Integers: $0, \pm 2, \pm 4, \pm 6, \ldots\ldots\ldots$

Odd Integers: $\pm 1, \pm 3, \pm 5, \pm 7, \ldots\ldots\ldots$

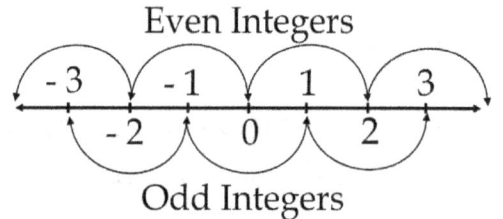

Ex8. Find the smallest positive integer n such that n/2 is a perfect square, n/3 is a perfect cube, and n/5 is a perfect fifth power.

Solution:

Since n is divisible by 2, 3, and 5, we may assume that it has the form $n = 2^a 3^b 5^c$. Then

$n/2 = 2^{a-1} 3^b 5^c$ $n/3 = 2^a 3^{b-1} 5^c$ $n/5 = 2^a 3^b 5^{c-1}$

Since n/2 must be a perfect square, $a - 1$, b, and c must all be even.

Since n/3 is a perfect cube, a, $b - 1$, and c must all be multiples of 3.

Since n/5 is a perfect fifth power, a, b, and $c - 1$ must all be multiples of 5.

The smallest values that satisfy these conditions are $a = 15$, $b = 10$, and $c = 6$.

Thus, $n = 2^{15} 3^{10} 5^6$ is the smallest such positive integer.

Ex9. If D = a² + b² + c² where a and b are consecutive, Integers and c = ab, then $\sqrt{D}$ is

(a) Always an even integer   (b) always an odd integer

(c) An integer, odd or even   (d) some times an irrational number

Solution:

Let b > a, hence b = a + 1, since these are consecutive integers, so D = a² + (a + 1)² + a² (a + 1)²

$$= a^4 + 2a^3 + 3a^2 + 2a + 1 = a^2\left(\left(a^2 + \frac{1}{a^2}\right) + 2\left(a + \frac{1}{a}\right) + 3\right) = a^2\left(\left(a + \frac{1}{a}\right)^2 + 2\left(a + \frac{1}{a}\right) + 1\right).$$

$$\rightarrow D = a^2\left(a + \frac{1}{a} + 1\right)^2 = (a(a+1) + 1)^2.$$ As, a(a+1)= even integer, so a(a+1)+1 = $\sqrt{D}$ = must be odd

integer always. Hence (B) is correct option.

Ex10. Four distinct integers p, q, r and s are chosen from the set {1, 2, 3, 4, 5,........, 16, 17}. Find the minimum possible value of $\frac{p}{q} + \frac{r}{s}$.

Solution:

The minimum is achieved when the numerator are as small as possible and denominator as large as possible, so {p, r} = {1, 2} and {q, s} = {16, 17}. Hence Minimum value of $\frac{p}{q} + \frac{r}{s} = \frac{1}{16} + \frac{2}{17} = \frac{49}{272}$.

Here you can see, $\frac{1}{16} + \frac{2}{17} < \frac{2}{16} + \frac{1}{17} \rightarrow \frac{1}{17} < \frac{1}{16}$.

CO – PRIME NUMBERS

The two natural numbers a and b are said to be co – prime if HCF (a, b) = 1. Like 4 and 5 are co-prime. 8 and 9 are co – prime. 5 and 7 are co – prime. These numbers are relatively primes.

TWIN PRIME NUMBERS

If the difference between two prime is two then they called twin primes.

Here are some twin primes pairs: (3, 5), (5, 7), (11, 13), (17, 19), (29, 31), (41, 43), (59, 61). You can get other too from the list of primes.

RATIONAL NUMBERS (Q)

If p and q are any two integers such that q ≠ 0 and H.C.F. (|p|, |q|) = 1, then a number p/q is called a rational number. For the case p = 0, q ≠ 0, the number 0/q = 0 is a rational number.

Here you can see 0 = 0 × 1 × 2 × 3 and 12 = 1× 2 × 2 × 3 so H.C.F. (0, 12) may be 1, 2, 3, 12, which should be unique. Hence we do not take H.C.F. of any number with 0.

IN OTHER WORDS, WE CAN SAY

a rational number is any number that can be expressed as the quotient or fraction p/q of two integers, a numerator p and a non-zero denominator q. Since q may be equal to 1, every integer is a rational number; hence '0' is also a rational number.

The set of rational numbers is denoted by Q, so Q = {p/q: p, q∈I, q ≠ 0 and p/q}.

Note:

❖ 2/4 is an equivalence rational number of 1/2.

❖ There exist infinite rational numbers between any two rational numbers.

❖ The rational numbers are either integers or mixed fraction. If it represented in decimals then it's has finite significant digits or recurring.

❖ Addition, subtraction and multiplication of two rational numbers is a rational number but division fails the property, as 2/0 = Not define.

❖ Every terminating decimal rational number can be written as $\dfrac{I}{2^n \times 5^m}$, where I, n, m $\in$ Z.

RATIONAL NUMBERS AND THEIR DECIMAL REPRESENTATION

To study the topic, let's have some decimal representation of rational numbers. Observe the table.

NO.	TERMINATING DECIMALS	NO.	NON-TERMINATING DECIMALS
$\dfrac{3}{2}$	1.5	$\dfrac{1}{2 \times 3}$	$0.1666666........ = 0.1\overline{6}$
$\dfrac{7}{2^3 5}$	0.175	$\dfrac{3}{5 \times 7}$	$0.0\overline{857142}$
$\dfrac{11}{2^4 5^3}$	0.0055	$\dfrac{1}{11}$	$0.\overline{09}$
$\dfrac{17}{5^2}$	0.68	$\dfrac{1}{2^2 \times 3}$	$0.08\overline{3}$
$\dfrac{41}{2^3}$	5.125	$\dfrac{1}{5 \times 3}$	$0.0\overline{6}$

THEOREM - 1:

Let x be a rational number whose decimal expansion terminates. Then x can be expressed in the form p/q. Where p and q are co - prime, and the prime factorisation of q is of the form $2^m \times 5^n$, where m, n are non-negative integers.

THEOREM - 2:

Let x = p/q be a rational number, such that the prime factorisation of q is of the form $2^m \times 5^n$, where m, n are non - negative integers. Then x has a decimal expansion which terminates.

THEOREM - 3:

Let x = p/q be a rational number, such that the prime factorisation of q is not of the form $2^m \times 5^n$, where n, m are non-negative integers. Then, x has a decimal expansion which is non – terminating, repeating (recurring).

Note: From above three theorems, we can concludes that every rational number is terminating or non – terminating, recurring.

Before moving on Irrational Number, we have a theorem on prime factorisation. Please have a look.

THEOREM: Let p be a prime number. If p divides a^2, then p divide a, where 'a' is a positive integer.

PROOF: Let the prime factorisation of a be as follow: a = $p_1 \times p_2 \times p_3 \times p_4 \times ..... \times p_n$, where $p_1, p_2, p_3, p_4, ..., p_n$ are primes, not necessarily distinct.

Therefore, $a^2 = (p_1 \times p_2 \times p_3 \times p_4 \times ..... \times p_n)(p_1 \times p_2 \times p_3 \times p_4 \times ..... \times p_n) = p_1{}^2 \times p_2{}^2 \times p_3{}^2 \times .... \times p_n{}^2$.

If p divide a^2, so p may be one of $p_1, p_2, p_3, p_4, ..., p_n \rightarrow$ p divide a.

IRRATIONAL NUMBERS (Q^c)

The first proof of the existence of irrational numbers is usually attributed to a Pythagorean, who probably discovered them while identifying sides of the pentagram. He started with an isosceles right angled triangle having sides length a, a, b ∈ N (relatively prime or having no common factor). Look, how he proceeded.....

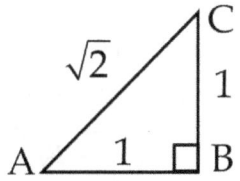

(i)   Ratio of sides is b : a which is a rational number, as a, b ∈ N.

(ii) $2a^2 = b^2$...(using Pythagoras theorem).

(iii) from (ii) we get, 'b' is an even integer. So let b = 2k, hence $a^2 = 2 k^2$.

(iv) from (iii) we get, 'a' to be an even integer → both 'a' and 'b' are even integers, which contradict our assumption of having no common factor, this means b : a must be a number other than rational. Hence, invented irrational numbers

Definition: Irrational number is a real number, that can't be express as the ratio of two integers.

Ex.   Proof that $\sqrt{2}$ is an irrational number.

Solution:

Let $\sqrt{2}$ is a rational number → $\sqrt{2} = p/q$, where p, q ∈ I and q ≠ 0 having no common factor.

On squaring both-side, we get $2 q^2 = p^2$ → 'p' is an even integer. Let p = 2 k, k∈I.

Hence, $2 q^2 = 4k^2$ → $q^2 = 2k^2$ → q is also even integer. Hence p and q both have a common factor 2, which contradict our assumption. So $\sqrt{2}$ is not a rational number but real.

Real numbers are composed of rational and irrational numbers; hence $\sqrt{2}$ is an irrational number.

Note: (i) $\sqrt{2} \approx 1.4142135623730950488016887242096980785696718753769480731766797973799$........

(ii) $\sqrt{2}^{\sqrt{2}^{\sqrt{2}^{\sqrt{2}^{......\infty}}}} = 2$. It's is assumed that $\sqrt{2}$ is the first irrational number discovered by Pythagoras.

(iii) Decimal expansion of every irrational number is non – terminating, non recursive.

It's unending decimal.

(iv) $\pi \neq \dfrac{22}{7}$ but $\pi \approx \dfrac{22}{7}$. Archimedes approximation $\dfrac{223}{71} < \pi < \dfrac{22}{7}$

(v)   Irrational numbers are infinite.

(vi) e, π, e + 1, 2 – π, e^{π}, π^e etc are some transcendental numbers, which can't be obtained as root of any non zero polynomial equation. Every transcendental number is irrational but converse is not true. as 1 - $\sqrt{2}$, $\sqrt{3}$ are not a transcendental number as it can be obtained from $x^2 – 2 x – 1 = 0$.

One question may arise here, that how Pythagoras comes to know about non-terminating, non-recursive nature of decimal expansion of $\sqrt{2}$? Well, the answer is not that simple as we understand! It uses paradoxes, computation, different experiments to prove the nature. People were used to calculate it by machine which gave unending non - recursive decimal results.

RADICALS/ SURDS & EXPONENTS ($\sqrt[n]{x}$)

Radicals, surds and exponents are similar and interrelated.

The expression $\sqrt[n]{x}$ is called a radical where Index 'n' called order of radical. The sign $\sqrt{}$ called radical and the number under it i.e. x is called radicand.

$$\sqrt[n]{x} = (x)^{\frac{1}{n}}$$

If we think of $(x)^{1/n}$ then 1/n is called the exponent of 'x' or power of 'x'.

If 'x' is a rational number and $n \in N$ such that $\sqrt[n]{x}$ is an irrational number, then $\sqrt[n]{x}$ is called the surd of the order 'n'. Let's differentiate them using some examples in note section.

Note: (i) $\sqrt{3}, \sqrt[3]{4}, \sqrt{7}, \sqrt[4]{6}, \sqrt{13}$ are pure surds and each are irrational numbers.

(ii) $5\sqrt{3}, 7\sqrt[3]{4}, 3\sqrt[4]{5}, 7 + \sqrt{3}$ is example of mixed surds.

(iii) $\sqrt{2} + \sqrt[3]{3}, \sqrt{2} - \sqrt{5}, \sqrt{3} - \sqrt{13}$ are example of compound surds but $5 - \sqrt{3}$ is not.

(iv) $\pi, \sqrt{5-\sqrt{3}}, \sqrt{e^\pi - 1}$ are irrationals but not surds.

(v) $\sqrt[3]{8} = 2, \sqrt{16} = 4$ are not surds but written under radicals.

(vi) $\sqrt{7+4\sqrt{3}} = 2+\sqrt{3}, \sqrt{5-2\sqrt{6}} = |\sqrt{2}-\sqrt{3}| = \sqrt{3}-\sqrt{2}$ are surds.

CRITICAL THINKING

❖ If we think of sign $\sqrt[n]{(...)}$ then it's a radical and have unique value or principle roots.

Ex. $\sqrt{4} = 2$ and $\sqrt[3]{(-8)} = -2$.

The fact that $x = a^{1/n}$ solve $x^n = a$ (where $n \in N$, $a \geq 0$) follow from nothing that

$$x^n = \underbrace{a^{1/n} \times a^{1/n} \times a^{1/n} \times ....\times a^{1/n}}_{n-times} = a^{\left(\frac{1}{n}+\frac{1}{n}+\frac{1}{n}+....+\frac{1}{n}\right)} = a^{\frac{n}{n}} = a^1 = a.$$

It behave like a real valued function $f:[0,\infty) \to [0,\infty)$, $f(x) = \sqrt[n]{x}$ where n = Even Natural No.

and $f: R \to R$, $f(x) = \sqrt[n]{x}$ when n = Odd Natural Number.

ie. $\sqrt{-4} = N.D.$

❖ If we think of Demoivre's Theorem

First Statement: If n = Integer then

$$re^{ix} = r(\cos(x) + i\sin(x))$$

$$\to (re^{ix})^n = r^n(\cos(x) + i\sin(x))^n = r^n(\cos(nx) + i\sin(nx))$$ i.e. unique value.

Second Statement: If n = Integer $\to 1/n \in$ Rational (Q) then

$(re^{ix})^{1/n} = r^{1/n}(\cos(x) + i\sin(x))^{1/n}$ has 'n' values and know as nth roots of $z = re^{ix}$ and calculate by

the formula $(r(\cos(x) + i\sin(x)))^{1/n} = r^{1/n}\left(\cos\left(\frac{2k\pi + x}{n}\right) + i\sin\left(\frac{2k\pi + x}{n}\right)\right)$ where $k \in \{0, 1, 2,.., n-1\}$

Ex. Cube root of 1 has three values $1, \omega, \omega^2$ where $\omega = \frac{-1+i\sqrt{3}}{2} = e^{i2\pi/3}$.

Similarly Square root of 4 has two values 2 and – 2 as per demoivre's but $\sqrt{4} = 2 \neq -2$ only since $\sqrt{()}$ is a radical sign and must have one value. So, writing $sqrt(1) \neq \sqrt{1}$.

Hope you were able to get the concept of surds. It's difficult to bind surds under single definition. You need to understand it by heart.

Laws of exponents / Surds (whenever defined) Holds good for a ≥ 0, b≥ 0

(i) $a^n b^n = (ab)^n \rightarrow \sqrt[n]{a} \times \sqrt[n]{b} = \sqrt[n]{ab}$.

(ii) $((a)^m)^n = ((a)^n)^m = a^{mn} \rightarrow \left(\sqrt[n]{a}\right)^n = \sqrt[n]{a^n} = a$.

(iii) $\sqrt[n]{\sqrt[m]{a}} = \sqrt[m]{\sqrt[n]{a}} = \sqrt[mn]{a}$.

(iv) $\sqrt[n]{a} = \sqrt[nm]{a^m}$

(v) $a^m \times a^n = a^{m+n}$

(vi) $a^m \div a^n = \dfrac{a^m}{a^n} = a^{m-n}$

(vii) $\left(a^m\right)^n = \left(a^n\right)^m = a^{mn}$

(viii) $\left(\dfrac{a}{b}\right)^{-m/n} = \left(\dfrac{b}{a}\right)^{m/n}$

(ix) $\sqrt[n]{\dfrac{a}{b}} = \dfrac{\sqrt[n]{a}}{\sqrt[n]{b}}$ whenever defined

(x) $\sqrt[n]{\sqrt[m]{\left(a^k\right)^m}} = \sqrt[n]{a^k} = \sqrt[nm]{a^{km}}$

(xi) $\left(\sqrt{a} + \sqrt{b}\right)^2 = a + b + 2\sqrt{ab}$

(xii) $\left(\sqrt{a} - \sqrt{b}\right)^2 = a + b - 2\sqrt{ab}$

You can see and observed that above rule work well if a > 0, n, m ∈ N but you need to be very careful in using these rules for a < 0. ie negative real numbers.

SOME FAILURE :

❖ $(a\,b)^x = (a)^x (b)^x$ fails if we think of $\{(-1)\times(-1)\}^{1/2} = (-1)^{1/2} \times (-1)^{1/2} \rightarrow 1 = i \times i \rightarrow 1 = -1$ where i = √-1.

❖ $i = \sqrt{-1} = (-1)^{1/2} = \left(\dfrac{1}{-1}\right)^{1/2} = \dfrac{(1)^{1/2}}{(-1)^{1/2}} = \dfrac{1}{i} = -i$.

So you needs to be very alert and precautious.

Ex1. Select the correct statement in perspective of complex numbers.

(A) $\sqrt{4} = \sqrt{(-2)^2} = -2$

(B) $\sqrt{a} \times \sqrt{b} = \sqrt{ab} \; \forall a, b \in R$

(C) $\sqrt{x^2} = x \; \forall x \in R$

(D) $\sqrt[3]{-8} = -2$ and $\sqrt[2]{0} = $N.D.

Solution:

(A) 4 = 2² = (- 2)² but √x is always a positive real number and defined for x ≥ 0. So √4 must be equal to 2. Even $\sqrt{x^2} = |x| \; \forall x \in R$ so $\sqrt{4} = \sqrt{(-2)^2} = |-2| = 2$. Hence options (A) and (C) are incorrect.

If we think of demoivre's theorem then it says √4 = ±2 but for real numbers √4 = 2.

(B) √2 × √3 = √(2×3)= √6, √2 × √- 3 = √(2× (- 3))= $\sqrt{-6}$ which is ≠ $-\sqrt{6}$.

$\sqrt{-2} \times \sqrt{-3} \neq \sqrt{(-2)\times(-3)}$ or √6 but $\sqrt{-2} \times \sqrt{-3} = -\sqrt{6}$.

Hence, $\sqrt{a} \times \sqrt{b} = \sqrt{ab}$ true sometime, not always ∴ option (B) is incorrect.

(D) $(-8)^{1/3} = \left((-2)^3\right)^{1/3} = -2$. $\sqrt[2]{0} = 0^{-2} = \dfrac{1}{0^2} = $ N.D. So, option (D) is correct.

Note: *Have your opinion and discuss the problem with your friends, teachers etc to get the concept.*

Ex2. Which one is an irrational number?

(A) $\sqrt{4+2\sqrt{3}} - \sqrt{4-2\sqrt{3}}$

(B) $\sqrt{1+(219567)(219569)}$

(C) $\sqrt{2+\sqrt{2+\sqrt{2+........\infty}}}$

(D) $\dfrac{1}{1+\sqrt{2}} + \dfrac{1}{\sqrt{2}+\sqrt{3}} + \dfrac{1}{\sqrt{3}+\sqrt{4}} + ... + \dfrac{1}{\sqrt{15}+\sqrt{14}}$

Solution:

(A) $\sqrt{4+2\sqrt{3}} = \sqrt{\left(1+\sqrt{3}\right)^2} = 1+\sqrt{3}$ and $\sqrt{4-2\sqrt{3}} = \sqrt{\left(1-\sqrt{3}\right)^2} = \left|\sqrt{3}-1\right| = \sqrt{3}-1$ so

$\sqrt{4+2\sqrt{3}} - \sqrt{4-2\sqrt{3}} = \left(1+\sqrt{3}\right) - \left(\sqrt{3}-1\right) = 2$

(B) $\sqrt{1+(n-1)(n+1)} = \sqrt{n^2} = |n|$ so we can say $\sqrt{1+(219567)(219569)} = 219568$

(C) $y = \sqrt{2+y} \rightarrow y^2 = 2+y \Rightarrow y = 2, -1$ but $y \neq 1$ so y=2.

(D) $\dfrac{1}{1+\sqrt{2}} + \dfrac{1}{\sqrt{3}+\sqrt{2}} + \dfrac{1}{\sqrt{4}+\sqrt{3}} + ... + \dfrac{1}{\sqrt{15}+\sqrt{14}}$

$\Rightarrow \left(\sqrt{2}-1\right) + \left(\sqrt{3}-\sqrt{2}\right) + \left(\sqrt{4}-\sqrt{3}\right) + ... + \left(\sqrt{15}-\sqrt{14}\right) = \sqrt{15}-1$. Hence, (D) is the correct option

COMPARISON OF IRRATIONALS

Comparison between irrationals, happen on the basis of their magnitude. Let's have examples to understand it clearly.

Ex3. Arrange the following surds in ascending order of their magnitudes.

(A) $\sqrt[5]{2}, \sqrt[3]{6}, \sqrt[4]{7}, \sqrt{5}$ (B) $\sqrt[3]{2}, \sqrt[6]{3}, \sqrt[9]{4}$ (C) $\sqrt{5}, \sqrt[3]{11}, 2\sqrt[6]{3}$ (D) $\sqrt[3]{2}, \sqrt{3}, \sqrt[4]{5}, \sqrt[6]{7}, \sqrt[12]{31}$

Solution:

(A) LCM {5, 3, 4, 2} = 60. So,

$\sqrt[5]{2} = (2)^{1/5} = (2)^{12/60} = \left(2^{12}\right)^{1/60}$ ....(1)

$\sqrt[3]{6} = (6)^{1/3} = (6)^{20/60} = \left(6^{20}\right)^{1/60}$ ....(2)

$\sqrt[4]{7} = (7)^{1/4} = (7)^{15/60} = \left(7^{15}\right)^{1/60}$ ....(3)

$\sqrt{5} = (5)^{1/2} = (5)^{30/60} = \left(5^{30}\right)^{1/60}$ ....(4)

From (1), (2), (3) and (4), you can say that $5^{30} > 6^{20} > 7^{15} > 2^{12} \rightarrow \sqrt[5]{2} < \sqrt[4]{7} < \sqrt[3]{6} < \sqrt{5}$.

(B) LCM {3, 6, 9} = 18. So,

$\sqrt[3]{2} = (2)^{1/3} = \left(2^6\right)^{1/18} = (64)^{1/18}$ ....(1)

$\sqrt[6]{3} = (3)^{1/6} = \left(3^3\right)^{1/18} = (27)^{1/18}$ ....(2)

$\sqrt[9]{4} = (4)^{1/9} = \left(4^2\right)^{1/18} = (16)^{1/18}$ ....(3)

16 < 27 < 64 so, $\sqrt[9]{4} < \sqrt[6]{3} < \sqrt[3]{2}$.

(C) LCM {2, 3, 6} = 6. So,

$$\sqrt{5} = (5)^{1/2} = \left(5^3\right)^{1/6} = (125)^{1/6} \quad\text{.........................(1)}$$

$$\sqrt[3]{11} = (11)^{1/3} = \left(11^2\right)^{1/6} = (121)^{1/6} \quad\text{....................(2)}$$

$$2\sqrt[6]{3} = 2(3)^{1/6} = \left(2^6 \times 3\right)^{1/6} = (192)^{1/6} \quad\text{................(3)}$$

121 < 125 < 192 $\rightarrow \sqrt[3]{11} < \sqrt{5} < 2\sqrt[6]{3}$.

(D) LCM {3, 2, 4, 6, 12} = 12. So, $\sqrt[3]{2}, \sqrt{3}, \sqrt[4]{5}, \sqrt[6]{7}, \sqrt[12]{31}$

$$\sqrt[3]{2} = (16)^{1/12}, \sqrt{3} = \left(3^6\right)^{1/12} = (729)^{1/12}, \sqrt[4]{5} = (5)^{1/4} = (125)^{1/12}, \sqrt[6]{7} = (49)^{1/12}, \sqrt[12]{31} = (31)^{1/12}$$

16 < 31 < 49 < 125 < 729 $\rightarrow \sqrt[3]{2} < \sqrt[12]{31} < \sqrt[6]{7} < \sqrt[4]{5} < \sqrt{3}$.

Ex4. Select the correct options

(A) $\sqrt{2} - 1 = \sqrt{3 - 2\sqrt{2}}$

(B) $2\sqrt[4]{2} > \sqrt{5}$

(C) $2\sqrt[5]{3} > 3\sqrt{2}$

(D) $1.\overline{3}$ can be written in $\dfrac{p}{q}, \{p, q \in I, q \neq 0\}$ form as $\dfrac{4}{3}$

Solution:

(A) As $(\sqrt{2} - 1)^2 = 3 - 2\sqrt{2}$ so, $\sqrt{3 - 2\sqrt{2}} = \sqrt{\left(1 - \sqrt{2}\right)^2} = \left|1 - \sqrt{2}\right| = \sqrt{2} - 1$.

(B) $2\sqrt[4]{2} = \left(2^4 \times 2\right)^{1/4} = (32)^{1/4}$ and $\sqrt{5} = \left(5^2\right)^{1/4} = 25^{1/4}$. Since, 25 < 32 so $\sqrt{5} < 2\sqrt[4]{2}$

(C) $2\sqrt[5]{3} = \left(2^5 \times 3\right)^{1/5} = (96)^{1/5} = \left(2^{10} \times 3^2\right)^{1/10}$ and $3\sqrt{2} = \left(3^2 \times 2\right)^{1/2} = (18)^{1/2} = \left(18^5\right)^{1/10} = \left(3^{10} \times 2^5\right)^{1/10}$.

Here, you can observe $3^{10} \times 2^5 > 2^{10} \times 3^2$. So, $2\sqrt[5]{3} < 3\sqrt{2}$.

(D) $x = 1.\overline{3} = 1.3333333333.......$

$\rightarrow 10x = 13.333333333......... \rightarrow 9x = 12$. Hence, x = 4/3.

Hence, options (A), (B) and (D) are correct.

Ex5. If $x = 1 - \sqrt{2}$ then find the value of $x^3 + (\pi - 2)x^2 - (2\pi + 1)x + \pi$.

Solution:

We can solve the problem by direct substitution but it will take time to solve. We have different strategy to solve the problem.

$\sqrt{2} = 1 - x$, now square both side, we get $x^2 - 2x - 1 = 0$.

Now divide $x^3 + (\pi - 2)x^2 - (2\pi + 1)x + \pi$ by $x^2 - 2x - 1$.

$$
\begin{array}{r}
x + \pi \\
x^2 - 2x - 1{\overline{\smash{\big)}\,x^3 + (\pi-2)x^2 - (2\pi+1)x + \pi}} \\
\underline{x^3 - 2x^2 - x} \\
\pi x^2 - 2\pi x + \pi \\
\underline{\pi x^2 - 2\pi x - \pi} \\
2\pi
\end{array}
$$

$$x^3 + (\pi - 2) x^2 - (2\pi + 1) x + \pi = (x^2 - 2x - 1)(x + \pi) + 2\pi = 2\pi \qquad (\text{as } x^2 - 2x - 1 = 0).$$

Ex6. Match the following

	COULUMN - I		COULUMN - II
A	The value of $\left(\sqrt{5}-2\right)\sqrt{1+4\sqrt{1+4\sqrt{9+4\sqrt{5}}}}$	P	2
B	The number of odd Natural divisors of $2^{10} \times 3^3 \times 5$, are?	Q	1
C	The value of $\left(\sqrt[2014]{2\sqrt{7}-3\sqrt{3}}\right)\times\left(\sqrt[4028]{55+12\sqrt{21}}\right)$ is?	R	4
D	Square root of $\dfrac{(0.75)^3}{1-0.75}+\left(0.75+(0.75)^2+1\right)$, is?	S	8

Solution:

(A) $9+4\sqrt{5}=\left(2+\sqrt{5}\right)^2 \to 1+4\sqrt{9+4\sqrt{5}}=1+4\left(2+\sqrt{5}\right)=9+4\sqrt{5}$.

$\to \sqrt{1+4\sqrt{1+4\sqrt{9+4\sqrt{5}}}}=\sqrt{9+4\sqrt{5}}=2+\sqrt{5}$ so, $\left(\sqrt{5}-2\right)\sqrt{1+4\sqrt{1+4\sqrt{9+4\sqrt{5}}}}=\left(\sqrt{5}-2\right)\left(2+\sqrt{5}\right)=1$

(B) Odd natural divisors of $2^{10} \times 3^3 \times 5$ can be obtained of the form $3^\alpha \times 5^\beta$. So, Total such divisors are $(1 + 3)(1 + 1) = 8$.

(C) $\left(2\sqrt{7}-3\sqrt{3}\right)^2=28+27-12\sqrt{21}=55-12\sqrt{21} \to \sqrt[2014]{2\sqrt{7}-3\sqrt{3}}=\sqrt[4028]{55-12\sqrt{21}}$

Hence, $\left(\sqrt[2014]{2\sqrt{7}-3\sqrt{3}}\right)\times\left(\sqrt[4028]{55+12\sqrt{21}}\right)=\left(\sqrt[4028]{55-12\sqrt{21}}\right)\times\left(\sqrt[4028]{55+12\sqrt{21}}\right)=\sqrt[4018]{55^2-12^2 \times 21}=1$

(D) as $1 - a^3 = (1 - a)(1 + a + a^2)$. Let $a = 0.75 \to$

$\dfrac{(0.75)^3}{1-0.75}+\left(0.75+(0.75)^2+1\right)=\dfrac{(0.75)^3}{1-0.75}+\dfrac{(1-0.75)\left(0.75+(0.75)^2+1\right)}{1-0.75}=\dfrac{0.75^3+1^3-0.75^3}{1-0.75}=4$

So, correct match is A→Q, B→S, C→Q, D→P.

Ex7. If $2^{5x} \div 2^x = \sqrt[5]{2^{20}}$ then find the value of x.

Solution:

$$2^{5x} \div 2^x = \sqrt[5]{2^{20}} \to 2^{5x-x}=\left(2^{20}\right)^{1/5} \to 2^{4x}=2^4 \to x=1.$$

Ex8. Find the sum of the real values of x satisfying $\dfrac{1}{\sqrt{x}}+\dfrac{1}{\sqrt{y}}=\dfrac{1}{\sqrt{20}}$.

Solution:

The only possible case is $\sqrt{x}$, $\sqrt{y}$ must be a multiple of $\sqrt{5}$ so

Let $\sqrt{x}=a\sqrt{5}$ and $\sqrt{y}=b\sqrt{5}$ where a, b ∈ N.

Hence, $\dfrac{1}{\sqrt{x}}+\dfrac{1}{\sqrt{y}}=\dfrac{1}{2\sqrt{5}} \to \dfrac{1}{a}+\dfrac{1}{b}=\dfrac{1}{2}$ as a, b ∈ N, to follow the equation, a > 2, b > 2 (must be true)

So, let a = 2 + p & b = 2 + q where p, q ∈ I.

$\frac{1}{a} + \frac{1}{b} = \frac{1}{2} \rightarrow 2(a + b) = ab \rightarrow 2(4 + p + q) = (2 + p)(2 + q) \rightarrow 4 = pq$.

Hence, p can take values 1, 2, 4 only → a = 3, 4, 6 → $\sqrt{x} = 3\sqrt{5}, 4\sqrt{5}, 6\sqrt{5}$

→ x = 45, 80, 180 → $\sum x = 45 + 80 + 180 = 305$.

Ex9. Let $y = \cfrac{1}{2 + \cfrac{1}{3 + \cfrac{1}{2 + \cfrac{1}{3 + ..}}}}$. What is the value of y?

(A) $\frac{\sqrt{13} + 3}{2}$ (B) $\frac{\sqrt{13} - 3}{2}$ (C) $\frac{\sqrt{15} + 3}{2}$ (D) $\frac{\sqrt{15} - 3}{2}$

Solution: As per question, $y = \cfrac{1}{2 + \cfrac{1}{3 + y}} \rightarrow y = \frac{3 + y}{7 + 2y} \rightarrow 2y^2 + 6y - 3 = 0$

→ $y = \frac{-3 \pm \sqrt{15}}{2}$. As y > 0 so $y = \frac{-3 + \sqrt{15}}{2}$. Hence (D) option is correct.

Ex10. If x = 5 + 2√6, find the value of

(i) $\sqrt{x} + \frac{1}{\sqrt{x}}$ (ii) $x^3 + \frac{1}{x^3}$

Solution:

(i) $5 + 2\sqrt{6} = (\sqrt{2} + \sqrt{3})^2 \rightarrow \sqrt{x} = \sqrt{2} + \sqrt{3}$.

$\frac{1}{\sqrt{x}} = \frac{1 \times (\sqrt{3} - \sqrt{2})}{(\sqrt{3} + \sqrt{2}) \times (\sqrt{3} - \sqrt{2})} = (\sqrt{3} - \sqrt{2})$. So $\sqrt{x} + \frac{1}{\sqrt{x}} = \sqrt{2} + \sqrt{3} + \sqrt{3} - \sqrt{2} = 2\sqrt{3}$.

(ii) $x = 5 + 2\sqrt{6} \rightarrow \frac{1}{x} = \frac{1}{5 + 2\sqrt{6}} \times \left(\frac{5 - 2\sqrt{6}}{5 - 2\sqrt{6}}\right) = 5 - 2\sqrt{6}$. So, x + 1/x = 10.

Now, $x^3 + \frac{1}{x^3} = \left(x + \frac{1}{x}\right)\left(x^2 - 1 + \frac{1}{x^2}\right) = \left(x + \frac{1}{x}\right)\left(\left(x + \frac{1}{x}\right)^2 - 3\right) = 10 \times (10^2 - 3) = 970$.

Ex11. If $\sqrt{17^2 + 17^2 + ... + 17^2 + 17^2} = 17^2 + 17^2 + 17^2$, then how many times should 17^2 appear under the square root sign for the equation above to be true?

Solution:

Let it appears n times so $17\sqrt{n} = 3 \times 17^2 \rightarrow \sqrt{n} = 51$ so $n = 51^2 = 2601$.

Ex12. The sum $S = \sqrt{1 + \frac{1}{1^2} + \frac{1}{2^2}} + \sqrt{1 + \frac{1}{2^2} + \frac{1}{3^2}} + ... + \sqrt{1 + \frac{1}{2015^2} + \frac{1}{2016^2}}$ is equal to

(A) 2016 – (1/2016) (B) 2015 – (1/2015) (C) 2015 – (1/2016) (D) 2016 – (1/2015)

Solution:

Let $T_r = \sqrt{1 + \dfrac{1}{r^2} + \dfrac{1}{(r+1)^2}}$ then $S = \displaystyle\sum_{r=1}^{2015} \sqrt{1 + \dfrac{1}{r^2} + \dfrac{1}{(r+1)^2}}$.

$$T_r = \sqrt{\dfrac{r^4 + 2r^3 + 3r^2 + 2r + 1}{r^2(r+1)^2}} = \sqrt{\left(\dfrac{r^2 + r + 1}{r(r+1)}\right)^2} = \dfrac{r^2 + r + 1}{r(r+1)}$$

$\rightarrow \dfrac{r(r+1)+1}{r(r+1)} = 1 + \dfrac{1}{r(r+1)} \rightarrow 1 + \dfrac{1}{r} - \dfrac{1}{r+1}$. So, $T_r = 1 + \dfrac{1}{r} - \dfrac{1}{r+1}$.

Hence sum $S = T_1 + T_2 + T_3 + ... + T_{2015}$

$$= \left(1 + \dfrac{1}{1} - \dfrac{1}{2}\right) + \left(1 + \dfrac{1}{2} - \dfrac{1}{3}\right) + \left(1 + \dfrac{1}{3} - \dfrac{1}{4}\right) + ... + \left(1 + \dfrac{1}{2015} - \dfrac{1}{2016}\right)$$

$= 2015 + 1 - 1/2016 = 2016 - 1/2016$. So, (A) is correct option.

Ex13. Evaluate $\displaystyle\lim_{x \to 0^+} \sqrt{x + \sqrt{x + \sqrt{x + \sqrt{x + ...}}}}$

Solution:

Let $y = \sqrt{x + \sqrt{x + \sqrt{x + \sqrt{x + ...}}}} \rightarrow y^2 = x + y \rightarrow y^2 - y - x = 0$ so $y = \dfrac{1 \pm \sqrt{1 + 4x}}{2}$ but $y > 0$ so

$y = \dfrac{1 + \sqrt{1 + 4x}}{2}$. Hence, $\displaystyle\lim_{x \to 0^+} \sqrt{x + \sqrt{x + \sqrt{x + \sqrt{x + ...}}}} = \displaystyle\lim_{x \to 0^+} \dfrac{1 + \sqrt{1 + 4x}}{2} = 1$.

Ex14. If $mx^m - nx^n = 0$, then what is the value of $\dfrac{1}{x^m + x^n} + \dfrac{1}{x^m - x^n}$ in terms of x^n?

(A) $\dfrac{2mn}{x^n(n^2 - m^2)}$ (B) $\dfrac{2mn}{x^n(n^2 + m^2)}$ (C) $\dfrac{2mn}{x^n(m^2 - n^2)}$ (D) none of these

Solution:

$mx^m - nx^n = 0 \rightarrow \dfrac{x^m}{x^n} = \dfrac{n}{m} \rightarrow x^{m-n} = \dfrac{n}{m}$. Now $\dfrac{1}{x^m + x^n} + \dfrac{1}{x^m - x^n} = \dfrac{2x^m}{x^{2m} - x^{2n}} = \dfrac{1}{x^n}\left(\dfrac{2x^{m-n}}{x^{2(m-n)} - 1}\right)$

$= \dfrac{1}{x^n}\left(\dfrac{2\dfrac{n}{m}}{\dfrac{n^2}{m^2} - 1}\right) = \dfrac{2nm}{x^n(n^2 - m^2)}$. Hence (A) option is correct.

Ex15. If $x = -0.5$, then which of the following has the smallest value?

(A) $2^{1/x}$ (B) $1/x$ (C) $1/x^2$ (D) $1/\sqrt{-x}$

Solution:

For $x = -0.5$, $2^{1/x} = 1/4$, $1/x = -2$, $1/x^2 = 4$, $1/\sqrt{-x} = \sqrt{2}$. So (B) is correct option.

You can observe that only $1/x$ is negative and other is positive values.

Ex16. If $\dfrac{a}{b} = \dfrac{1}{3}, \dfrac{b}{c} = 2, \dfrac{c}{d} = \dfrac{1}{2}, \dfrac{d}{e} = 3$ and $\dfrac{e}{f} = \dfrac{1}{4}$, then the value of $\dfrac{abc}{def}$?

(A) 3/8   (B) 27/8   (C) 3/4   (D) 27/4

Solution:

$$\dfrac{a}{d} = \dfrac{a}{b} \times \dfrac{b}{c} \times \dfrac{c}{d} = \dfrac{1}{3} \times 2 \times \dfrac{1}{2} = \dfrac{1}{3}, \quad \dfrac{b}{e} = \dfrac{b}{c} \times \dfrac{c}{d} \times \dfrac{d}{e} = 2 \times \dfrac{1}{2} \times 3 = 3 \text{ and } \dfrac{c}{f} = \dfrac{c}{d} \times \dfrac{d}{e} \times \dfrac{e}{f} = \dfrac{1}{2} \times 3 \times \dfrac{1}{4} = \dfrac{3}{8}$$

$\therefore \dfrac{abc}{def} = \dfrac{1}{3} \times 3 \times \dfrac{3}{8} = \dfrac{3}{8}$. Hence (A) is correct option.

Ex17. Let $x = \sqrt{4 + \sqrt{4 - \sqrt{4 + \sqrt{4 - \ldots \ldots \text{infinity}}}}}$. Then x equals

(A) 3   (B) $\dfrac{\sqrt{13} - 1}{2}$ (C) $\dfrac{\sqrt{13} + 1}{2}$ (D) $\sqrt{13}$

Solution:

$x = \sqrt{4 + \sqrt{4 - x}} \to x^2 = 4 + \sqrt{(4 - x)} \to x^2 - 4 = \sqrt{(4 - x)} \to (x^2 - 4)^2 = 4 - x.$

$\to (x^2 - 4)^2 + x - 4 + x^2 - x^2 = 0 \to (x^2 - 4)^2 + (x^2 - 4) + x(1 - x) = 0 \ldots\ldots\ldots\ldots\ldots(1)$

So let $x^2 - 4 = y$, the equation becomes $y^2 + y + x(1 - x) = 0$

$\to y = \dfrac{-1 \pm \sqrt{1 - 4x(1 - x)}}{2} \to 2y = -1 \pm \sqrt{4x^2 - 4x + 1} = -1 \pm (2x - 1) \to y = x - 1 \text{ and } - x$

Hence equation (1) becomes, $(y - x + 1)(y + x) = 0$ .i.e. $(x^2 - x - 3)(x^2 + x - 4) = 0$

$\to x = \dfrac{1 \pm \sqrt{13}}{2}, \dfrac{-1 \pm \sqrt{17}}{2}$, as per question $x > 2$ so $x = \dfrac{1 + \sqrt{13}}{2}$ is the only answer, Hence (C) option is correct.

Ex18. Mrs Chandra is excited to learn about the distributive law, and thinks that it applies to every possible operation. As such, he claims that $\sqrt{a^2 + b^2} = \sqrt{a^2} + \sqrt{b^2}$. How many ordered pairs (a, b) of integers are there, such that $-10 \le a \le 10$, $-10 \le b \le 10$ and $\sqrt{a^2 + b^2} = \sqrt{a^2} + \sqrt{b^2}$

Solution:

On squaring both side the equation $\sqrt{a^2 + b^2} = \sqrt{a^2} + \sqrt{b^2}$ we get $a^2 + b^2 = a^2 + b^2 + 2|ab| \to ab = 0$.

Now, If a = 0 then b has 21 choices i.e. $b \in \{-10, -9, -8, \ldots\ldots\ldots, 8, 9, 10\}$ similarly, if b = 0 then a has 21 choices. So, total number of such ordered pair (a, b) = 21 + 21 - 1 = 41.

Can you tell me that why i have remove 1? $\because$ we counted (0, 0) twice. So, Answer = 41.

Ex19. If N > 1 then value of $\sqrt[3]{N\sqrt[3]{N\sqrt[3]{N}}}$ = ...

(A) $N^{1/27}$ (B) $N^{1/9}$ (C) $N^{13/27}$ (D) N

Solution:

$\sqrt[3]{N} = N^{1/3} \to \sqrt[3]{N\sqrt[3]{N}} = \sqrt[3]{N \times N^{1/3}} = \sqrt[3]{N^{4/3}} = N^{4/9}$. So, $\sqrt[3]{N\sqrt[3]{N\sqrt[3]{N}}} = \sqrt[3]{N \times N^{4/9}} = N^{13/27}$

Hence, (C) is correct option.

Ex20. The value of $\sqrt{x+2\sqrt{x-1}}+\sqrt{x-2\sqrt{x-1}}$ is equal to 2 for x equals

    (A) 3/2   (B) 2/3   (C) 5/4   (D) 7/5

Solution:

$\sqrt{x+2\sqrt{x-1}}+\sqrt{x-2\sqrt{x-1}}=2$ on squaring both side we get

$x+2\sqrt{x-1}+x-2\sqrt{x-1}+2\sqrt{x^2-4(x-1)}=4 \rightarrow 2x+2\sqrt{(x-2)^2}=4$

$\rightarrow |x-2|=2-x$ true $\forall\ x \le 2$ but $x \ge 1$ for real roots. Hence, options A, C & D are correct options.

I know you mind need refreshment, so we have a mind refreshers problem. Just solve it. Study or learning should not be by force, it should be choice. Understand it better and develop interest in it.

PUZZLE # 5

Here basic shape represents digits {0, 1, 2, 3, 4, 5, 6, 7, 8, 9}. Find the value of '?'.

Till now we have learn something. You will forget the concepts if you haven't applied it. So, solve the given exercise for better understanding. ☺ ☺ ☺

EXERCISE - 1

1. The value of $\sqrt{\left|\sqrt{3}-\sqrt{4+\sqrt{5}+\sqrt{17-4\sqrt{15}}}\right|}$ =

   (A) 1   (B) $\sqrt{3}$ (C) $\sqrt{2\sqrt{3}-1}$ (D) $\sqrt{2-\sqrt{3}}$

2. If $R = \dfrac{30^{65} - 29^{65}}{30^{64} + 29^{64}}$ then

   (A) $0 < R \le 0.1$ (B) $0.1 < R \le 0.5$ (C) $0.5 < R \le 1.0$ (D) $R > 1.0$

3. A new sequence is obtained from the sequence of the positive integers $\{1, 2, 3, 4,...\}$ by deleting all the perfect squares. Then the 2003rd term of the new sequence is?

   (A) 2046   (B) 2047   (C) 2048   (D) 2049

4. Let x and y be positive integers such that x is prime and y is composite. Then,

   (A) y – x cannot be an even integer   (B) xy cannot be an even integer.

   (C) $(x + y)/x$ can't be an even integer (D) none of these

5. The value of $\sqrt{\dfrac{1}{2}\sqrt{\dfrac{1}{2}\sqrt{\dfrac{1}{2}\sqrt{\dfrac{1}{2}}}}.......\infty}$ is

   (A) 0   (B) 3/2   (C) 1/2   (D) 1

6. The value of $\sqrt{1+2\sqrt{1+2\sqrt{1+2\sqrt{1+}}}..........}$ is

   (A) 1 - $\sqrt{2}$ (B) $1 + \sqrt{2}$ (C) $1 + \sqrt{3}$ (D) none of these

7. The greatest number among $2^{1/2}, 3^{1/3}, 8^{1/8}$ & $9^{1/9}$, is

   (A) $2^{1/2}$ (B) $3^{1/3}$ (C) $8^{1/8}$ (D) $9^{1/9}$

8. How many pairs of natural numbers are there so that difference of their squares is 60?

   (A) 4   (B) 3   (C) 2   (D) none of these

9. If the sum of three consecutive odd natural numbers is a perfect square between 200 and 400, then the root of this sum is

   (A) 15   (B) 16   (C) 18   (D) 19

10. If the mean of three numbers a, b, c is 3, then $\sqrt[3]{\left(7^{a+b-c}\right)\left(7^{b+c-a}\right)\left(7^{c+a-b}\right)}$ equals

    (A) $7^{1/3}$ (B) $7^{2/3}$ (C) 7^3 (D) 7

11. Select the correct option(s) for x, y, z, w, $a \in R - \{0\}$.

    (A) If $x > y \to \dfrac{1}{x} < \dfrac{1}{y}$.   (B) If $x > y$ and $z > w \to xz > yz$

    (C) If $x > y \to ax > ay$ (D) $ax = ay \to \cancel{a}x = \cancel{a}y \to x = y$ only if $a \ne 0$

REAL NUMBERS (R)

A French philosopher, Mathematician & scientist Rene Descartes suggested a field of all rational and irrational numbers called Real Numbers or simply the real and denoted by 'R'. A real number is a value that represents a quantity along a continuous line. It contains, Rational Numbers such as the integers and fractions, Irrational Numbers such as $\sqrt{2}$, $2 + \sqrt{3}$, transcendental numbers e, π, etc. A real number can be thought of a point on an infinite long line called Number Line as shown in figure.

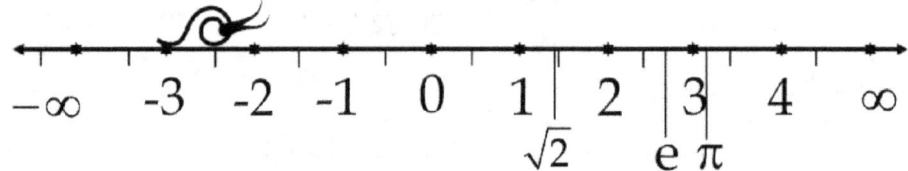

The number line generated due to the nature of roots of a real polynomial. Like $x^2 + 1 = 0$ and $x^2 - 1 = 0$. The thinking also generate an another field i.e. Complex Numbers Field where real numbers are subset of it. Here i am having the axiomatic approach of real number, please look at it.

AXIOMATIC APPROACH

Let $\mathbb{R}$ denote the set of all real numbers. Then:

❖ The set $\mathbb{R}$ is a field, meaning that addition and multiplication are defined and have the usual properties.

❖ The field $\mathbb{R}$ is ordered, meaning that there is a total order $\geq$ such that, for all real numbers x, y & z:

  ➢ If $x \geq y$ then $x + z \geq y + z$.

  ➢ If $x \geq 0$ and $y \geq 0$ then $xy \geq 0$.

❖ The order is Dedekind-complete; that is: every non-empty subset S of $\mathbb{R}$ with an upper bound in $\mathbb{R}$ has a least upper bound (also called supremum) in $\mathbb{R}$.

Here, you can prove that ∞ & $-\infty$ are not real, they are just a thinking of infinite long number line. These are not following addition and multiplication as $2 + 3 = 5$ or $2 \times 3 = 6$ but $\infty + \infty = \infty$ and $\infty \times \infty = \infty$.

COMPLEX NUMBERS (C)

If we solve the equation $x^2 + 1 = 0$, there is no any real root. This equation leads to generate the thinking of new number field i.e. complex number field. It's defines the roots of $x^2 + 1 = 0$ as $x = \pm\sqrt{-1} = \pm i$ (iota).

$i = \sqrt{-1}$, $i^2 = -1$, $i^3 = -i$, $i^4 = 1$, $i^{4k} = 1$, $i^{4k+1} = i$, $i^{4k+2} = -1$, $i^{4k+3} = -i$, where $k \in I$.

A complex number is represented by $z = a + i b$, where a, b $\in R$ and a = real part of z or Re (z) and b = imaginary part of z or Im (z).

FUNDAMENTAL THEOREM OF ALGEBRA

Statement: Every non – constant single variable polynomial equation with complex coefficients has at-least one complex root and if its degree is 'n' then it has exactly 'n' complex roots.

It includes real polynomial having real coefficients, as every real is a complex number with imaginary part 0.

Ex1. Represent the following in a + i b form.

  (i) $2 + \sqrt{-3}$ (ii) $\dfrac{1}{5 - \sqrt{-7}}$ (iii) $\dfrac{1}{2 + \sqrt{-3}} + 3 - \sqrt{-3}$ (iv) $\dfrac{5 + 3i}{1 + i}$

Solutions:

  (i) $2 + \sqrt{-3} = 2 + \sqrt{-1}\sqrt{3} = 2 + i\sqrt{3}$

(ii) $\dfrac{1}{5-\sqrt{-7}}=\dfrac{1}{5-i\sqrt{7}}=\dfrac{1}{5-i\sqrt{7}}\times\dfrac{5+i\sqrt{7}}{5+i\sqrt{7}}=\dfrac{5+i\sqrt{7}}{5^2-7i^2}=\dfrac{5+i\sqrt{7}}{32}=\dfrac{5}{32}+i\dfrac{\sqrt{7}}{32}$

(iii) $\dfrac{1}{2+\sqrt{-3}}=\dfrac{1}{2+i\sqrt{3}}\times\dfrac{2-i\sqrt{3}}{2-i\sqrt{3}}=\dfrac{2-i\sqrt{3}}{7}=\dfrac{2}{7}-i\dfrac{\sqrt{3}}{7}$, so $\dfrac{1}{2+\sqrt{-3}}+3-\sqrt{-3}=\dfrac{23}{7}-i\dfrac{8\sqrt{3}}{7}$

(iv) $\dfrac{5+3i}{1+i}=\dfrac{5+3i}{1+i}\times\dfrac{1-i}{1-i}=\dfrac{5+3i-5i-3i^2}{1^2-i^2}=\dfrac{8-2i}{2}=4-i$

PERFECT NUMBERS

The factors of 6 are 1, 2, 3 and 6. Also, 1+2+3+6 = 12 = 2 × 6. We find that the sum of the factors of 6 is twice the number 6. All the factors of 28 are 1, 2, 4, 7, 14 and 28. Adding these we have, 1 + 2 + 4 + 7 + 14 + 28 = 56 = 2 × 28.

The sum of the factors of 28 is equal to twice the number 28.

A number for which sum of all its factors is equal to twice the number is called a perfect number.

The numbers 6 and 28 are perfect numbers. Is 10 a perfect number? (No, as 1 + 2 + 5 + 10 ≠ 2×10)

NUMBER SYSTEM HIERARCHY

❖ Prime numbers set is subsets of Natural numbers (N)

❖ Real numbers are composition of rational and irrational numbers. i.e. $Q^C \cup Q = R$ and $Q^C \cap Q = \emptyset$.

❖ Complex numbers are the largest set till known.

❖ There is no upper bound for these numbers, they are infinite in number.

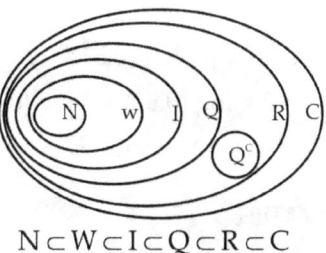

$$N \subset W \subset I \subset Q \subset R \subset C$$

Ex2. If a, b, c are three positive real numbers, then $\dfrac{a+c}{b+c}$ is

(A) Always smaller than a/b (B) Always greater than a/b

(C) Greater than a/b only if a > b (D) greater than a/b only if a < b

Solution:

Let $\dfrac{a+c}{b+c}>\dfrac{a}{b}\to ab+bc>ab+ac\to bc>ac\to b>a$. Hence, (D) option is correct.

Ex3. If a, b, c, d are positive real numbers such that $\dfrac{a}{3}=\dfrac{a+b}{4}=\dfrac{a+b+c}{5}=\dfrac{a+b+c+d}{6}$, then $\dfrac{a}{b+2c+3d}$ is

(A) 1/2 (B) 1 (C) 2 (D) Not determinable

Solution:

Let $\dfrac{a}{3}=\dfrac{a+b}{4}=\dfrac{a+b+c}{5}=\dfrac{a+b+c+d}{6}=k\to a=3k, a+b=4k\to b=k, a+b+c=5k\to c=k,$

$a+b+c+d=6k\to d=k$. So, $\dfrac{a}{b+2c+3d}=\dfrac{3k}{k+2k+3k}=\dfrac{1}{2}$. Hence, option (A) is correct.

Ex4. Let a, b, c, d be real numbers such that $|a - b| = 2$, $|b - c| = 3$, $|c - d| = 4$. Then the sum of all possible values of $|a - d|$ is

    (A) 9   (B) 18   (C) 24   (D) 30

Solution:

$|a - b| = 2 \Rightarrow a - b = \pm 2$ and $|a - c| = 3 \Rightarrow b - c = \pm 3$ and $|c - d| = 4 \Rightarrow c - d = \pm 4$. So, possible values of a − d are $\pm 9, \pm 5, \pm 3, \pm 1 \rightarrow |a - d| = 9, 5, 3, 1$. So Sum = 18. Option (B) is correct.

Ex5. If a, b are natural numbers such that $2013 + a^2 = b^2$, then the minimum possible value of ab is

    (A) 671   (B) 668   (C) 658   (D) 645

Solution:

$2013 + a^2 = b^2 \rightarrow b^2 - a^2 = 2013 \rightarrow (b - a)(b + a) = 11 \times 3 \times 61$. For minimum value of ab

b − a = 33 and a + b = 61 → b = 47 and a = 14. Hence $ab|_{MIN} = 14 \times 47 = 658$. Option (C) is correct.

Ex6. If $\log(0.57) = \bar{1}.756$, then the value of $\log(57) + \log(0.57)^3 + \log\sqrt{0.57}$ is:

    (A) 0.902   (B) $\bar{2}.146$ (C) 1.902   (D) $\bar{1}.146$

Solution:

$\log(57) + \log(0.57)^3 + \log\sqrt{0.57} = \log(0.57 \times 10^2) + \log(0.57)^3 + \log(0.57)^{1/2}$

$= 2 + \log(0.57) + 3\log(0.57) + 0.5\log(0.57) = 2 + 4.5\log(0.57) = 2 + 4.5(-1 + 0.756) = 0.902$

Hence (A) option is correct.

SUMMATION

This is the topic, which most student hate at start, I am going to explain it in details so that your fear factor becomes zero. Hence, first understands the meaning of it, only then we can play with the summation type problems.

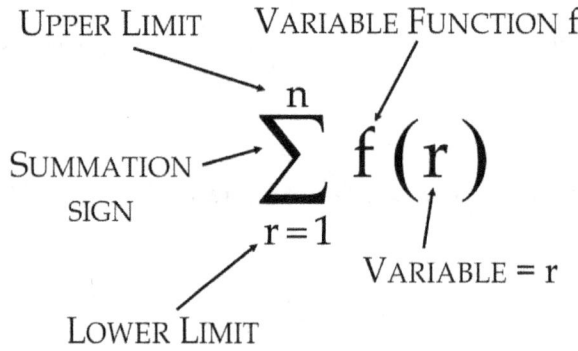

Ex1. $\displaystyle\sum_{r=1}^{5} T_r$ in this case lower limit = 1 and upper limit = 5 and Variable function is T_r.

Its meaning is as shown, $\displaystyle\sum_{r=1}^{5} T_r = T_1 + T_2 + T_3 + T_4 + T_5$. (Look, how r is changing.)

Now let say $T_r = (2r + 1)$ then $\displaystyle\sum_{r=1}^{5}(2r+1) = (2\cdot1+1) + (2\cdot2+1) + (2\cdot3+1) + (2\cdot4+1) + (2\cdot5+1)$.

$= 2\,(1+2+3+4+5)+(1+1+1+1+1) = 2\times\dfrac{5\times6}{2}+5=35$. Here we are using the formula

$1+2+3+4+.....+n = \dfrac{n(n+1)}{2}$. Let's have another example.

Ex2. $\displaystyle\sum_{m=2}^{10} m^2 = 2^2+3^2+4^2+.........+9^2+10^2.$

Observe the expression, variable m is starting from 2 and end at 10 and variable is increasing by 1 unit each time.

$$\sum_{m=2}^{10} m^2 = \left(1^2+2^2+3^2+.....+10^2\right)-1^2 = \dfrac{10\times11\times21}{6}-1 = 385-1=384.$$

❖ Here, we are using the formula $1^2+2^2+3^2+.....+n^2 = \dfrac{n\times(n+1)\times(2n+1)}{6}$.

FORMULA RELATED TO SUMMATION

(I) $\displaystyle\sum_{r=1}^{n} r = 1+2+3+......+n = \dfrac{n(n+1)}{2}.$ **(II)** $\displaystyle\sum_{r=1}^{n} r^2 = 1^2+2^2+.....+n^2 = \dfrac{n\times(n+1)\times(2n+1)}{6}.$

(III) $\displaystyle\sum_{r=1}^{n} r^3 = 1^3+2^3+3^3+.....+n^3 = \left(\dfrac{n(n+1)}{2}\right)^2.$ **(IV)** $\displaystyle\sum_{r=1}^{n} k = nk$ where k is any constant.

Let's prove them one by one.

PROOF:

(I) $\displaystyle\sum_{r=1}^{n} r = 1+2+3+......+n = \dfrac{n(n+1)}{2}.$

Let S = 1 + 2 + 3 + 4 +......+ n, then

$$
\begin{aligned}
S &= 1 \;+\; 2 \;+\; 3 \;+\; 4 \;+......+(n-1)+\; n\\
+S &= \; n \;+(n-1)+(n-2)+(n-3)+......+\; 2 \;+\; 1\\
\hline
2S &= \underbrace{(n+1)+(n+1)+(n+1)+(n+1)+......+(n+1)+(n+1)}_{n\ times}=n(n+1)
\end{aligned}
$$

$\Rightarrow S = \dfrac{n(n+1)}{2}$

(II) $\displaystyle\sum_{r=1}^{n} r^2 = 1^2+2^2+.....+n^2 = \dfrac{n\times(n+1)\times(2n+1)}{6}$

To prove the formula we have to think in different way. As we know

$(k+1)^3 = k^3 + 3k^2 + 3k + 1$. Apply it for first n natural number and add them, we get as shown

$2^3 = 1^3 + 3\cdot1^2 + 3\cdot1 + 1$

$3^3 = 2^3 + 3\cdot2^2 + 3\cdot2 + 1$

$4^3 = 3^3 + 3\cdot3^2 + 3\cdot3 + 1$

....

....

$+\ (n+1)^3 = n^3 + 3\cdot n^2 + 3\cdot n + 1$

$\overline{(n+1)^3 = 1^3 + 3\cdot\left(1^2+2^2+3^2+....+n^2\right)+3\cdot\left(1+2+3+......+n\right)+\underbrace{(1+1+1+1...+1)}_{n\ times}}$

You need to observe the sum, I apply cancellation also, be focused for a while.

$$(n+1)^3 = 1^3 + 3 \cdot \left(1^2 + 2^2 + 3^2 + \dots + n^2\right) + 3 \cdot \left(1 + 2 + 3 + \dots + n\right) + \underbrace{\left(1 + 1 + 1 + 1 \dots + 1\right)}_{n \text{ times}}$$

$$\to n^3 + 3n^2 + 3n + 1 = 1 + 3\sum_{r=1}^{n} r^2 + 3\,(n(n+1)/2) + n \to 3\sum_{r=1}^{n} r^2 = n^3 + 3n^2 - 3\,(n(n+1)/2) + 2n.$$

$$\sum_{r=1}^{n} r^2 = \frac{1}{3}\left\{ n(n^2 + 3n + 2) - 3\frac{n(n+1)}{2} \right\} \to \sum_{r=1}^{n} r^2 = \frac{1}{3}\left\{ \frac{2n(n+1)(n+2)}{2} - \frac{3n(n+1)}{2} \right\}$$

$$\sum_{r=1}^{n} r^2 = \frac{1}{6}\left\{ n(n+1)(2n+4-3) \right\} = \frac{n(n+1)(n+1)(2n+1)}{6}\,.$$

Hope, you enjoy the proof! ☺ ☺ ☺

(III) $\displaystyle\sum_{r=1}^{n} r^3 = 1^3 + 2^3 + 3^3 + \dots + n^3 = \left(\frac{n(n+1)}{2}\right)^2$

Many mathematicians, proved this formula but I am writing simple one given by Charles Wheatstone (1854).

$$\sum_{r=1}^{n} r^3 = 1^3 + 2^3 + 3^3 + \dots + n^3 = 1 + 8 + 27 + 64 + \dots + n^3\,.$$

Let's observe the series $1 + 3 + 5 + 7 + 9 + 11 + 13 + \dots + (2n - 1) + \dots$ (A)

OBSERVATION · I: $\underbrace{\underbrace{\underbrace{\underbrace{1}_{1^2} + 3 + 5}_{2^2} + 7 + 9 + 11}_{3^2} + 13}_{4^2} + \dots +$

OBSERVATION – II: $\underbrace{1}_{1^3} + \underbrace{3 + 5}_{2^3} + \underbrace{7 + 9 + 11}_{3^3} + \underbrace{13 + 15 + 17 + 19}_{4^3} + \dots +$

Similarly, The starting number of $n^3 = (\{1 + 2 + 3 + \dots + (n - 1)\} + 1)^{TH}$ term of series (A)

$= ((n - 1)\,n/2 + 1)^{th}$ term $= 1 + (n - 1)n = n^2 - n + 1$.

Similarly, last term of $n^3 = (1 + 2 + 3 + \dots + n)^{th}$ term of series (A) $= (n^2 + n)/2$ th term $= n^2 + n - 1$. Hence, Series (A) can be written as

$$\underbrace{1}_{1^3} + \underbrace{3 + 5}_{2^3} + \underbrace{7 + 9 + 11}_{3^3} + \underbrace{13 + 15 + 17 + 19}_{4^3} + \dots + \underbrace{\left(n^2 - n + 1\right) + \left(n^2 - n\right) + \dots + \left(n^2 + n - 1\right)}_{n^3}$$

Now, let's move on proof part.

$$\to \sum_{r=1}^{n} r^3 = \underbrace{1}_{1^3} + \underbrace{3 + 5}_{2^3} + \underbrace{7 + 9 + 11}_{3^3} + \underbrace{13 + 15 + 17 + 19}_{4^3} + \dots + \underbrace{\left(n^2 - n + 1\right) + \left(n^2 - n\right) + \dots + \left(n^2 + n - 1\right)}_{n^3}$$

$$\to \sum_{r=1}^{n} r^3 = \underbrace{\underbrace{\underbrace{\underbrace{1}_{1^2} + 3 + 5}_{2^2} + 7 + 9 + 11}_{3^2} + 13}_{4^2} + \dots + \left(n^2 + n - 1\right)}_{\left(\frac{n(n+1)}{2}\right)^2}$$

So, $\displaystyle\sum_{r=1}^{n} r^3 = (1+2+3+....+n)^2 = \left(\dfrac{n(n+1)}{2}\right)^2$.

(IV) $\displaystyle\sum_{r=1}^{n} k = \underbrace{k+k+k+......+k}_{n \text{ times}} = nk$

Let's refresh your mind and see, how much focused are you?

IQ TEST# 1

Find the number of circular rings in totality. Don't touch the figure to count.

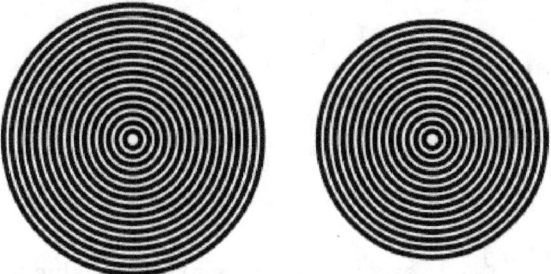

You may visit Mathsarc Education YouTube Channel to see 'Circle Dance'

Ex3. $\displaystyle\sum_{r=1}^{n} 2 = \underbrace{2+2+2+2+......+2}_{n-\text{times}} = 2n$

(V) $\displaystyle\sum_{r=1}^{n}(k_1 r + k_2) = k_1\sum_{r=1}^{n} r + \sum_{r=1}^{n} k_2$ or in general $\displaystyle\sum_{r=1}^{n}(k_1 f(r) + k_2 g(r)) = k_1\sum_{r=1}^{n} f(r) + k_2\sum_{r=1}^{n} g(r)$.

Ex4. $\displaystyle\sum_{r=1}^{n}(5r+2) = (5\cdot 1+2)+(5\cdot 2+2)+(5\cdot 3+2)+(5\cdot 4+2)+.......(5\cdot n+2)$.

$= 5\cdot 1 + 5\cdot 2 + 5\cdot 3 + ... + 5\cdot n + \underbrace{2+2+2+2+......+2}_{n-\text{times}}$.

$= 5(1+2+3+4+......+n) + \underbrace{2+2+2+2+......+2}_{n-\text{times}} = 5\sum_{r=1}^{n} r + \sum_{r=1}^{n} 2 = 5\times\dfrac{n(n+1)}{2} + 2n$.

Note: *Observe, how we write* $\displaystyle\sum_{r=1}^{n}(5r+2) = 5\sum_{r=1}^{n} r + \sum_{r=1}^{n} 2$.

Now answer the next example i.e. formula and remember the result.

Ex5. Show that $\displaystyle\sum_{r=1}^{n}(f(r)\times g(r)) \neq \left(\sum_{r=1}^{n} f(r)\right)\times\left(\sum_{r=1}^{n} g(r)\right)$. verify it by taking any suitable example

Solution:

Let us take an example as $\displaystyle\sum_{r=1}^{3} r(r+1)$, here $f(r) = r$ and $g(r) = r+1$.

So $\displaystyle\sum_{r=1}^{3} r(r+1) = 1\cdot 2 + 2\cdot 3 + 3\cdot 4 = 20$ & $\displaystyle\sum_{r=1}^{3} r = 1+2+3 = 6$, $\displaystyle\sum_{r=1}^{3}(r+1) = 2+3+4 = 9$, Clearly $20 \neq 6\times 9$.

Hence, here we can verify that $\sum_{r=1}^{n}(f(r) \times g(r)) \neq \left(\sum_{r=1}^{n} f(r)\right) \times \left(\sum_{r=1}^{n} g(r)\right)$.

Ex6. Find the value of $\sum_{r=1}^{100}\{r \times (r!)\}$.

Solution:

$$\sum_{r=1}^{100} r \times (r!) = \sum_{r=1}^{100}(r+1-1) \times (r!) = \sum_{r=1}^{100}((r+1)(r!)-(r!)) \to \sum_{r=1}^{100}\{(r+1)!-r!\}$$

$$= \left.\begin{matrix} 2! & - & 1! \\ 3! & - & 2! \\ 4! & - & 3! \\ \ldots\ldots\ldots \\ \ldots\ldots\ldots \\ 100! & - & 99! \\ 101! & - & 100! \end{matrix}\right\} = 101! - 1.$$

Ex7. For $\dfrac{2^2+4^2+6^2+\ldots\ldots+(2n)^2}{1^2+3^2+5^2+\ldots\ldots+(2n-1)^2}$ to exceed 1.01, the maximum value of n is

(A) 99   (B) 100   (C) 101   (D) 150

Solution:

$$\frac{2^2+4^2+6^2+\ldots\ldots+(2n)^2}{1^2+3^2+5^2+\ldots\ldots+(2n-1)^2} = \frac{2^2\left(1^2+2^2+3^2+\ldots\ldots+n^2\right)}{\left(1^2+2^2+3^2+\ldots\ldots+(2n)^2\right)-2^2\left(1^2+2^2+3^2+\ldots\ldots+n^2\right)}$$

$$= \frac{4 \times \dfrac{n(n+1)(2n+1)}{6}}{\dfrac{2n(2n+1)(4n+1)}{6}-4 \times \dfrac{n(n+1)(2n+1)}{6}} = \frac{2 \times (n+1)}{(4n+1)-2 \times (n+1)} = \frac{2(n+1)}{2n-1}. \text{ As per question}$$

$$\frac{2(n+1)}{2n-1} > 1.01 \to \frac{2n+2}{2n-1} > \frac{101}{100} \to \frac{3}{2n-1} > \frac{1}{100} \to 2n-1 < 300 \to n < 150.5, \text{ so } n\mid_{\text{MAX}} = 150.$$

Ex8. The sum $S = (1^2 - 1 + 1)(1!) + (2^2 - 2 + 1)(2!) + \ldots\ldots + (n^2 - n + 1)(n!)$ is

(A) $(n+2)!$ (B) $(n-1)((n+1)!) + 1$ (C) $(n+2)! - 1$ (D) $n((n+1)!) - 1$

Solution:

Let $T_r = (r^2 - r + 1)(r!) = (r^2 - 1 - r + 2)(r!) = ((r-1)(r+1) - (r-2))(r!) = (r-1)((r+1)!) - (r-2)(r!)$

So $T_r = (r-1)((r+1)!) - (r-2)(r!)$, True $\forall\, r \geq 2$. $S = T_1 + T_2 + T_3 + \ldots\ldots + T_n$

$$S = \left.\begin{matrix} T_1 + 1 \times 3! - 0 \times 2! \\ +2 \times 4! - 1 \times 3! \\ +3 \times 5! - 2 \times 4! \\ \ldots\ldots\ldots\ldots\ldots \\ \ldots\ldots\ldots\ldots\ldots \\ +(n-1) \times ((n+1)!) - (n-2) \times (n!) \end{matrix}\right\} = T_1 + (n-1) \times (n+1)! = 1 + (n-1) \times (n+1)! \therefore \text{(B) option is correct.}$$

Ex9. Find the value of $x = \sum\limits_{i=1}^{10} \sum\limits_{j=1}^{3} a_{ij}$, where $a_{ij} = \begin{cases} 2i & \text{if } i = j \\ i & \text{if } i < j \\ j & \text{if } j < i \end{cases}$.

Solution:

Here, you can observe, we have double summation which based on condition.

$$x = \sum_{i=1}^{10} \sum_{j=1}^{3} a_{ij} = \sum_{i=1}^{10} (a_{i1} + a_{i2} + a_{i3}) = \sum_{i=1}^{10} a_{i1} + \sum_{i=1}^{10} a_{i2} + \sum_{i=1}^{10} a_{i3}$$

$$\sum_{i=1}^{10} a_{i1} = a_{11} + a_{21} + a_{31} + ... + a_{(10)1} = 2 \times 1 + \underbrace{1 + 1 + 1 + ... + 1}_{9 \text{ times}} = 11.$$

$$\sum_{i=1}^{10} a_{i2} = a_{12} + a_{22} + a_{32} + ... + a_{(10)2} = 1 + 2 \times 2 + \underbrace{2 + 2 + 2 + ... + 2}_{8 \text{ times}} = 21.$$

$$\sum_{i=1}^{10} a_{i3} = a_{13} + a_{23} + a_{33} + ... + a_{(10)3} = 1 + 2 + 2 \times 3 + \underbrace{3 + 3 + ... + 3}_{7 \text{ times}} = 30.$$

Hence, $x = 11 + 21 + 30 = 62$.

Ex10. Evaluate $\sum\limits_{1 \le i < j \le n}' i \cdot j$.

Solution:

Let $S = \sum\limits_{1 \le i < j \le n}' i \cdot j = 1 \times 2 + 1 \times 3 + 1 \times 4 + ... + 1 \times n + 2 \times 3 + 2 \times 4 + ... + 2 \times n + 3 \times 4 + ... + (n-1) \times n.$

Observe the dark part of figure. You can easily find that sum of these numbers = 2S.

If we think of sum of

1st Row = $1(1 + 2 + 3 + 4 + ... + n)$

2nd Row = $2(1 + 2 + 3 + 4 + ... + n)$

3rd Row = $3(1 + 2 + 3 + 4 + ... + n)$

....

nth Row = $n(1 + 2 + 3 + 4 + ... + n)$

Sum of all the elements in the table

= $(1 + 2 + 3 + 4 + ... + n) \times (1 + 2 + 3 + 4 + ... + n)$

1x1	1x2	1x3	1x4	.....	......	.....	1xn
2x1	2x2	2x3	2x4	....	....	...	2xn
3x1	3x2	3x3	.....	....	...	...	3xn
......	....	.....	.....	......	......	...	
....	...	...		...	...	...	....
......	......	....	....	....	......		
n-1x1	n-1x2	...	....	...	....	......	n-1xn
nx1	nx2	....	.....	....	....	.....	nxn

= $(1 + 2 + 3 + 4 + ... + n)^2 = 2S + \sum\limits_{r=1}^{n} r^2.$

$\rightarrow 2S = \left(\dfrac{n(n+1)}{2}\right)^2 - \dfrac{n(n+1)(2n+1)}{6} \rightarrow S = \dfrac{n^2(n+1)^2}{8} - \dfrac{n(n+1)(2n+1)}{12}$

$= \dfrac{n(n+1)}{4}\left(\dfrac{n^2+n}{2} - \dfrac{2n+1}{3}\right) = \dfrac{n(n+1)}{4}\left(\dfrac{3n^2-n-2}{6}\right) \rightarrow S = \dfrac{n(n+1)(n-1)(3n+2)}{24}.$

NUMBER BASES

(I) THE DECIMAL (BASE - 10) SCALE

Any natural number n can be written in the form $n = a_k 10^k + a_{k-1} 10^{k-1} + a_{k-2} 10^{k-2} + \ldots\ldots + a_1 10 + a_0$ where $1 \le a_k \le 9, 0 \le a_i \le 9, i \ge 1$.

Ex1. Let n = 67402, then $67402 = 6 \times 10^4 + 7 \times 10^3 + 4 \times 10^2 + 0 \times 10 + 2$.

(II) Binary (Base - 2) Scale

Ex2. Express the number $(35)_{10}$ to Binary system.

Solution:

2 | 35
2 | 17.......1
2 | 8.........1
2 | 4.........0 . Observe the division and remainders. $(35)_{10} = (100011)_2 = 1 \times 2^5 + 1 \times 2^1 + 1 \times 2^0 = 35$.
2 | 2.........0
2 | 1.........0
2 | 0.........1

Ex3. Consider all four digit numbers for which the first two digits are equal and the last two digits are also equal. How many such numbers are perfect square?

(A) 0   (B) 1   (C) 2   (D) 4

Solution:

Let the four digit number be N = aabb where $a \ne 0$.

$N = 1000a + 100a + 10b + b = 1100a + 11b = 11 \times (100a + b) \rightarrow 100a + b$ must be a multiple of 11.

$100a + b = a0b$ is divisible by 11 if a + b is a multiple of 11. So possible pairs of (a, b) are (2, 9), (3, 8), (4, 7), (5, 6), (6, 5), (7, 4), (8, 3), (9, 2). Out of these only (7, 4) $\equiv 704 = 11 \times 64$ gave the perfect square.

So, N = 7744 is the only such four digit number. Hence (B) is correct option.

Ex4. The digits of a three-digit number A are written in the reverse order to form another three-digit number B. If B > A and B – A is perfectly divisible by 7, then which of the following is necessarily true?

(A) 100 < A < 299   (B) 106 < A < 305   (C) 112 < A < 311   (D) 118 < A < 317

Solution:

Let A = abc, $a \ne 0$. $\rightarrow$ B = cba.

Now, B > A $\rightarrow$ 100c + 10b + a > 100a + 10b + c $\rightarrow$ 99(c - a) > 0 so c > a

B – A = 99(c - a) is multiple of 7 so c – a must be a multiple of 7, hence possible ordered pair (c, a) are (8, 1), (9, 2) $\rightarrow$ A may be 1b8 or 2b9. Hence (B) is an appropriate answer.

Ex5. Number S is equal to the square of the sum of the digits of a 2 digit number D. If the difference between S and D is 27, then D is

(A) 32   (B) 54   (C) 64   (D) 52

Solution:

Let D = ab, then $S = (a + b)^2 \rightarrow |S - D| = 27$. So $|(a + b)^2 - (10a + b)| = 27$.

$|a^2 - 10a + b^2 - b + 2ab| = 27$ ....(1)

At the moment we need a hit and trial method to get number D.

So, from the given options, only (B) option satisfy the relation (1).

To get answer of such kind of questions, it's better to go from options

Ex6. If m = 777777.......7777 is a 99 digit number and n = 99999......9999 is 77 digit number then the sum of the digits in the product m × n is _____

(A) 890   (B) 891   (C) 892   (D) 893

Solution:

m = 7777......777777 → 99 times.

n = 99999......99999 (77 times 9) = $10^{77} - 1$. So, $n \times m = \underbrace{777.....7777}_{99 \text{ times}} \times (10^{77} - 1)$

$$= - \begin{array}{l} \underbrace{777777......777}_{99 \text{ digits}} \underbrace{0000......000000}_{77 \text{ digits}} \\ \underbrace{777..7}_{22 \text{ digits}} \underbrace{7777......777777}_{77 \text{ digits}} \\ \hline \underbrace{77....77}_{76 \text{ digits}} 6 \underbrace{99...99}_{22 \text{ digits}} \underbrace{22222..........22}_{76 \text{ digits}} 3 \end{array}$$

→ sum of digits = $7 \times 76 + 6 + 9 \times 22 + 2 \times 76 + 3 = 891$. Answer = (B).

Ex7. What are all the two-digit positive integers in which the difference between the integer and the product of its two digits is 12?

Solution:

Let such an integer be n = ab (a two digit number) i.e. 10a + b, where a, b are digits.

As per question, on solving 10a + b − ab =12 for a, we obtain $a = \dfrac{12 - b}{10 - b} = 1 + \dfrac{2}{10 - b}$.

Since 'a' is an integer, 10 − b must be a positive integer that divides 2. This gives b = 8, a = 2 or b = 9, a = 3. Thus, 28 and 39 are the only such integers.

Ex8. Let S be the set of all rational numbers r, 0 < r < 1, that have a repeating decimal expansion of the form

$$0.abcabcabc...... = 0.\overline{abc}$$

Where the digits a, b, c are not necessarily distinct. To write the elements of S as fractions in lowest terms, how many different numerators are required?

Solution:

Observe that 0.abcabcabc . . . = abc / 999, and that 999 = $3^3 \times 37$. If abc is neither divisible by 3 nor by 37, the fraction is already in lowest terms. By Inclusion - Exclusion there are

$999 - \left(\dfrac{999}{3} + \dfrac{999}{37}\right) + \dfrac{999}{3 \times 37} = 648$ such fractions. Also, fractions of the form $\dfrac{s}{37}$ where s is divisible by 3 but not by 37 are in S. There are 12 fractions of this kind (with s = 3, 6, 9, 12, . . . , 36). We do not consider fractions of the form $\dfrac{p}{3^t}, t \le 3$ with p divisible by 37 but not by 3, because these fractions are > 1 and hence not in S. The total number of distinct numerators in the set of reduced fractions is thus 640 + 12 = 660.

EXERCISE – 2

1. The value of $1 + 1 \times 1! + 2 \times 2! + 3 \times 3! + \ldots\ldots + n \times n!$ is
   (A) n^n (B) $(n!)^2$ (C) $(n + 1)!$ (D) $n! + n(n + 1)$

2. Determine the square root of $\underbrace{111\ldots\ldots111}\underbrace{222\ldots\ldots222}5$.
   $\overset{2010 \text{ times}}{} \quad \overset{}{2011 \text{ times}}$

   (A) $10^{2010} - 5$ (B) $\dfrac{10^{2011} + 5}{3}$ (C) $\dfrac{10^{2010} + 5}{3}$ (D) $10^{2010} + 5$

3. Magician R. Chandra claims to know two positive whole numbers that multiply to 1000, neither of which contains the digit 0. What is the sum of these 2 numbers?
   (A) 125   (B) 133   (C) 136   (D) none

4. If $x! = \dfrac{(7!)!}{7!}$, then the value of x is _____?

   (A) 5013   (B) 5038   (C) 5039   (D) 5040

5. If $\dfrac{1}{a} + \dfrac{1}{b} = \dfrac{1}{1000}$ then how many distinct ordered pairs of positive integers (a, b) are there which satisfy the given equation?
   (A) 35   (B) 36   (C) 42   (D) 49

6. What are the last two digits of 7^{2008}?
   (A) 21   (B) 61   (C) 01   (D) 41

7. Total number of integer pairs (x, y) satisfying the equation $x + y = xy$ is _____
   (A) 0   (B) 1   (C) 2   (D) None of these

8. Suppose n is an integer such that the sum of the digits of n is 2, and $10^{10} < n < 10^{11}$. The number of different values for n is
   (A) 11   (B) 10   (C) 9   (D) 8

9. Let T be the set of integers {3, 11, 19, 27,….., 451, 459, 467} and S be a subset of T such that the sum of no two elements of S is 470. The maximum possible number of elements in S is
   (A) 32   (B) 28   (C) 29   (D) 30

10. The sum of all positive integers n, such that $\dfrac{(n+1)^2}{n+7}$ is an integer.
    (A) 45   (B) 42   (C) 36   (D) 47

11. The base – 3 representation of x is 121122111122211112222_3. What is the first digit (on the left) of the base – 9 representation of x?
    (A) 2   (B) 4   (C) 5   (D) 3

I know your mind need relax, you can do it by walking in corridor, or having some extra curriculum activities or have an IQ Test.

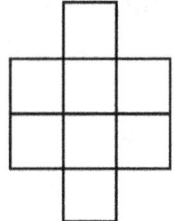

IQ TEST # 2

Arrange the numbers 1, 2, 3, 4, 5, 6, 7, & 8 in the grid as shown in figure so that no two consecutive integers are adjacent

(Vertically, Horizontally, and Diagonally).

DIVISIBILITY TEST

DIVISIBILITY BY 2:

a number is divisible by 2 if it has any of the digits 0, 2, 4, 6 or 8 in its ones place or unit place.

Ex.   4, 12, 30, 18, 102, etc., are all divisible by 2.

DIVISIBILITY BY 3:

A number is divisible by 3 if the sum of digits in the number is divisible by 3.

Ex.   The number 3792 is divisible by 3 since 3 + 7 + 9 + 2 = 21, which is divisible by 3. Other number divisible by 3 are 21, 27, 36, 54, 219 etc.

DIVISIBILITY BY 4:

a number with 3 or more digits is divisible by 4 if the number formed by its last two digits (i.e. ones and tens) is divisible by 4.

Ex.   The number 2616 is divisible by 4 since 16 is divisible by 4.

DIVISIBILITY BY 5:

A number is divisible by 5 if the unit's digit in the number is 0 or 5.

Ex.   13520, 7805, 7945, 8484765880, 640, 745, etc., are all divisible by 5.

DIVISIBILITY BY 6:

A number is divisible by 6 if the number is even and sum of its digits is divisible by 3 or in other words "if a number is divisible by 2 and 3 both then it is divisible by 6 also".

Ex.   The number 4518 is divisible by 6 since it is even and sum of its digits 4 + 5 + 1 + 8 = 18 is divisible by 3.

DIVISIBILITY BY 7:

A number of the form $10x + y$ is divisible by 7 if and only if $x - 2y$ is divisible by 7. In other words, the unit digit of the given number is doubled and then it is subtracted from the number obtained after omitting the unit digit. If the obtained number is divisible by 7, then the given number is also divisible by 7.

Ex.   Consider the number 448. On doubling the unit digit 8 of 448 we get 16, then, 44 - 16 = 28. Since 28 is divisible by 7, 448 is divisible by 7.

DIVISIBILITY BY 8:

a number with 4 or more digits is divisible by 8, if the number formed by the last three digits is divisible by 8.

Ex.   Number 41784 is divisible by 8 as the number formed by last three digits, i.e. 784 is divisible by 8. Other number divisible by 8 are 9216, 8216, 7216, 10216, 9995216 etc.

DIVISIBILITY BY 9:

A number is divisible by 9 if the sum of its digits is divisible by 9.

For example, the number 19044 is divisible by 9 as the sum of its digits $1 + 9 + 0 + 4 + 4 = 18$ is divisible by 9. There are other numbers like 4608, 837234, 48476547, 5283 that are also divisible by 9.

DIVISIBILITY BY 10:

A number is divisible by 10, if it unit place is zero.

Ex.   The last digit of 580 is zero, therefore, 580 is divisible by 10. Other numbers divisible by 10 are 20, 340, 650, 204569370, 3456390 etc.

DIVISIBILITY BY 11:

A number is divisible by 11 if the difference of the sum of the digits at odd places and sum of the digits at even places is either zero or divisible by 11.

Ex.   The number 38797, the sum of the digits at odd places is $3 + 7 + 7 = 17$ and the sum of the digits at even places is $8 + 9 = 17$. The difference is $17 - 17 = 0$, so the number is divisible by 11. The numbers 308, 1331 and 61809 are also divisible by 11.

DIVISIBILITY BY 12:

A number is divisible by 12 if it is divisible by 3 and 4.

Ex.   The number 72, this number is divisible by both 3 and 4 hence it's divisible by $3 \times 4 = 12$.

DIVISIBILITY BY 25:

A number is divisible by 25 if the number formed by the last two digits is divisible by 25 or the last two digits are zero.

Ex.   The number 13675 is divisible by 25 as the number formed by the last two digits is 75 which is divisible by 25.

DIVISIBILITY BY 125:

A number is divisible by 125 if the number formed by the last three digits is divisible by 125 or the last three digits are zero.

Ex.   The number 5250 is divisible by 125 as 250 is divisible by 125.

DIVISIBILITY BY 18:

An even number satisfying the divisibility test of 9 is divisible by 18.

DIVISIBILITY BY PRODUCT OF PRIMES:

A natural number 'N' is divisible by $P_1 \times P_2 \times P_3 \times ..... \times P_n$ (where $P_1, P_2, P_3, ..., P_n$ are primes) if N is divisible by all the primes $P_1, P_2, P_3, ... \& P_n$.

Ex.   Let $N = 60 = 2^2 \times 3 \times 5$, N is divisible by 2, 3 & 5 hence it will be divisible by $2 \times 3 \times 5 = 30$.

NOTE:

❖   Product of n consecutive natural numbers is divisible by n! (i.e n factorial)

   Ex $4 \times 5 \times 6 \times 7$ is divisible by $4! = 4 \times 3 \times 2 \times 1 = 24$.

❖   If a number is divisible by another number then it is divisible by each of the factors of that number.

❖   If two given numbers are divisible by a number, then their sum and difference is also divisible by that number.

Let's have another IQ test to distract from divisibility test to check retaining power of your brain.

IQ TEST # 3

What is the angle between Minute and Hour Hand at time 03:10 A.M.?

Have a verbal test to judge your learning.

VERBAL ABILITY TEST - I

1. The number divisible by 2, 3 & 5 is___
   (A) 47866391   (B) 4763890   (C) 9846730   (D) 984662010

2. Which of the following is prime number?
   (A) 899   (B) 997   (C) 1003   (D) 1147

3. The digit K for which the natural number 86739K2 is divisible by 8?
   (A) 1   (B) 3   (C) 5   (D) 7

4. Select the number divisible by 7.
   (A) 8567492   (B) 78675940   (C) 497057921   (D) 67486371

5. Select the number divisible by 11.
   (A) 784653782   (B) 758737550   (C) 9678592710   (D) 8484674563

6. If $n \in N$, then $n^3 - n$ is always divisible by
   (A) 5   (B) 4   (C) 6   (D) 7

7. If $3249 \times 1*5 = 438615$, then *=?
   (A) 1   (B) 2   (C) 3   (D) 6

8. If $2^{2010} - 2^{2009} - 2^{2008} + 2^{2007} = k \times 2^{2007}$, then k equals....
   (A) 1   (B) 2   (C) 3   (D) 0

9. Mr. Ramesh Chandra wishes to write down a list of different positive integers less than or equal to 10 such that for each pair of adjacent numbers one number is divisible by the other number. What is the length of the longest list of numbers that Mr. R. Chandra could write down?
   (A) 6   (B) 7   (C) 8   (D) 9

10. Select the correct option
    (A) Unit digit of 3786913^{20918} is 3   (B) Unit place of 45602^{26765} is 2
    (C) Number of zero in the end of 20! is 5   (D) 8484984345 is divisible by 125

PUZZLE # 6 (GOLD)

There are three boxes in a table. One of the box contains Gold and the other two are empty. A printed message contains in each box. One of the messages is true and the other two are lies.

   (i) The first box says '*The Gold is not here*'.
   (ii) The Second box says '*The Gold is not here*'.
   (iii) The Third box says '*The Gold is in the Second box*'.

Which box has the Gold?

ANSWER KEY – VERBAL TEST – I

1.	D	2.	B	3.	A, C	4.	B
5.	A	6.	C	7.	C	8.	C
9.	D Hint: 6, 3, 9, 1, 5, 10, 2, 4, 8.			10.	B		

MODULAR ARITHMETIC

Euclid's Division lemma states that for given positive integers 'a' and 'b', there exist unique integers 'q' and 'r'
satisfying $a = b \times q + r$, $0 \leq r < b$.

In another words, if we divide a natural number 'a' by another natural number 'b' then we will get
remainder 'r' such that $0 \leq r < b$ or 'r' may be 0, 1, 2, 3, 4, ..., b – 2, b – 1.

Ex. If we divide 23 by 5 then as per Euclid division lemma $23 = 5 \times 4 + 3$, where $0 \leq 3 < 5$ so remainder
is 3

NOTE: If we divide 23 by 5 we get remainder 3. The same thing can be written as $23 \equiv 3(\bmod 5)$.

EXPERIMENT # 1

Let's have an experiment to understand remainder in depth.

We are dividing a number $N = 8 + 21 + 43 + 87 + 321 + 95$ by 7 to obtain a remainder. Here $N = 575$. If we
divide it by 7 then we will get remainder '1'. Look and observe the following.

$8 \equiv 1 \,(\bmod 7)$, $21 \equiv 0(\bmod 7)$, $43 \equiv 1 \,(\bmod 7)$, $87 \equiv 3(\bmod 7)$, $321 \equiv 6(\bmod 7)$, $95 \equiv 4(\bmod 7)$.

Now let sum the remainders as: $1 + 0 + 1 + 3 + 6 + 4 = 15$. If we again divide 15 by 7 we will get
remainder '1'. Hence remainder = '1' for the case.

Even you can do like this: $1 + (0 + 1 + 3 + 6 + 4) = 1 + 14$ and $1 \equiv 1 \,(\bmod 7)$, $14 \equiv 0 \,(\bmod 7)$ so remainder is
$0 + 1 = 1$.

THEOREM #1

If $a \equiv b \,(\bmod m)$, then $a + c \equiv (b + c) \,(\bmod m)$ and $ac \equiv bc \,(\bmod m)$, where c is any integer.

THEOREM #2

If $a \equiv b \,(\bmod m)$ and $c = d \,(\bmod m)$, then

    (i) $a + c \equiv (b + d) \,(\bmod m)$ (ii) $a - c \equiv (b - d) \,(\bmod m)$

    (iii) $ac \equiv bd \,(\bmod m)$

THEOREM # 3

If $a \equiv b \,(\bmod m)$, then $a^k = b^k \,(\bmod m) \,\forall\, k \in N$.

EXPERIMENT # 2

Let $N = 253 \times 17 \times 15 = 64515$, if we divide it by usual method from 7 then we will get remainder 3.

Now observe, $253 \equiv 1(\bmod 7)$, $17 \equiv 3(\bmod 7)$, $15 \equiv 1 \,(\bmod 7)$. Hence, $253 \times 17 \times 15 = 1 \times 3 \times 1 \,(\bmod 7)$, so
remainder is 3.

CONCEPTS:

    (i) $a^n + b^n$ is divisible by $(a + b)$ if n is an odd natural number.

    (ii) $a^n - b^n$ is divisible by $a - b \,\forall\, n \in N$.

    (iii) As per binomial theorem, we can prove $(1 + x)^n - 1$ is divisible by x.

    (iv) Using binomial theorem we can say, $(1 + x)^n - nx - 1$ is divisible by x^2.

Ex1. If $x = (16^3 + 17^3 + 18^3 + 19^3)$, then x divided by 70 leaves a remainder of

    (A) 0   (B) 1   (C) 69   (D) 35

Solution:

$16^3 + 19^3$ is divisible by $(16 + 19) = 35$.

$17^3 + 18^3$ is divisible by $(17 + 18) = 35$.

$16^3 + 17^3 + 18^3 + 19^3$ is even so divisible by 2. Hence $16^3 + 17^3 + 18^3 + 19^3$ is divisible by 70.

So, (A) is correct option.

Ex2. The number of distinct primes dividing 12! + 13! + 14! Is

(A) 5　　　　　　　(B) 6　　　　　　　(C) 7　　　　　　　(D) 8

Solution:

12! + 13! + 14! = 12! $\times$(1 + 13 + 13 $\times$ 14) = 12! $\times 14^2$. Hence distinct prime divisors are 2, 3, 5, 7 & 11.

Option (A) is correct.

Ex3. Let n! = 1×2×3×.....×n for n ≥ 1. If p = 1! + (2×2!) + (3×3!) + ...+ (10×10!), then p + 2 when divided by 11! Leaves a remainder of

(A) 10　　　　　　　(B) 0　　　　　　　(C) 7　　　　　　　(D) 1

Solution:

r $\times$ r! = (r + 1 – 1) $\times$ r! = (r + 1)! – r!. So

p = (2! – 1!) + (3! – 2!) + (4! – 3!) + (5! – 4!) +.......+(11! – 10!) = 11! – 1!.

Hence p + 2 = 11! + 1. So desire remainder = 1. (D) Option is correct.

Ex4. If n∈N then find the remainder when $37^{n+2} + 16^{n+1} + 30^n$ is divided by 7.

Solution:

$37 \equiv 2$ (mod 7) i.e. if we divide 37 by 7 we get remainder 2. So $37^{n+2} \equiv 4 \times 2^n$ (mod 7)

Similarly, $16^{n+1} \equiv 2 \times 2^n$ (mod 7) and $30^n \equiv 2^n$ (mod 7), hence sum of all the remainders = $4 \times 2^n + 2 \times 2^n + 2^n = 7 \times 2^n$, it's divisible by 7, so remainder is '0'.

Ex5. Find the smallest Natural number x > 15 such that if be divide x by Numbers 77 or 286 we get the remainder 15.

Solution:

77 = 7 $\times$ 11 and 286 = 2 $\times$ 11 $\times$ 13

Hence, least Natural Number, multiple of 77 and 286 = 2×11×13×7 = 2002, so number x = 2002 + 15

x = 2017.

Ex6. The remainder when $1^{1997} + 2^{1997} + 3^{1997} + ...+ 1996^{1997}$ is divided by 1997, is

(A) 0　　　　　　　(B) 1　　　　　　　(C) 197　　　　　　　(D) 1996

Solution:

As per concept $a^n + b^n$ is divisible by (a + b) ∀ n ∈ odd naturals.

Hence, $(1^{1997} + 1996^{1997}) + (2^{1997} + 1995^{1997}) + (3^{1997} + 1994^{1997}) +.........+ (998^{1997} + 999^{1997})$ is divisible by 1997. So, remainder = 0. Hence, (A) is correct option.

Ex7. What is the remainder when $1^{2013} + 2^{2013} + 3^{2013} + 4^{2013} + \ldots 2012^{2013} + 2013^{2013}$ is divided by 2014?

Solution:

As we know that if $n \in$ Odd Natural numbers then $a^n + b^n$ is divisible by $(a + b)$. You can see the formula derived from sum of n terms of a Geometric Progression as

$$a^{n-1} - a^{n-2}b + a^{n-3}b^2 - a^{n-4}b^3 + \ldots + (-1)^{n-1}b^{n-1} = \frac{a^n + b^n}{a+b}.$$

So, $a^n + b^n = (a + b)(a^{n-1} - a^{n-2}b + a^{n-3}b^2 - a^{n-4}b^3 + \ldots + (-1)^{n-1}b^{n-1})$.

Now $1^{2013} + 2013^{2013}, 2^{2013} + 2012^{2013}, 3^{2013} + 2011^{2013} \ldots$ Must be divisible by 2014, but there will remain a number 1007^{2013} which will give us a remainder.

$$1007^{2013} = \left(1007^{2012}\right)(1006 + 1) = \underbrace{\left(1007^{2012} \times 1006\right)}_{\text{divisible by 2014}} + 1007^{2012}$$

Similarly, $1007^{2012} = \left(1007^{2011}\right)(1006+1) = \underbrace{\left(1007^{2011} \times 1006\right)}_{\text{divisible by 2014}} + 1007^{2011}$. If we continue the process then

will get a situation $1007^2 = (1007)(1006 + 1)$. Hence remainder is 1007.

Even we can apply Chinese remainder theorem or a result.

$1007^{2013} \equiv 1 \pmod 2$ and $1007^{2013} \equiv 0 \pmod{1007} \rightarrow 1007^{2013} \equiv 1007 \pmod{2014}$.

Ex8. The largest non – negative integer k such that 24^k divides 13! Is

(A) 2   (B) 3   (C) 4   (D) 5

Solution:

$13! = 1 \times 2 \times 3 \times 4 \times 5 \times 6 \times 7 \times 8 \times 9 \times 10 \times 11 \times 12 \times 13 = 2^{10} \times 3^5 \times 5^2 \times 7 \times 11 \times 13 = 24^3 \times (2 \times 3^2 \times 5^2 \times 7 \times 11 \times 13)$

$\therefore$ Maximum k can be 3. Hence Option (B) is correct.

Ex9. When a natural number x is divided by 5, the remainder is 2. When a natural number y is divided by 5, the remainder is 4. The remainder is z when x + y is divided by 5. The value of $\frac{2z-5}{3}$ is

(A) -1   (B) 1   (C) -2   (D) 2

Solution:

As per question $x = 5k_1 + 2$, $y = 5k_2 + 4$, where $k_1, k_2 \in I$. $x + y = 5(k_1 + k_2) + 6$. So when x + y is divided by 5, remainder will be '1'. So z = 1. $\therefore \frac{2z-5}{3} = -1$ Hence, Option (A) is correct.

Ex10. The remainder when 2^{2003} is divided by 100.

(A) 16   (B) 13   (C) 8   (D) 64

Solution:

$2^{2003} = 2^3 \times (2^{10})^{200} \rightarrow 2^{2003} \equiv 8 \times (24)^{200} \pmod{100}$.

$$24^{200} = (25-1)^{200} = \underbrace{{}^{200}C_0(25)^{200} - {}^{200}C_1(25)^{199} + {}^{200}C_2(25)^{198} - \ldots - {}^{200}C_{199}25}_{\text{a positive Number}} + {}^{200}C_{200}. \text{ So}$$

$8 \times 24^{200} = 100 \times (\text{a Natural Number}) + 8$. As $24^{200} > 0 \rightarrow 8 \times (24)^{200} \equiv 8 \pmod{100}$, hence I option is correct.

Ex11. The remainder of $2005^{2002} + 2002^{2005}$ when divided by 200 is _____

    (A) 0   (B) 71   (C) 27   (D) 57

Solution:

$2005^{2002} + 2002^{2005} \equiv r \pmod{200}$ is equivalent to $5^{2002} + 2^{2005} \equiv r \pmod{200}$.

In 5^{2002}, last 3 digits = 625 and $2^{2005} = 32 \times (2^{10})^{200} \equiv 32 \times (24)^{200} \bmod (200)$.

Now, $24^{200} > 0$ so, $(25 - 1)^{200} = 25k + 1$, where $k \in N$ (using Binomial Expansion)

Hence $32 \times (24)^{200} = 200 I + 32$, where $I \in$ Integers $\rightarrow 32 \times (24)^{200} \equiv 32 \pmod{200}$

$\therefore 2005^{2002} + 2002^{2005} \equiv (25 + 32) \pmod{200}$ so, remainder = 57. Hence, (D) option is correct.

HIGHEST COMMON FACTOR (H C F) or (G C D)

HCF of two natural numbers 'a' and 'b' is the largest natural number 'c' that divide both 'a' and 'b'. Its an application of Euclid division.

Ex1. 1, 2, 3, 4, 6, 12 are natural divisors of number a = 12.

    1, 2, 3, 6, 9, 18 are natural divisors of number b = 18.

    Here, you can observe, 2, 3, 6 are common divisors of 12 and 18 both and 6 is the highest one, hence HCF (12, 18) = 6.

Working Rule:

$12 = 2^2 \times 3^1$ and $18 = 2 \times 3^2$ so HCF $(12, 18) = 2^p \times 3^q$ where p and q are the lowest exponents of common primes factors. i.e. p = 1 and q = 1. So, HCF $(12, 18) = 2^1 \times 3^1 = 6$.

General Note:

Let $a = p_1^{\alpha_1} p_2^{\alpha_2} p_3^{\alpha_3} \ldots\ldots p_n^{\alpha_n}$ & $b = p_1^{\beta_1} p_2^{\beta_2} p_3^{\beta_3} \ldots\ldots p_n^{\beta_n}$, where a, b $\in N$ and p_1, p_2,...p_n are different primes, then, $\text{HCF}(a, b) = p_1^{\text{MIN}\{\alpha_1, \beta_1\}} p_2^{\text{MIN}\{\alpha_2, \beta_2\}} p_3^{\text{MIN}\{\alpha_3, \beta_3\}} \ldots\ldots p_n^{\text{MIN}\{\alpha_n, \beta_n\}}$.

Ex2. Find the HCF of 120 and 150.

Solution:

$120 = 2^3 \times 3^1 \times 5^1$ and $150 = 2^1 \times 3^1 \times 5^2$ so, HCF $(120, 150) = 2^1 \times 3^1 \times 5^1 = 30$.

Ex3. Find HCF (60, 75) using Euclid Division.

Solution:

    STEP 1: $75 = 60 \times 1 + 15$

    STEP 2: $60 = 15 \times 4 + 0$ ....$\rightarrow$ process stop after getting 0. So, HCF (60, 75) = 15.

Ex4. Find the GCD of 2016! + 1 and 2017! + 1.

Solution:

    2016! + 1 and 2017! + 1 are two prime numbers. Hence, GCD (2016! + 1, 2017! + 1) = 1.

Ex5. Find HCF (48, 60, 75).

Solution:

Prime factorizations of these numbers are as follow:

$48 = 2^4 \times 3$

$60 = 2^2 \times 3 \times 5$

$75 = 3 \times 5^2$

so, HCF $(48, 60, 75) = 2^0 \times 3^1 \times 5^0 = 3$.

Ex6. Find the value of $\sum\limits_{x=1}^{2015} \gcd(x, 2015)$.

Solution:

Prime factorisation of $2015 = 5 \times 13 \times 31$

Let A_n = set of natural numbers $\in [1, 2015]$ which are divisible by n, hence

$A_5 = \{5, 10, 15, 20, 25, ..., 2015\} \rightarrow n(A_5) = 403$

$A_{13} = \{13, 26, 39, ..., 2015\} \rightarrow n(A_{13}) = 155$

$A_{31} = \{31, 62, 93, ..., 2015\} \rightarrow n(A_{31}) = 65$

$A_5 \cap A_{13} = \{65, 130, ..., 2015\} \rightarrow n(A_5 \cap A_{13}) = 31$

$A_5 \cap A_{31} = \{155, 310, 465, ..., 2015\} \rightarrow n(A_5 \cap A_{31}) = 13$

$A_{13} \cap A_{31} = \{403, 806, ..., 2015\} \rightarrow n(A_{13} \cap A_{31}) = 5$

$A_5 \cap A_{13} \cap A_{31} = \{2015\} \rightarrow n(A_5 \cap A_{13} \cap A_{31}) = 1$.

$N(A_5 \cup A_{13} \cup A_{31}) = 403 + 155 + 65 - 31 - 13 - 5 + 1 = 575$.

Let $a \in \{1, 2, 3, 4, ..., 2015\}$ and $a \notin A_5 \cup A_{13} \cup A_{31}$, then $\gcd(a, 2015) = 1$.

Number of such element a = 2015 − 575 = 1440.................(1)

Now look at the figure as shown and observe it.

Figure is showing the number of elements in different regions.

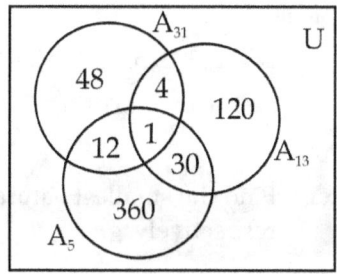

So, sum of the remaining gcd(x, 2015) apart from (1) is

$= 360 \times 5 + 120 \times 13 + 48 \times 31 + 4 \times 403 + 12 \times 155 + 30 \times 65 + 1 \times 2015$

$= 12285$.

Hence $\sum\limits_{x=1}^{2015} \gcd(x, 2015) = 12285 + 1440 = 13725$

LOWEST COMMON MULTIPLE (L C M)

If we multiply a number 'a' by counting numbers N = {1, 2, 3, 4, 5,.....} then we will get multiple of 'a'.

MULTIPLE OF 3

3 → 6 → 9 → 12 → 15 → 18

0 1 2 3 4 5 6 7 8 9 10 11 12 13 14 15 16 17 18 19 20

5 → 10 → 15 → 20

MULTIPLE OF 5

The multiple of 3 are 3, 6, 9, 12, 15, 18, 21,........

Similarly multiple of 5 are 5, 10, 15, 20, 25,.......

Common multiple of 3 and 6 are 15, 30, 45, 60,........

Lowest or least common multiple = 15. Look at the figure and analyse.

Definition:

LCM of two numbers a and b usually denoted as LCM (a, b) is the smallest number divisible by both a and b.

Ex1. Find LCM (30, 40).

Solution:

Multiple of 30 are 30, 60, 90, 120, 150, 180, ...

Multiple of 40 are 40, 80, 120, 160, 200, ...

So, lowest common multiple = 120. Hence, LCM (30, 40) = 120.

PRIME FACTORISATION METHOD

$30 = 2 \times 3 \times 5$

$40 = 2^3 \times 5$

Let 'a' be the LCM (30, 40) then its divisible by 30 as well as 40. So the number must be a = $2^3 \times 3 \times 5$.

= 120.

WORKING RULE:

Let $a = p_1^{\alpha_1} \times p_2^{\alpha_2} \times p_3^{\alpha_3} \times ..... \times p_n^{\alpha_n}$ and $b = p_1^{\beta_1} \times p_2^{\beta_2} \times p_3^{\beta_3} \times ..... \times p_n^{\beta_n}$ where $p_1, p_2, p_3, ..., p_n$ are primes, then $LCM(a,b) = p_1^{MAX\{\alpha_1, \beta_1\}} \times p_2^{MAX\{\alpha_2, \beta_2\}} \times p_3^{MAX\{\alpha_3, \beta_3\}} \times .... \times p_n^{MAX\{\alpha_n, \beta_n\}}$.

Ex2. If we divide a natural number 'n' by 2, 3, 5, 7, 6 it will leave remainder '1' then find its least value.

Solution

LCM (2, 3, 5, 7, 6) = $2 \times 3 \times 5 \times 7$ which is divisible by 2, 3, 5, 7 & 6.

So least value of n = $2 \times 3 \times 5 \times 7 + 1 = 211$.

Ex3. Find the smallest natural number 'N' leave remainder 5, 4 and 3 on division by 6, 5 and 4 respectively.

Solution:

N = $6k_1 + 5 = 5k_2 + 4 = 4k_3 + 3$ where $k_1, k_2, k_3 \in I$

N = $6k_1 + 6 - 1 = 5k_2 + 5 - 1 = 4k_3 + 4 - 1$. Here you need to observe the pattern.

So, N = LCM (6, 5, 4) – 1 = 60 – 1 = 59.

Ex4. What is the smallest positive integer n such that LCM (n, 30) = 180?

Solution:

$180 = 2^2 \times 3^2 \times 5$.

$30 = 2 \times 3 \times 5$. For smallest n, it must include the factors 2^2 and 3^2, so n = $2^2 \times 3^2 = 36$.

Ex5. Three cats start racing on circular track from a point P. They take time 5, 8 and 12 minute to complete the one round of the track. After, how much time they all will meet together first time, to finish the Race?

Solution:

Race will complete after LCM (5, 8, 12) minutes i.e. 120 Minutes.

Ex6. Find the LCM of following pairs

(i) 4 & - 2   (ii) π and e   (iii) 2 and π (iv) – 6 and – 15

Solution:

(i) Multiple of 4 = {4, 8, 12, 16,........}

Multiple of – 2 = {- 2, - 4, - 6, - 8, - 10, ...}

So common multiple = $\emptyset$ i.e. LCM (4, - 2) = Does not exist.

(ii) LCM (π, e) = Not define or Does not exist (LCM calculated in like quantities)

(iii) LCM (π, 2) = Not define or Does not exist

(iv) LCM (- 6, - 15) = - 30.

Note: *LCM is subject to disputed case, it works well in natural numbers. I am writing our thinking.*

EXPERIMENT # 3 (LCM & HCF)

Consider a number $a = p_1^{\alpha_1} \times p_2^{\alpha_2} \times p_3^{\alpha_3} \times ..... \times p_n^{\alpha_n}$ and $b = p_1^{\beta_1} \times p_2^{\beta_2} \times p_3^{\beta_3} \times ..... \times p_n^{\beta_n}$.

Hence, $LCM(a,b) = p_1^{MAX\{\alpha_1, \beta_1\}} \times p_2^{MAX\{\alpha_2, \beta_2\}} \times p_3^{MAX\{\alpha_3, \beta_3\}} \times .... \times p_n^{MAX\{\alpha_n, \beta_n\}}$

$HCF(a, b) = p_1^{MIN\{\alpha_1, \beta_1\}} p_2^{MIN\{\alpha_2, \beta_2\}} p_3^{MIN\{\alpha_3, \beta_3\}} .... p_n^{MIN\{\alpha_n, \beta_n\}}$

$HCF(a, b) \times LCM(a, b) = p_1^{\alpha_1} \times p_1^{\beta_1} \times p_2^{\alpha_2} \times p_2^{\beta_2} \times p_3^{\alpha_3} \times p_3^{\beta_3} \times .... \times p_n^{\alpha_n} \times p_n^{\beta_n}$

$= p_1^{\alpha_1} \times p_2^{\alpha_2} \times p_3^{\alpha_3} \times .... \times p_n^{\alpha_n} \times p_1^{\beta_1} \times p_2^{\beta_2} \times p_3^{\beta_3} \times .... \times p_n^{\beta_n} = a \times b$

Ex7. Let a = $2^3 \times 3$ and b = $2^4 \times 5$ then LCM (a, b) = $2^4 \times 3 \times 5$, HCF (a, b) = 2^3.

HCF (a, b) × LCM (a, b) = $(2^3) \times (2^4 \times 3 \times 5) = 2^{4+3} \times 3^{1+0} \times 5^{1+0} = (2^3 \times 3^1) \times (2^4 \times 5^1) = a \times b$.

NOTE:

(i) HCF (a, b) × LCM (a, b) = a × b

(ii) $LCM(p,q,r) = \dfrac{p \cdot q \cdot r \cdot HCF(p,q,r)}{HCF(p,q) \cdot HCF(p,r) \cdot HCF(q,r)}$ $\forall$ p, q, r $\in$ N.

(iii) $HCF(p,q,r) = \dfrac{p \cdot q \cdot r \cdot LCM(p,q,r)}{LCM(p,q) \cdot LCM(p,r) \cdot LCM(q,r)}$ $\forall$ p, q, r $\in$ N.

(iv) $LCM\left(\dfrac{a}{b}, \dfrac{c}{d}, \dfrac{e}{f}\right) = \dfrac{LCM(a,c,e)}{HCF(b,d,f)}$ Whenever defined.

(v) $HCF\left(\dfrac{a}{b}, \dfrac{c}{d}, \dfrac{e}{f}\right) = \dfrac{HCF(a,c,e)}{LCM(b,d,f)}$ Whenever defined.

Ex8. The LCM of two numbers is 2079 and their HCF is 27. If one of the numbers is 189, find the other number.

Solution:

HCF (a, b) × LCM (a, b) = a × b → 2079 × 27 = 189 × b. So, b = 297.

Ex9. Find the number of ordered pair (x, y) where x, y ∈ N such that LCM (x, y) = $2^3 × 3^2$.

Solution:

Lets distribute 2 and 3 among x and y.

If x = 2^3 then y has 4 choices {2^0, 2^1, 2^2, 2^3}.

If y = 2^3 then x has 4 choices {2^0, 2^1, 2^2, 2^3}.

So total ways of distribution of 2 are (4 + 4 – 1) = 7.

Similarly, total ways of distribution of 3 = (3 + 3 – 1) = 5.

Hence, total ordered pairs (x, y) = 7×5 = 35.

PRACTICE PROBLEM:

Q1. Find the number of ordered triplets (x, y, z), where x, y, z ∈ N such that LCM(x, y, z) = $2^3 × 3^3$?

Solution:

Since we are taking LCM so we can assume $x = 2^a × 3^d$, $y = 2^b × 3^e$ & $z = 2^c × 3^f$ where a, b, c, d, e, f, g ∈ W. As per question max {a, b, c} = 3 and max {d, e, f} = 3.

Number of ways in which max {a, b, c} = 3 is $4^3 – 3^3$ = 37.

Similarly, number of cases for max {d, e, f} = 3 is 37.

Hence numbers of required ordered triplets = 37 × 37 = 1369.

Ex10. Find the 2 digits natural numbers which gives remainder 3, 3 and 7 on dividing by 5, 7 and 11 respectively.

Solution:

Let the number be N = $5k_1$ + 3 = $7k_2$ + 3 = $11k_3$ + 7 where k_1, k_2, k_3 ∈ I.

LCM (5, 7) = 35 so 2 digits required number may be 35+3 = 38 or 70 + 3 = 73 and only 73 will satisfy the divisibility norm of 11. Hence, 73 is the answer.

Ex11. LCM of two numbers x & y is 720 and the LCM of numbers 12x and 5y is also 720. The number y is

(A) 180   (B) 144   (C) 120   (D) 90

Solution:

720 = $2^4 × 3^2 × 5$.

LCM{x, y} = LCM {12x, 5y} = LCM {2^2.3 x, 5y} = 720 →x = $2^2 × 3 × 5$ and y = $2^4 × 3^2$. Hence y = 144.

Ex12. What is the greatest number that will divide 38, 45 and 52 and leave 2, 3 and 4 respectively as remainders?

Solution.

Let the required number be 'N'

If 38, 45 and 52 and leave 2, 3 and 4 respectively as remainders then

38 – 2 = 36, 45 – 3 = 42 and 52 – 4 = 48 will be divisible by N.

→ N = HCF {36, 42, 48} = 6. So, 6 is the required answer.

Ex13. Find the greatest number which will divide 410, 751 and 1030 so as to leave remainder 7 in each case.

Sol. Let N is the required number

→ 410 – 7 = 403, 751 – 7 = 744 and 1030 – 7 = 1023 are divisible by N

For greatest value of

N = HCF {403, 746, 1023}

Concept: If N = HCF {a, b, c} then N will divide $k_1 a - k_2 b$, $l_1 b - l_2 c$, e.t.c.

Here 403 × 2 – 746 = 62 = 2 × 31 → N can't be greater than 62 and N is a factor of 2 × 31.

But 1023 is odd so not divisible by 2 hence 31 may be the HCF.

If you divide 403, 746 and 1023 you will get 31 as highest common factor.

So, N = HCF {403, 746, 1023} = 31.

Ex14. Find the greatest number which is such that when 76, 151 and 226 are divided by it, the remainders are alike. Find also the common remainder.

Sol. Let the greatest number be N then

(i) $76 = k_1 N + r$ (ii) $151 = k_2 N + r$ (iii) $226 = k_3 N + r$ where $k_i \in$ Integers

From (i), (ii) and (iii) we get

$151 - 76 = 75 = (k_2 - k_1) N$ → $N_{max} \le 75$ and N_{max} is a factor of 75.

Now N will divide 151 – 76 = 75, 226 – 151 = 75 and 226 – 76 = 150

→ N = HCF {75, 75, 150} = 75. So, N = 75 is the required number.

Now common remainder r = 1 as 76 = 1×75 + 1.

ALGEBRAIC EXPRESSIONS AND THEIR IDENTITIES

Here we are going to have some important formula related to the topic.

Let $n \in N$, a, b and c $\in$ R or C then

(i) $(a + b)^2 = a^2 + b^2 + 2ab$

(ii) $(a - b)^2 = a^2 + b^2 - 2ab$

(iii) $(a + b)^2 + (a - b)^2 = 2 (a^2 + b^2)$

(iv) $(a + b)^2 - (a - b)^2 = 4ab$

(v) $(x + a)(x + b) = x^2 + (a + b)x + ab$

(vi) $(x + a)(x + b)(x + c) = x^3 + (a + b + c)x^2 + (ab + bc + ac)x + abc$

(vii) $(a + b)^3 = a^3 + b^3 + 3ab(a + b) = a^3 + 3a^2b + 3ab^2 + b^3$

(viii) $(a - b)^3 = a^3 - b^3 - 3ab(a - b) = a^3 - 3a^2b + 3ab^2 - b^3$

(ix) $(a+b)^n = {}^nC_0a^nb^0 + {}^nC_1a^{n-1}b^1 + {}^nC_2a^{n-2}b^2 + \cdots + {}^nC_2a^0b^n$, where ${}^nC_r = \dfrac{n!}{(n-r)! \times r!}$

(x) $a^2 - b^2 = (a - b)(a + b)$

(xi) $a^4 - b^4 = (a - b)(a + b)(a^2 + b^2)$

(xii) $a^3 - b^3 = (a - b)(a^2 + ab + b^2)$

(xiii) $a^3 + b^3 = (a + b)(a^2 - ab + b^2)$

(xiv) $a^n - b^n = (a - b)(a^{n-1} + a^{n-2}b + a^{n-3}b^2 + \ldots + b^{n-1})$

(xv) $a^n + b^n = (a + b)(a^{n-1} - a^{n-2}b + a^{n-3}b^2 - \ldots + (-1)^{n-1}b^{n-1})$ $\forall$ $n \in$ Odd Natural numbers.

(xvi) $(a + b + c)^2 = a^2 + b^2 + c^2 + 2(ab + bc + ca)$

(xvii) $a^3 + b^3 + c^3 - 3abc = (a + b + c)(a^2 + b^2 + c^2 - ab - bc - ca)$

　　　If $a + b + c = 0$ then $a^3 + b^3 + c^3 = 3abc$

(xviii) $a^2 + b^2 + c^2 - ab - bc - ca = ((a - b)^2 + (b - c)^2 + (c - a)^2)/2$

(xix) $x^4 + x^2 + 1 = (x^2 + x + 1)(x^2 - x + 1)$

(xx) $ab + bc + ca = abc\left(\dfrac{1}{a} + \dfrac{1}{b} + \dfrac{1}{c}\right)$ $\forall$ $a, b, c \in R - \{0\}$

Ex1. If $x + \dfrac{1}{x} = 4$ then find the value of following

(i) $x^2 + \dfrac{1}{x^2}$ 　　　(ii) $x^3 + \dfrac{1}{x^3}$ 　　　(iii) $\sqrt{x} + \dfrac{1}{\sqrt{x}}$

Solution:

(i) $x^2 + \dfrac{1}{x^2} = \left(x + \dfrac{1}{x}\right)^2 - 2 \to x^2 + \dfrac{1}{x^2} = 4^2 - 2 = 14$

(ii) $x^3 + \dfrac{1}{x^3} = \left(x + \dfrac{1}{x}\right)^3 - 3\left(x + \dfrac{1}{x}\right) = 4^3 - 3 \times 4 = 52$

(iii) let $\sqrt{x} + \dfrac{1}{\sqrt{x}} = t \to x + \dfrac{1}{x} + 2 = t^2 \to 4 + 2 = t^2 \to t = \pm\sqrt{6}$ but $t > 0$ so $t = \sqrt{6}$

Ex2. The number of distinct prime divisors of the number $512^3 - 253^3 - 259^3$ is

(A) 4 　　　(B) 5 　　　(C) 6 　　　(D) 7

Solution:

As $a^3 + b^3 + c^3 - 3abc = (a + b + c)(a^2 + b^2 + c^2 - ab - bc - ca)$

If $a + b + c = 0$, then $a^3 + b^3 + c^3 - 3abc = 0$. $\to a^3 + b^3 + c^3 = 3abc$.

Hence, $512^3 - 253^3 - 259^3 = 3(512)(-253)(-259) = 3 \times 2^9 \times 11 \times 23 \times 7 \times 37$. So, distinct prime numbers are 2, 3, 7, 11, 23 & 37. Therefore I is the correct option.

Ex3. If $\sqrt{a} + \sqrt{b} - \sqrt{c} = 0$, then value of $(a + b - c)^2$ is:
(A) 2ab 　　　(B) 2bc 　　　(C) 4ab 　　　(D) 4ac

Solution:

$\sqrt{a} + \sqrt{b} - \sqrt{c} = 0 \rightarrow \sqrt{a} + \sqrt{b} = \sqrt{c}$. Now square both side, we get $a + b + 2\sqrt{(ab)} = c$.

$\rightarrow (a + b - c) = -2\sqrt{(ab)} \rightarrow (a + b - c)^2 = 4ab$. I is correct option.

Ex4. If $x + \dfrac{1}{x} = a$, $x^2 + \dfrac{1}{x^3} = b$, then $x^3 + \dfrac{1}{x^2}$ is –

(A) $a^3 + a^2 - 3a - 2 - b$ (B) $a^3 - a^2 - 3a + 4 - b$ (C) $a^3 - a^2 + 3a - 6 - b$ (D) $a^3 + a^2 + 3a - 16 - b$

Solution:

As $x + \dfrac{1}{x} = a$, $x^2 + \dfrac{1}{x^3} = b$

$\left(x + \dfrac{1}{x}\right)^2 = a^2 \rightarrow x^2 + \dfrac{1}{x^2} + 2 = a^2 \quad\ldots\ldots\ldots\ldots(1)$ and $\left(x + \dfrac{1}{x}\right)^3 = a^3 \rightarrow x^3 + \dfrac{1}{x^3} + 3 \cdot x \cdot \dfrac{1}{x}\left(x + \dfrac{1}{x}\right) = a^3 \quad\ldots..(2)$

On adding (1) and (2) we get, $a^2 + a^3 = x^2 + \dfrac{1}{x^2} + 2 + x^3 + \dfrac{1}{x^3} + 3 \cdot x \cdot \dfrac{1}{x}\left(x + \dfrac{1}{x}\right)$

$\rightarrow a^2 + a^3 = x^2 + \dfrac{1}{x^3} + 2 + x^3 + \dfrac{1}{x^2} + 3\left(x + \dfrac{1}{x}\right) \rightarrow a^2 + a^3 = b + 2 + x^3 + \dfrac{1}{x^2} + 3a$

$\rightarrow x^3 + \dfrac{1}{x^2} = a^3 + a^2 - 3a - 2 - b$. Hence option (A) is correct.

Ex5. If $2x + y = 6$ and $xy = 5$ then find the value of $8x^3 + y^3$.

Solution:

$2x + y = 6$, Cube both side we get $(2x + y)^3 = 6^3$

$\rightarrow 8x^3 + y^3 + 3(2x)y(2x + y) = 6^3 \rightarrow 8x^3 + y^3 + 6xy(2x + y) = 216$

$\rightarrow 8x^3 + y^3 = 216 - 6 \times 5(6) = 36$.

Ex6. If $x + y + z = 10$ and $xy + yz + zx = 20$, then find the value of $x^3 + y^3 + z^3 - 3xyz$.

Solution:

$x + y + z = 10 \rightarrow (x + y + z)^2 = 100$

$x^2 + y^2 + z^2 + 2(xy + yz + zx) = 100 \rightarrow x^2 + y^2 + z^2 = 100 - 2 \times 20 = 60$.

Now, $x^3 + y^3 + z^3 - 3xyz = (x + y + z)(x^2 + y^2 + z^2 - (xy + yz + zx)) = 10(60 - 20) = 400$.

Ex7. If $x = \sqrt{5 - 2\sqrt{6}}$, then find the value of $x^4 + \dfrac{1}{x^4}$.

Solution:

$x = \sqrt{5 - 2\sqrt{6}} \rightarrow x = \sqrt{3} - \sqrt{2}$ so $\dfrac{1}{x} = \dfrac{1}{\sqrt{3} - \sqrt{2}} = \sqrt{3} + \sqrt{2}$

Hence, $x + \dfrac{1}{x} = 2\sqrt{3}$. Now, $x^4 + \dfrac{1}{x^4} = \left(x^2 + \dfrac{1}{x^2}\right)^2 - 2 = \left(\left(x + \dfrac{1}{x}\right)^2 - 2\right)^2 - 2 = (12 - 2)^2 - 2 = 98$.

Ex8. If $x + y = 1$, $x^3 + y^3 = 4$, then find the value of $x^2 + y^2$

Solution:

$x^3 + y^3 = (x + y)(x^2 + y^2 - xy)$ ....(1)

Now, $(x + y)^2 = x^2 + y^2 + 2xy \rightarrow xy = (1 - (x^2 + y^2))/2$ .....(2)

Now by (1) and (2) we can say

$x^3 + y^3 = (x + y)(x^2 + y^2 - (1 - (x^2 + y^2))/2) \rightarrow x^3 + y^3 = (x+y)\left(\frac{3}{2}(x^2 + y^2) - \frac{1}{2}\right)$

$2(x^3 + y^3) = (x + y)(3(x^2 + y^2) - 1) \rightarrow 8 = 3(x^2 + y^2) - 1 \rightarrow x^2 + y^2 = 3$.

Ex9. If $x > 0$ and $x^4 + \dfrac{1}{x^4} = 47$ then find the value of $x^3 + \dfrac{1}{x^3}$

Solution:

$x^3 + \dfrac{1}{x^3} = \left(x + \dfrac{1}{x}\right)\left(x^2 + \dfrac{1}{x^2} - 1\right)$ ....(1)

$x^4 + \dfrac{1}{x^4} = \left(x^2 + \dfrac{1}{x^2}\right)^2 - 2 = \left(\left(x + \dfrac{1}{x}\right)^2 - 2\right)^2 - 2$ ......(2)

Let $x + \dfrac{1}{x} = t > 0$ as $x > 0$. So from (2) we can get, $47 = (t^2 - 2)^2 - 2$

$\rightarrow t^2 - 2 = \pm 7 \rightarrow t^2 = 2 \pm 7$, but $t^2 > 0$, hence $t^2 = 9$ and $t = 3$ as $t > 0 \rightarrow x + \dfrac{1}{x} = 3 \rightarrow x^2 + \dfrac{1}{x^2} = 7$.

So, from (1) $x^3 + \dfrac{1}{x^3} = 3(7 - 1) = 18$.

Ex10. Solve the system of equations

$x^2 + xy + y^2 = 19$ and $x^2 - xy + y^2 = 49$

Solution:

If we add these two equations we get $2(x^2 + y^2) = 68 \rightarrow x^2 + y^2 = 34$ ....(1)

If we subtract one equation from other we get, $2xy = -30 \rightarrow xy = -15$ ....(2)

Now, we know that $(x + y)^2 = x^2 + y^2 + 2xy$

So from eq (1) and (2) we get $(x + y)^2 = 34 - 30 = 4 \rightarrow x + y = \pm 2$ ....(3)

Similarly, $(x - y)^2 = x^2 + y^2 - 2xy = 34 + 30 = 64 \rightarrow x - y = \pm 8$ ....(4)

Now, there are 4 cases.

Case (I) $x + y = 2$ and $x - y = 8 \rightarrow x = 5$ and $y = -3$

Case (II) $x + y = -2$ and $x - y = 8 \rightarrow x = 3$ and $y = -5$

Case (III) $x + y = 2$ and $x - y = -8 \rightarrow x = -3$ and $y = 5$

Case (IV) $x + y = -2$ and $x - y = -8 \rightarrow x = -5$ and $y = 3$

So, solutions (x, y) are $(5, -3)$, $(3, -5)$, $(-3, 5)$ and $(-5, 3)$.

Ex11. How many different solutions (x, y, z) does the system of equations

$xy + yz + zx = 12$

$xyz = 2 + x + y + z$

have if x, y and z are positive real numbers?

(A) 0   (B) 1   (C) 3   (D) Infinite

Solution:

Let $xyz = a^3$ where $a \in R^+$.

Now, Apply AM. $\geq$ GM. On $xy + yz + zx$, we get $xy + yz + zx \geq 3 a^2$.

Similarly, apply AM. $\geq$ GM. On $x + y + z$, we get $x + y + z \geq 3 a$.

So, $12 = xy + yz + zx \geq 3 a^2$ and $a^3 = 2 + x + y + z \geq 2 + 3a$

Therefore we get, $a^2 \leq 4$ and $a^3 \geq 2 + 3a \rightarrow a^3 - 3a - 2 \geq 0 \rightarrow (a - 2)(a + 1)^2 \geq 0 \rightarrow a \in [2, \infty)$.

Hence, both in-equations $a^2 \leq 4$ and $(a - 2)(a + 1)^2 \geq 0$ valid only if $a = 2$.

Equality holds only if $x = y = z = a = 2$, i.e. $(2, 2, 2)$ is the only solution. Option (B) is correct.

Ex12. If $a^2 + 2b = 7$, $b^2 + 4c = -7$ and $c^2 + 6a = -14$ where $a, b, c \in R$, then the value of $a^2 + b^2 + c^2$ is.

Solution:

If we add all the three given equation, we get $a^2 + 2b + b^2 + 4c + c^2 + 6a = -14$

$\rightarrow a^2 + 6a + b^2 + 2b + c^2 + 4c + 3^2 + 1^2 + 2^2 = 0$

$\rightarrow (a + 3)^2 + (b + 1)^2 + (c + 2)^2 = 0$, is possible only if $a + 3 = 0$, $b + 1 = 0$ and $c + 2 = 0$ as $a, b, c \in R$.

$\rightarrow a = -3$, $b = -1$, $c = -2$. Hence, $a^2 + b^2 + c^2 = 14$.

❖ *Here we have verbal ability test. Please do not use pen and paper. It will enhance your thinking and mental calculation power. Rest is your choice!*

VERBAL ABILITY TEST – II

1. If $\dfrac{a}{b} = \dfrac{c}{d}$, where b, d ≠ 0, then $\dfrac{a^2 + c^2}{b^2 + d^2}$ equals

    (A) 1   (B) 2   (C) 2a/b   (D) a^2/b^2

2. Given 2a3 + 326 = 5b9. If 5b9 is divisible by 9 then the digits a + b =

    (A) 2   (B) 4   (C) 6   (D) 9

3. Let A = 1 – x and let B = $(1 + x)(1 + x^2)(1 + x^4)$. If x = 0.1 then AB equals....

    (A) 0.999999   (B) 0.99999999   (C) 1.00000001   (D) 1.000001

4. Which of the following is a perfect square?

    (A) $4^4 5^5 6^6$ (B) $4^4 5^6 6^5$ (C) $4^5 5^4 6^6$ (D) $4^6 5^4 6^5$

5. If a, b and c are digits for which

    $$
    \begin{array}{r}
    7\ \ a\ \ 2 \\
    -\ 4\ \ 8\ \ b \\
    \hline
    c\ \ 7\ \ 3
    \end{array}
    $$

    Then a + b + c equals

    (A) 14   (B) 15   (C) 16   (D) 17

6. If a + b + c = 0 and $a^3 + b^3 + c^3 = 27$, find abc.

    (A) 9   (B) 0   (C) 3   (D) $3\sqrt[3]{9}$

7. What is the value of 2012 + 201.2 + 20.12 + 2.012?

    (A) 2235.332   (B) 2236.322   (C) 2326.232   (D) 2235.532

8. If $a * b = \dfrac{ab}{a+b}$, then what is the value of $(2 * 3) * 4$?

    (A) 24/5   (B) 6/7   (C) 3/8   (D) 12/13

9. The numbers 2, 3, 4, 5, 6, 7, 8 are to be placed, one per square, in the diagram shown such that the four numbers in the horizontal row add up to 21 and the four numbers in the vertical column add up to 21. Which number should replace x?

    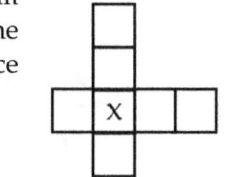

    (A) 5   (B) 6

    (C) 7   (D) 8

10. Tens digit of 2015^2 – 2015 is

    (A) 1   (B) 2   (C) 3   (D) 4

11. Mr. Chandra places numbers on a 3 × 3 grid according to Chandra's principle:

    For any three numbers in a horizontal, vertical, or diagonal line, the middle number is always the average (mean) of its two neighbours.

    Determine the values of x and y when the grid is completed according to Chandra's principle.

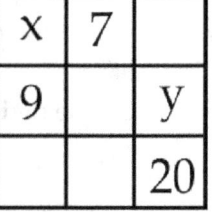

x	7	
9		y
		20

    (A) x = 15, y = 4

    (B) x = 6, y = 13

    (C) x = 4, y = 15

    (D) x = 13, y = 6

12. 20152015 × 20152015 + 20152015 + 20152016 is the square of

    (A) 20152016   (B) 20152017   (C) 20162015   (D) 20162016

13. Consider all 3-digit numbers N = 100a + 10b + c where 0 < a < b < c < 10. The smallest prime number that does not divide N is:

    (A) 5   (B) 7   (C) 11   (D) 17

14. The fraction $\dfrac{2015}{199}$ can be written in the form $u + \cfrac{1}{v + \cfrac{1}{w + \cfrac{1}{x}}}$ where u, v, w and x are positive integers, then x is

    (A) 25   (B) 24   (C) 1   (D) 7

15. Given that a, b, c and d are positive integers with $a < 2b$, $b < 3c$, $c < 4d$ and $d < 2$, what is the largest possible value of a?

    (A) 48   (B) 45   (C) 47   (D) 15

ANSWER KEY- VERBAL ABILITY TEST – II

1. D	2. C	3. B	4. C
5. D	6. A	7. A	8. D
9. C	10. A	11. C	12. A
13. C	14. B	15. D	☺☺☺

PLAY WITH NUMBERS (CRYPTOGRAMS)

Here we are going to replace digits with letters or with any symbol and doing sum, multiplication, addition, subs traction and division etc. It's like coding – decoding game. Just solve them and enjoy the mathematics. Answers are given below.

Ex1. Find the digits in place of letters for the following.

```
      3   A            1   A              A   B
    + 2   5          ×     A            ×     3
    ─────────        ─────────          ─────────
      B   2            9   A            C   A   B

      (I)              (II)               (III)
```

Ex2. Solve the following Cryptograms.

```
              C                      1  3                   A  2                         9  5
  (I)     +  C  C        (II)     +  Δ  7      (III)         A  A       (IV)    –   1  A
          ─────────               ─────────            +    1  A                ─────────
             D  4                  1  0  0             ─────────                   A  8
                                                          7  8

             A  B                   A  B                   4  A                     5  A
  (V)     ×        9     (VI)    ×   A  B      (VII)   ×    2  1     (VIII)  ×   1  B
          ─────────               ─────────            ─────────                ─────────
          C  2  3                 6  A  B              B  4  5                   8  0  0
```

Ex3. Solve the following puzzles.

```
         A  B  A  B  C             1  B  B              A  B  C  A  D
  (I)  + A  B  D  D  B    (II)   +    2  3     (III)            ×     7
       ─────────────────         ─────────             ─────────────────
         C  A  B  D  C             2  E  E              G  A  A  F  E
```

Ex4. Solve the following divisions

```
                AAB5                                                1 F G
            9) A02A6                                          AB) 4 C D E
               –9                                                –2 8
               ──                                                ─────
               A2                                                 H5 6
              –9                                                 –I J K
              ──                                                 ─────
               BA                                                 L M N
              –2C                                                –P Q R
              ───                                                ─────
               D 6                                                  0
              –DE
              ───
                A

  (I)                                            (II)
```

```
                   1ABCDE
Ex5.  Solve the Cryptogram ×      3 .
                   ──────────
                   ABCDE1
```

ANSWER KEY · CRYPTOGRAMS

Ex1: (I) A = 7 and B = 6   (II) A = 6

(III) A = 5 , B = 0, C = 1

Ex2: (I) C = 7, D = 8   (II) Δ = 8   (III) A = 3   (IV) A = 7

(V) A =4, B =7, C = 4   (VI) A = 2, B = 5   (VII) A = 5, B = 9   (VIII) A = 0, B = 6

Ex3: (I) A = 1, B = 0, C = 2, D = 9

(II) B = 8, E = 1 or B = 7, E = 0

(III) A = 1, B = 3, C = 0, D = 8, E = 6, F = 2, G = 9

Ex4: (I) A = 1, B = 3, C = 7, D = 4, E = 5

(II) A = 2, B = E = R = N= 8, C = 3, D = G = Q = M = 6, F = 5, H = P= I = L = 1, J = 4, K = 0.

Ex5. A = 4, B = 2, C = 8, D = 5, E = 7.

FAST CALCULATION

Knowing the behavior of numbers in different situation gave, feeling of a magic. It may be a magic for normal person but it's like a musical tone for a mathematician, that he remember the nature of digits. He remembers the very basic properties, formulae and different rules that he/ she generate during learning, experiments, hits and trials etc. even you can observe some pattern in numbers and generate some results for upcoming generations. If you want to be magician then do not skip any part. Do experiments, gave surprise to your family members, friends, teachers etc. Let's start with very basic thing.

Remember this table for square and cube and square roots. (Take time to remember, write 3 – 4 times)

Number (N)	N^2	N^3	$\sqrt{N}$
1	1	1	1
2	4	8	1.4142
3	9	27	1.7320
4	16	64	2
5	25	125	2.2360
6	36	216	2.4494
7	47	343	2.6457
8	64	512	2.8284
9	81	729	3
10	100	1000	3.1622
11	121	1331	3.3166
12	144	1728	3.4641
13	169	2197	3.6055
14	196	2744	3.7416
15	225	3375	3.8729
16	256	4096	4
17	289	4913	4.1231
18	324	5832	4.2426
19	361	6859	4.3588
20	400	8000	4.4721

Note: Observe the pattern of decimals in square roots between two integers.

SQUARE OF TWO DIGIT NUMBERS

To get square of any two digit number (Let N = ab = 73), then follow the following steps.

Step 1:   write a^2 and b^2 (Ex: 7^2 = 49 and 3^2 = 9 so write as 09, now your expression should be '4909')

Step 2:   now multiply a and b and write the value of 2×a×b (2× 7 × 3 = 42)

$$
\begin{array}{cccc}
4 & 9 & 0 & 9 \\
\end{array}
$$

Step 3:   leave, unit place and add as shown here (Ex: $\underline{+\ \ 4\ \ 2\ \ *}$)

$$
\begin{array}{cccc}
5 & 3 & 2 & 9 \\
\end{array}
$$

The number 5329 is square of 73. ie. 73^2 = 5329

Do experiments with different numbers till, get it into habit

Here you have question, why?

Answer is very simple. N = ab = 10 a + b

Now, $N^2 = (10 a + b)^2 = 100 a^2 + b^2 + 10 \times (2 \times a \times b)$. I hope, you are smart enough to get the concept.

Ex: $73^2 = 100 \times 7^2 + 3^2 + 10 \times (2 \times 7 \times 3) = 4900 + 09 + 42 \times 10 = 4909 + 420 = 5329$.

SQUARE OF TWO DIGIT NUMBER HAVING UNIT DIGIT 5

To find the square of two digit number 'a5', do the following steps. (Ex: take N = 65)

Step 1:   multiply a × (a + 1) (Ex. 6 × 7 = 42)

Step 2:   write 25 in the end of a(a + 1) (Ex. 4225)

Step 3:   4225 is square of 65. ie. 65^2 = 4225.

Do experiments with different numbers till, get it into habit

OBSERVATIONS

(I)   there are 2n non – perfect square between n^2 and $(n + 1)^2$.

(II)   The sum of first n odd natural numbers is n^2.   (III)

$$1 = 1^2$$
$$1 + 3 = 2^2$$
$$1 + 3 + 5 = 3^2$$
$$1 + 3 + 5 + 7 = 4^2$$
$$1 + 3 + 5 + 7 + 9 = 5^2$$

$$1^2 = 1$$
$$11^2 = 121$$
$$111^2 = 12321$$
$$1111^2 = 1234321$$
$$11111^2 = 123454321$$

(IV)   We can express square of any odd natural number as sum of two consecutive positive integers.

Ex. $3^2 = 9 = 4 + 5$, $7^2 = 49 = 24 + 25$, $11^2 = 121 = 60 + 61$

For being extraordinary! You have to do extraordinary things, which let other to surprise!

Here I am writing table of some important numbers, please memorize and rewrite them. Use them in your calculation to remain in your mind for longer duration. If you apply them in daily life then they will remain forever.

TABLE									
12	13	14	15	16	17	18	19	21	23
24	26	28	30	32	34	36	38	42	46
36	39	42	45	48	51	54	57	63	69
48	52	56	60	64	68	72	76	84	92
60	65	70	75	80	85	90	95	105	115
72	78	84	90	96	102	108	114	126	138
84	91	98	105	112	119	126	133	147	161
96	104	112	120	128	136	144	152	168	184
108	117	126	135	144	153	162	171	189	207
120	130	140	150	160	170	180	190	210	230
24	25	26	27	28	29	31	37	43	45
48	50	52	54	56	58	62	74	86	90
72	75	78	81	84	87	93	111	129	135
96	100	104	108	112	116	124	148	172	180
120	125	130	135	140	145	155	185	215	225
144	150	156	162	168	174	186	222	258	270
168	175	182	189	196	203	217	259	301	315
192	200	208	216	224	232	248	296	344	360
216	225	234	243	252	261	279	333	387	405
240	250	260	270	280	290	310	370	430	450

MULTIPLICATION (SHORT CUT METHODS)

The shortcuts are situational. The same method may be longer for another example, so you need to understand, that which method is good in given situation to get fast result. I am performing some experiments, so you need to observe them and apply to get better understanding.

Ex1. Find the value of following.

    (i)   367456 × 99   (ii) 367854 × 999   (iii) 989034 ×9999

Solution:

    (i)   367456 × 99 = 367456 × (100 - 1) = 36745600 – 367456 = 36378144

    (ii) 367854 × 999 = 367854 ×(1000 - 1) = 367854000 – 367854 = 367486146

    (iii) 989034 ×9999 = 9890340000 – 989034 = 9889350966

Ex2. Evaluate the following

    (i)   26548 × 125   (ii) 63562 × 25   (iii) 3678546 × 625

Solution:

    (i) $26548 \times 125 = 26548 \times \dfrac{125 \times 8}{8} = \dfrac{26548}{8} \times 1000 = 3318.5 \times 1000 = 3318500$

So, instead of multiplying with 125, just divide by 8 and then multiply by 1000.

    (ii) $63562 \times 25 = 63562 \times \dfrac{25 \times 4}{4} = \dfrac{63562}{4} \times 100 = 15890.5 \times 100 = 1589050$

Instead of multiplying by 25 just divide the number by 4 and then multiply by 100.

(iii) $3678546 \times 625 = 3678546 \times \dfrac{625 \times 16}{16} = \dfrac{3678546}{16} \times 10000 = 229909.125 \times 10000 = 2299091250$.

Instead of multiplying by 625 just divide the number by 16 and then multiply by 10000.

You can generate other results too.

Ex3. Find the value of following

(i) 37689×11 (ii) 98467×101 (iii) 67456×1001

Solution:

Here, 11, 101, 1001 are of the form $10^k + 1$, $k \in N$.

(i) $37689 \times 11 = 37689 \times (10 + 1) = 376890 + 37689 = 414579$

(ii) $98467 \times 101 = 98467 \times (100 + 1) = 9846700 + 98467 = 9945167$

(iii) $67456 \times 1001 = 67456 \times (1000 + 1) = 67456000 + 67456 = 67523456$

Ex4. Calculate the value of following

(i) $3675 \times 234 + 3675 \times 766$ (ii) $678 \times 294 - 678 \times 144$

(iii) $84 \times 274 + 84 \times 13$ (iv) $7846 \times 846 - 7846 \times 646$

Solution:

The distributive properties of numbers are

(A) $a \times b + a \times c = a \times (b + c)$ (B) $a \times b - a \times c = a \times (b - c)$

(i) $3675 \times 234 + 3675 \times 766 = 3675 \times (234 + 766) = 3675 \times 1000 = 3675000$

(ii) $678 \times 294 - 678 \times 144 = 678 \times (294 - 144) = 678 \times 150 = 678 \times 300/2 = 339 \times 300 = 101700$

(iii) $84 \times 274 + 84 \times 13 = 84 \times (274 + 13) = 84 \times 287 = 24108$

(iv) $7846 \times 846 - 7846 \times 646 = 7846 \times (846 - 646) = 7846 \times 200 = 1569200$

Ex5. Evaluate the following

(i) 289×271 (ii) 802×304 (iii) 871×2001

Solution:

(i) $289 \times 271 = (290 - 1) \times (270 + 1) = 78300 - 270 + 290 - 1 = 78319$

(ii) $802 \times 304 = (800 + 2) \times (300 + 4) = 240000 + 600 + 3200 + 8 = 243808$

(iii) $871 \times 2001 = (870 + 1) \times (2000 + 1) = 1740000 + 2000 + 870 + 1 = 1742871$

VEDIC MATHS (MULTIPLICATION)

This is the powerful tool to multiply two numbers. You need to do little bit practice to get the method.

Ex6. Evaluate the following

(i) 672×24 (ii) 193×216 (iii) 9201×312

Solution:

First understand the steps of working and observe the method. Do experiments on numbers to get better understanding.

(i) $672 \times 24 = 16128$

STEP: 1- Just multiply the unit digits and write them as shown ie. $2 \times 4 = 8$, so write it as 08 instead of 8. Here 0 will be carry. Write a separator too.

$$\begin{array}{r} 6\,7\,2 \\ \uparrow \\ 2\,4 \\ \hline |08 \end{array}$$

STEP: 2 - Next multiplication will be as shown

$$672$$
$$\times \quad 28 + 04 = 32$$
$$24$$
$$\overline{32}$$

STEP: 3 – This time multiplication will shift toward left as shown.

$$672$$
$$\times \quad 24+14=38$$
$$24$$
$$\overline{38}$$

STEP: 4 – Now multiplication reduces

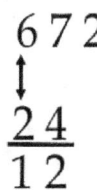

$$672$$
$$\updownarrow$$
$$24$$
$$\overline{12}$$

STEPS OF WORKING

(A)   (B)   (C)

(A)
$$672$$
$$\updownarrow$$
$$24$$
$$\overline{|08}$$

(B)
$$672$$
$$\times$$
$$24 \quad \boxed{28 + 04 = 32}$$
$$\overline{32|08}$$

$$\boxed{24 + 14 = 38}$$

(C)
$$672$$
$$\times$$
$$24$$
$$\overline{38|32|08}$$

$$1\;2\;\boxed{2|3}\boxed{8|3}\boxed{2|0}\,8$$
$$1 \quad 6 \quad 1\;2\;8$$

$$672 \times 24 = 16128$$

(D)
$$672$$
$$\updownarrow$$
$$24$$
$$\overline{12|38|32|08}$$

At last you will get product as 12 | 38 | 32 | 08.

Now, have 8 at unit place and add carry '0' to 2, so 0 + 2 = 2, 8 + 3 = 11, have 1 and carry forward 1 so, 2 + 3 + 1 = 6. Observe the steps of working part.

(ii)   193 × 216

Now, I am going to explain it in single figure. Just observe and analyze the method.

STEPS OF WORKING

$$193 \quad 193 \quad 193 \quad 193 \quad 193$$
$$\updownarrow \quad\; \times \quad\; \times\!\!\times \quad\; \times \quad\; \updownarrow$$
$$216 \quad 216 \quad 216 \quad 216 \quad 216$$

$$02|19|21|57|18 = 4\,1\,6\,8\,8$$

(iii)   9201 × 312

STEPS OF WORKING

$$9201 \quad 9201 \quad 9201 \quad 9201 \quad 9201 \quad 9201$$
$$\updownarrow \quad\;\; \times \quad\;\; \times\!\!\times \quad\; \times\!\!\times \quad\;\; \times \quad\;\; \updownarrow$$
$$312 \quad\; 312 \quad\; 312 \quad\;\; 312 \quad\;\; 312 \quad\; 312$$

$$27|15|20|07|01|02 = 2\,8\,7\,0\,7\,1\,2$$

I hope! You got the method. Solve more multiplication and enjoy two line method of multiplication. ☺ ☺ ☺

APPROXIMATION (A Solid Tool)

Ex1. Find the approximate value of $\sqrt[3]{65}$.

Solution:

As we know $f'(a) = \lim\limits_{h \to 0} \dfrac{f(a+h)-f(a)}{h}$, so we can approximate value as

$f(a+h) \approx f(a) + h\,f'(a)$ ....(1)

Let $f(x) = x^{1/3}$, then $f'(x) = \dfrac{1}{3}x^{-2/3} \rightarrow f'(64) = 1/48$

So, $f(65) = \sqrt[3]{65} = f(64+1) \approx f(64) + 1 \times f'(64) = 4 + 1/48 = 4.0208333$

So, $\sqrt[3]{65} \approx \mathbf{40208333}$ but as per calculator its $\sqrt[3]{65} = \mathbf{4.0207257}$.

You can observe the difference. It's very less.

Ex2. If side of a cube is 8.004 cm, then it's approximate volume is

(A) 512.8 cm^3 (B) 512.96 cm^3 (C) 512.768 cm^3 (D) 512.850 cm^3

Solution:

Let $f(x) = x^3 \rightarrow f(8.004) = 8.004^3$

$f(a+h) \approx f(a) + h\,f'(a)$

$f(8.004) \approx f(8) + 0.004 \times f'(8)$

$f'(x) = 3\,x^2 \rightarrow f'(8) = 192$, so

$f(8.004) \approx 512 + 0.004 \times 192 = 512.768$ so, $f(8.004) \approx 512.768$. Option (C) is correct.

Ex3. Find $(1.0002)^{3000} \approx$

(A) 1.2 (B) 1.4 (C) 1.6 (D) 1.8

Solution:

As per Binomial theorem for any index

$$(1+x)^n = 1 + nx + \frac{n(n-1)}{2!}x^2 + \frac{n(n-1)(n-2)}{3!}x^3 + \frac{n(n-1)(n-2)(n-3)}{4!}x^4 + \ldots\ldots$$

If x is very – very small then we can write $(1+x)^n \approx 1 + nx$. Since $x^2 \rightarrow 0$

So, $(1.0002)^{3000} = (1 + 0.0002)^{3000} \approx 1 + 0.0002 \times 3000 = 1.6$, Hence option (C) is correct.

Ex4. If $1° = 0.017^C$, $\sqrt{2} \approx 1.4142$ then $\sin(46°) \approx$ ......

(A) 0.7294 (B) 0.7191 (C) 0.7394 (D) 0.8

Solution:

Let $f(x) = \sin(x) \rightarrow f'(x) = \cos x$

As per $f(a+h) \approx f(a) + h\,f'(a)$, a and h both are in radian so can't apply directly in degree.

$\sin(46°) = \sin(45° + 1°) = \sin(\pi/4 + 0.017) \approx \sin(\pi/4) + 0.017 \times \cos(\pi/4)$

$\sin(46°) \approx 1/\sqrt{2} + 0.017 \times 1/\sqrt{2} = (1/\sqrt{2}) \times 1.017 = 0.7191$, so Option (B) is correct.

ARITHEMATIC OPERATIONS

It is a common need to simplify the expressions formulated according to the statements of the problems relating to practical life. To do this, it is essential to follow in sequence the mathematical operations given by the term "BODMAS".

BODMAS

Each letter of the word BODMAS stands as follows:

B for Bracket: [{(–)}]

There are four brackets, namely, – bar, ( ), { } and [ ]. They are removed, strictly in the order - , ( ), { } & [ ].

O for Of: of

D for Division: ÷

M for Multiplication: ×

A for Addition: +

S for Subtraction: –

The order of various operations in exercises involving brackets and fractions must be performed strictly according to the order of the letters of the word BODMAS.

Note: Here, $-\overline{5-7} = -(-2) = 2$.

Ex1. Simplify $3 - \left[50 \div 5 \text{ of } 2 + \left\{ 12 - \left(7 - \dfrac{\overline{9}}{2} - \dfrac{7}{3} \right) \right\} \right]$

Solution:

Given expression $3 - \left[50 \div 5 \text{ of } 2 + \left\{ 12 - \left(7 - \dfrac{\overline{9}}{2} - \dfrac{7}{3} \right) \right\} \right]$

$= 3 - \left[50 \div 5 \text{ of } 2 + \left\{ 12 - \left(7 - \dfrac{13}{6} \right) \right\} \right]$

$= 3 - \left[50 \div 5 \text{ of } 2 + \left\{ 12 - \dfrac{29}{6} \right\} \right]$

$= 3 - \left[50 \div 5 \text{ of } 2 + \dfrac{43}{6} \right]$

$= 3 - \left[50 \div 5 \times 2 + \dfrac{43}{6} \right]$

$= 3 - \left[50 \div 10 + \dfrac{43}{6} \right]$

$= 3 - \left[5 + \dfrac{43}{6} \right]$

$= 3 - \dfrac{73}{6} = -\dfrac{55}{6}$

Ex2. Simplify $2 - 4 \div 2 \times 2 + 6 - 2\,[60 \div 5\,\{\overline{6 \times 4} \div 2 + (7 \text{ of } 3 - \overline{22 - 1})\}]$.

Solution:

The given expression $2 - 4 \div 2 \times 2 + 6 - 2\,[60 \div 5\,\{\overline{6 \times 4} \div 2 + (7 \text{ of } 3 - \overline{22 - 1})\}]$

$= 2 - 4 \div 2 \times 2 + 6 - 2\,[60 \div 5\,\{\overline{6 \times 4} \div 2 + (7 \text{ of } 3 - 21)\}]$

$= 2 - 4 \div 2 \times 2 + 6 - 2\,[60 \div 5\,\{\overline{6 \times 4} \div 2 + (21 - 21)\}]$

$= 2 - 4 \div 2 \times 2 + 6 - 2\,[60 \div 5\,\{\overline{6 \times 4} \div 2\}]$

$= 2 - 4 \div 2 \times 2 + 6 - 2\,[60 \div 5\,\{24 \div 2\}]$

$= 2 - 4 \div 2 \times 2 + 6 - 2 [60 \div 5 \{12\}]$

$= 2 - 4 \div 2 \times 2 + 6 - 2 [60 \div 60]$

$= 2 - 4 \div 2 \times 2 + 6 - 2$

$= 2 - 2 \times 2 + 6 - 2$

$= 2 - 4 + 6 - 2$

$= 8 - 6$

$= 2$

PRACTICE PROBLEM

Simplify the following

(i) $8\dfrac{1}{2} - \left[3\dfrac{1}{5} \div 4\dfrac{1}{2} \text{ of } 5\dfrac{1}{3} + \left\{ 11 - \left(3 - 1\overline{\dfrac{1}{4} - \dfrac{5}{8}} \right) \right\} \right]$ Answer: $-\dfrac{31}{120}$

(ii) $5\dfrac{1}{3} - \left\{ 4\dfrac{1}{3} - \left(3\dfrac{1}{3} - 2\overline{\dfrac{1}{3} - \dfrac{1}{3}} \right) \right\}$ Answer: $2\dfrac{1}{3}$

EXERCISE – 3 (LADDULAL'S UNDERSTANDING)

1. If a and b are two integers and $b > 0$, then there exist two integers q and r such that
    (A) $b = aq + r$, where $0 < r < b$ 　　　(B) $b = rq + a$, where $0 < r < b$
    (C) $a = bq + r$, where $0 \leq r < b$ 　　　(D) None of these

2. The value of $0.32 \times 0.32 + 0.64 \times 0.68 + 0.68 \times 0.68$ is
    (A) 0.99 　　　(B) 0.98 　　　(C) 1.02 　　　(D) 1

3. The value of $(2.35)^3 + 1.95 \times (2.35)^2 + 7.05 \times (0.65)^2 + (0.65)^3$, is
    (A) 26 　　　(B) 27 　　　(C) 26.5 　　　(D) 27.5

4. $\dfrac{0.62 \times 0.62 \times 0.62 - 0.41 \times 0.41 \times 0.41}{0.62 \times 0.62 + 0.62 \times 0.41 + 0.41 \times 0.41}$
    (A) 0.21 　　　(B) 1.03 　　　(C) 0.41 　　　(D) 0.62

5. The value of $3 \div \left[(8-5) \div \left\{ (4-2) \div \left(2 + \dfrac{8}{13} \right) \right\} \right] = ?$
    (A) $\dfrac{33}{71}$ 　　　(B) $\dfrac{55}{17}$ 　　　(C) $\dfrac{13}{17}$ 　　　(D) $\dfrac{31}{17}$

6. The value of $(20 \div 5) \div 2 + (16 \div 8) \times 2 + (10 \div 5) \times (3 \div 2) = ?$
    (A) 9 　　　(B) 12 　　　(C) 15 　　　(D) 18

7. If $\dfrac{a}{a+b} = \dfrac{17}{23}$, then value of $\dfrac{a+b}{a-b}$ is equals to

   (A) $\dfrac{11}{23}$ (B) $\dfrac{17}{32}$ (C) $\dfrac{23}{11}$ (D) $\dfrac{23}{17}$

8. The square root of 104976

   (A) 324 (B) 424 (C) 326 (D) None of these

9. If $\sqrt{0.03 \times 0.3 \times a} = 0.03 \times 0.3 \times \sqrt{b}$, then value of a/b is

   (A) 0.009 (B) 0.03 (C) 0.09 (D) None of these

10. If $\dfrac{\sqrt{7}-1}{\sqrt{7}+1} - \dfrac{\sqrt{7}+1}{\sqrt{7}-1} = a + b\sqrt{7}$, then select the wrong statement

    (A) $a > b$ (B) $b^a = 1$ (C) $b = 0$ (D) $|b| > a$

11. If $\left(\dfrac{\left(1-\left(\dfrac{a}{b}\right)^{-2}\right)a^2}{\left(\sqrt{a}-\sqrt{b}\right)^2 + 2\sqrt{ab}} \right) = 1$ and $a + b = 5$ where $a, b \in R^+$, then value of $a^b - b^a$ is?

    (A) 2 (B) 1 (C) 0 (D) 3

12. Select the correct statement

    (A) $6 - \sqrt{35} > \dfrac{1}{10}$

    (B) There is positive integral solution (x, y) of $x^2 - y^2 = 1$, where $x, y \in I^+$

    (C) $n! > 2^n \ \forall \ n \geq 4$

    (D) The sum of the interior angles of any n - gon is $180(n - 1)$ degrees

13. Express $0.3\overline{4} + 0.3\overline{4}$ as a single decimal.

    (A) $0.67\overline{88}$ (B) $0.68\overline{78}$ (C) $0.6\overline{89}$ (D) $0.68\overline{7}$

14. If $2^{-m} \times \dfrac{1}{2^m} = \dfrac{1}{4}$, then $\dfrac{1}{14}\left[\left(4^m\right)^{1/2} + \left(\dfrac{1}{5^m}\right)^{-1} \right] = ?$

    (A) $1/2$ (B) 2 (C) 4 (D) $-\frac{1}{4}$

15. The smallest among the surds $\sqrt{10} - \sqrt{5}$, $\sqrt{19} - \sqrt{14}$, $\sqrt{22} - \sqrt{17}$ and $\sqrt{8} - \sqrt{3}$ is

    (A) $\sqrt{10} - \sqrt{5}$ (B) $\sqrt{19} - \sqrt{14}$ (C) $\sqrt{22} - \sqrt{17}$ (D) $\sqrt{8} - \sqrt{3}$

16. If $\sum_{k=4}^{143} \dfrac{1}{\sqrt{k}+\sqrt{k+1}} = a - \sqrt{b}$, where a, b $\in$ I then a and b are respectively

   (A) 10 and 0   (B) - 10 and 4   (C) 10 and 4   (D) - 10 and 0

17. If a $\in$ (0, 1) then select the wrong option

   (A) $a < \sqrt{a}$ (B) $\dfrac{1}{a^4} > a$ (C) $a < \dfrac{1}{a}$ (D) $a^2 < a^3$

18. Which one is wrong statement?

   (A) $\dfrac{7^3}{5^4}$ is a non terminating repeating decimal

   (B) Sum of two irrationals may be rational

   (C) If 7 divides a natural number 'a²' then it will divide 'a' as well

   (D) LCM (6, 14) × HCF (6, 14) = 21

19. If a, b and c are rational numbers such that a + b$\sqrt{2}$ + c $\sqrt{5}$ = 0, then
   (A) b = c = 0   (B) a = 0
   (C) a = b = c = 0   (D) at least one of a, b or c is zero

20. If $\sqrt{\dfrac{6+2\sqrt{3}}{33-19\sqrt{3}}} = a + b\sqrt{3}$, then a + b =?

   (A) 6   (B) 8   (C) 10   (D) 12

ROUGH SPACE

Here are the IIT JEE Foundation Sample Test Papers for class VIII, IX and X standard students! Please solve them honestly. It will grow and test your understanding power! Keep smiling; it's a need of life! ☺ ☺ ☺

MATHSARC EDUCATION

A learning place to fulfill your dream of success!

IIT JEE FOUNDATION   SAMPLE TEST PAPER

Mathsarc Education
A learning place to fulfill your dream of success!

BRAHMASTRA

Name:_____   MM : 70   Time : 1 Hr

SECTION - I
Single Correct Answer Type

This section contains **10 multiple choice questions**. Each question has four choices (A), (B), (C) and (D) out of which **ONLY ONE is correct.** Marking (+3, -1)

1. If a + b = 4 and ab = 3, then the value of $a^3 + b^3$, is
    (A) 64   (B) 36   (C) 28   (D) 26

2. If $m = 3^{n-1}$ & $3^{4n-1} = 27$, what is the value of m + n?
    (A) 3   (B) 2   (C) 0   (D) 1

3. Which of the following is equal to $\dfrac{1 \cdot 3 \cdot 5 \cdot 7 \cdots (2n-1)}{2 \cdot 4 \cdot 6 \cdot 8 \cdots (2n)}$?

    (A) 2^n (B) $(2n-1)! \div (n-1)!$

    (C) $(2n)! \div n!$ (D) $(2n)! \div \left(2^n (n!)\right)^2$

4. A special coin is designed having faces as '1' and '2'. Now this coin is tossed with two dices. What is the probability to get sum divisible by '3'
    (A) $\dfrac{1}{2}$ (B) $\dfrac{1}{3}$ (C) $\dfrac{2}{3}$ (D) $\dfrac{3}{4}$

5. If $\sqrt{2} - \sqrt{3} + \sqrt{k}$ is rational, then k =
    (A) $\sqrt{3} - \sqrt{2}$ (B) $5 - 2\sqrt{6}$ (C) $-5 - 2\sqrt{6}$ (D) 5

ROUGH WORK

6. A regular octagon is formed by cutting congruent isosceles right angle triangle from the corner of a square. If length of square is 1 unit then length of octagon is?

(A) $\dfrac{\sqrt{2}-1}{2}$ (B) $\sqrt{2}-1$ (C) $\dfrac{\sqrt{5}-1}{4}$ (D) $\dfrac{\sqrt{5}-1}{3}$

7. The sum of the digits of the number $2^{2000}5^{2002}$ when express in the base 10, is

(A) 6 (B) 7 (C) 9 (D) 5

COMPREHENSION (8 - 10)

Consider the system of linear equations $L_1 \equiv a_1x + b_1y + c_1 = 0$ & $L_2 \equiv a_2x + b_2y + c_2 = 0$

8. if $\dfrac{a_1}{a_2} = \dfrac{b_1}{b_2} \neq \dfrac{c_1}{c_2}$, then system has

(A) one Solution (B) Zero Solution (C) Infinite Solutions (D) None of These

9. If $a_1 + 2b_1 + c_1 > 0$ & $3a_1 - 2b_1 + c_1 < 0$ then points

(A) (1, 2) & (3, -2) are same side of line L_1 (B) (1, 2) & (3, -2) are same side of line L_2

(C) (1, 2) & (3, -2) are opposite side of line L_1 (D) (1, 2) & (3, -2) are opposite side of line L_2

10. Consider the Graph of Lines L_1 & L_2 as shown in figure
Scale: 1 square is of 1 unit. Select the wrong option

(A) $c_1(2a_1 + c_1) > 0$

(B) $c_2(2a_2 + 3b_2 + c_2) < 0$

(C) $2a_1 + 3b_1 + c_1 = 0$

(D) $2a_2 + 3b_2 + c_2 = 0$

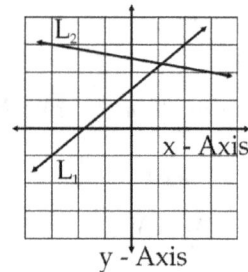

ROUGH WORK

SECTION – II
(Multiple Correct Answer(s) Type)

This section contains **5 multiple choice questions.** Each question has four choices (A), (B), (C) and (D) out of which **ONE OR MORE** may be correct. Marking (+4, 0)

11. If $\alpha, \beta \in R$ are roots of the equation $3^{(x+2)} + 3^{-x} = 10$, then

    (A) $|\alpha + \beta| = 2$ (B) $|\alpha - \beta| = 2$

    (C) if $\alpha < \beta$ then $\alpha^\beta + \beta = 1$ (D) if $\alpha < \beta$ then $\alpha + \beta = 1$

12. Consider a grid as shown in the figure, then
    (A) Number of squares in the grid = 20
    (B) Number of rectangle (squares excluded) = 40
    (C) Number of shortest way to go from A to B on grid is 7!

    (D) Number of shortest way to go from A to B on grid is $\dfrac{7!}{4! \times 3!}$.

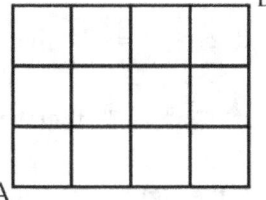

13. Select the correct statements (m, n are two Natural Numbers)

    (A) 143 is a prime number   (B) $\pi = \dfrac{22}{7}$

    (C) $\sqrt{3}$ is an irrational number   (D) $\dfrac{1}{2^m \cdot 5^n}$ is terminating rational number

14. If $N = \sqrt{3 - 2\sqrt{2}}$, Then
    (A) $N - \sqrt{2}$ is an irrational Number
    (B) $N - \sqrt{2}$ is a rational Number
    (C) $N - \sqrt{3}$ is a rational number
    (D) if $N = p + q\sqrt{r}$, where p, q, r are integers, Then $p + q + r = 2$.

15. Consider a natural number $N = 2^3 \times 3^2 \times 5$, then
    (A) Number of natural divisors = 6   (B) L.C.M. of N & $2 \times 3 \times 5^2 = 2^3 \times 3^2 \times 5^2$
    (C) H.C.F. of N & $2 \times 3 \times 5^2 = 2 \times 3 \times 5$ (D) Number of odd natural divisors of N = 6

ROUGH WORK

SECTION -III
(One Integer Value Correct Type)

This section contains **5 questions. Each question,** when worked out will result in **one integer from 0 to 9.** (*Both inclusive*) Marking (+4, 0)

16. An n digit number is a positive number with exactly n digits. Nine hundred distinct n-digit numbers are to be formed using only the three digits 2, 5 and 7. The smallest value of n for which this is possible, is?

17. If $2^{5x} \div 2^x = \sqrt[5]{2^{20}}$ then x equals?

18. If $a^m \, a^n = a^{mn}$, then value of m(n - 2) + n(m - 2) is

19. If the eight digit number 2575d568 is divisible by 54 and 87, the value of digit "d" is?

20. If a, b, c are three positive real number and $x \neq 0$, $x \neq 1$, then $\left(\dfrac{x^a}{x^b}\right)^{\frac{1}{ab}} \cdot \left(\dfrac{x^b}{x^c}\right)^{\frac{1}{bc}} \left(\dfrac{x^c}{x^a}\right)^{\frac{1}{ca}} =$

ROUGH WORK

******** ☺ ☺ ☺ ☺ ☺ ********

SOME SOLVED EXAMPLES

Ex1. There is a prime number p such that 16p + 1 is the cube of a positive integer. Find p.

Solution:

Let $16p + 1 = a^3 \rightarrow 16p = a^3 - 1 \rightarrow 16p = (a - 1)(a^2 + a + 1)$. Now we need to think different situations

Since $a^2 + a + 1 = a (a + 1) + 1$ would always be odd, thus we know that 16 must divide a - 1.

If $a - 1 = 16p$, then we have $a^2 + a + 1 = 1$ which gives us a = 0, which we reject. Thus, a – 1 = 16 and $a^2 + a + 1 = p \rightarrow a = 17$. Now that we have a = 17, we get $a^2 + a + 1 = 289 + 17 + 1 = 307$.

Ex2. Let a, b, c, d and e be consecutive positive integers such that b + c + d is a perfect square and a + b + c + d + e is a perfect cube. Find the smallest possible value of c.

Solution:

$b + c + d = 3c$ ....a perfect square $\rightarrow$ c is a multiple of 3^{2k+1} k∈W.

$a + b + c + d + e = 5c$ .... a perfect cube $\rightarrow$ c is multiple of 5^{3k+2}, k∈ W. for minimum c must be multiple of $5^2 = 25$. For both to be true, $c\,|_{min} = 3^3 \times 5^2 = 675$.

Ex3. N has 6 divisors inclusive of 1 and N. The product of five of them is 784. What is the missing divisor?

Solution:

784 = 2x2x4x7x7, so prime factors are 2 and 7, now N has 6 divisors which can be expressed as (1+2)(1+1) so one prime number has power 2 and other has 1 but 784 = 2x2x4x7x7 so $N = 2^2 \times 7$.

Its divisors are 1, 2, 4, 7, 2×7, 4×7. You can observe that missing number is 4×7 = 28.

Ex4. In the real number system, the equation $\sqrt{x+3-4\sqrt{x-1}} +\sqrt{x+8-6\sqrt{x-1}} =1$ has:

(A) No solution   (B) Exactly two distinct solutions

(C) Exactly four distinct solutions   (D) Infinitely many solutions

Solution:

$\sqrt{x+3-4\sqrt{x-1}} +\sqrt{x+8-6\sqrt{x-1}} =1$. Let $\sqrt{(x - 1)} = t \rightarrow t > 0$. Hence, $\sqrt{t^2 +4-4t}+\sqrt{t^2 +9-6t} =1 \rightarrow$ $|t - 2| + |t - 3| = 1$, which is the sum of distance of points 't' on real number line from 2 and 3 is 1 $\rightarrow t\in [2, 3]$, hence infinite many solutions are there. Option (D) is correct.

Ex5. The least positive integer n for which $\sqrt{n+1} -\sqrt{n-1} <0.2$ is

(A) 24   (B) 25   (C) 26   (D) 27

Solution:

$\sqrt{n+1} -\sqrt{n-1} <0.2 \rightarrow 0<5\sqrt{n+1} <1+5\sqrt{n-1}$.

On square both side, we get $25 (n + 1) < 1 + 25(n - 1) + 10 \sqrt{(n - 1)}$.

$\rightarrow 0 < 49 < 10 \sqrt{(n - 1)}$. Now square both side, we get $2401 < 100 (n - 1) \rightarrow 2501 < 100$ n. So n > 25.01, Hence, $n\,|_{MIN} = 26$. Option (C) is correct.

Ex6. If n is the smallest natural number such that n + 2n + 3n +.......+ 99n is a perfect square, then the number of digits in n^2 is

(A) 1   (B) 2   (C) 3   (D) more than 3

Solution:

n + 2n + 3n +.......+ 99n = n (1 + 2 + 3 + 4 +.......+ 99) = n × 99 × 50 = 9 × 25 × 2 × 11 × n. As per condition, n = 2 × 11 = 22. Hence n^2 = 484, so option (C) is correct.

Ex7. Number of distinct real roots of $\left(x^2 - 7x + 11\right)^{x^2 - 11x + 30} = 1$, is

(A) 4   (B) 5   (C) 6   (D) 0

Solution:

First we need to understand the various situations in which we can get $a^b = 1$.

(1) $a^0 = 1$ (2) $1^{(\text{any real})} = 1$ (3) $(-1)^{\text{Even}}$ (4) $0^0 = 1$

Case (1) apply $a^0 = 1$

$x^2 - 11x + 30 = 0 \rightarrow x = 5, 6$.

Case (2) apply $1^{(\text{any real})} = 1$

$x^2 - 7x + 11 = 1 \rightarrow x^2 - 7x + 10 = 0 \rightarrow x = 2, 5$.

Case (3) apply $(-1)^{\text{even}}$

$x^2 - 7x + 11 = -1 \rightarrow x^2 - 7x + 12 = 0$. So x = 3, 4 for both values 3, 4, $x^2 - 11x + 30 \in$ Even.

Hence x = 3 & 4 are solutions.

Case (4) apply $0^0 = 1$

In this case there is no any x for which both $x^2 - 7x + 11$ & $x^2 - 11x + 30$ become 0 simultaneously.

Even, we have considered this case in case (1). Hence, option (B) is correct

Ex8. Show that $\sqrt[3]{ax^2 + by^2 + cz^2} = \sqrt[3]{a} + \sqrt[3]{b} + \sqrt[3]{c}$ if $ax^3 = by^3 = cz^3$ and $\dfrac{1}{x} + \dfrac{1}{y} + \dfrac{1}{z} = 1$.

Solution:

$$\sqrt[3]{ax^2 + by^2 + cz^2} = \sqrt[3]{\frac{ax^3}{x} + \frac{by^3}{y} + \frac{cz^3}{z}} = \sqrt[3]{ax^3\left(\frac{1}{x} + \frac{1}{y} + \frac{1}{z}\right)} = \sqrt[3]{by^3\left(\frac{1}{x} + \frac{1}{y} + \frac{1}{z}\right)} = \sqrt[3]{cz^3\left(\frac{1}{x} + \frac{1}{y} + \frac{1}{z}\right)}$$

$$\rightarrow \sqrt[3]{ax^2 + by^2 + cz^2} = \sqrt[3]{ax^3\left(\frac{1}{x} + \frac{1}{y} + \frac{1}{z}\right)} = x\sqrt[3]{a} \quad \text{............... (1)}$$

$$\rightarrow \sqrt[3]{ax^2 + by^2 + cz^2} = \sqrt[3]{by^3\left(\frac{1}{x} + \frac{1}{y} + \frac{1}{z}\right)} = y\sqrt[3]{b} \quad \text{...............(2)}$$

$$\rightarrow \sqrt[3]{ax^2 + by^2 + cz^2} = \sqrt[3]{cz^3\left(\frac{1}{x} + \frac{1}{y} + \frac{1}{z}\right)} = z\sqrt[3]{c} \quad \text{...............(3)}$$

Now add (1)/x, (2)/y and (3)/z, we get

$$\frac{\sqrt[3]{ax^2 + by^2 + cz^2}}{x} + \frac{\sqrt[3]{ax^2 + by^2 + cz^2}}{y} + \frac{\sqrt[3]{ax^2 + by^2 + cz^2}}{z} = \sqrt[3]{a} + \sqrt[3]{b} + \sqrt[3]{c}$$

$$\rightarrow \sqrt[3]{ax^2 + by^2 + cz^2}\left(\frac{1}{x} + \frac{1}{y} + \frac{1}{z}\right) = \sqrt[3]{a} + \sqrt[3]{b} + \sqrt[3]{c} \rightarrow \sqrt[3]{ax^2 + by^2 + cz^2} = \sqrt[3]{a} + \sqrt[3]{b} + \sqrt[3]{c}.$$

Ex9. Find the value of $a^3 + b^3 + c^3 - 2abc$ if $a = \dfrac{1}{\sqrt{3} + \sqrt{2}}$, $b = \dfrac{1}{\sqrt{2} + 1}$ and $c = 1 - \sqrt{3}$.

Solution:

As $a = \dfrac{1}{\sqrt{3} + \sqrt{2}} = \sqrt{3} - \sqrt{2}$, $b = \dfrac{1}{\sqrt{2} + 1} = \sqrt{2} - 1$, so $a + b + c = 0$

$\rightarrow a^3 + b^3 + c^3 - 3abc = 0 \rightarrow a^3 + b^3 + c^3 - 2abc = abc = \left(\sqrt{3} - \sqrt{2}\right)\left(\sqrt{2} - 1\right)\left(1 - \sqrt{3}\right)$

$\rightarrow \left(\sqrt{6} - 2 - \sqrt{3} + \sqrt{2}\right)\left(1 - \sqrt{3}\right) = \sqrt{6} - 2 - \sqrt{3} + \sqrt{2} - 3\sqrt{2} + 2\sqrt{3} + 3 - \sqrt{6} = 1 - 2\sqrt{2} + \sqrt{3}$

Ex10. If $x^{1/3} + y^{1/3} + z^{1/3} = 0$, then show that $(x + y + z)^3 = 27\ xyz$

Solution:

As per question $x^{1/3} + y^{1/3} + z^{1/3} = 0 \rightarrow x^{1/3} + y^{1/3} = -z^{1/3}$

Now, cube on both sides, we get $(x^{1/3} + y^{1/3})^3 = (-z^{1/3})^3$

$\rightarrow x + y + 3(xy)^{1/3} (x^{1/3} + y^{1/3}) = -z \rightarrow x + y + z + 3(xy)^{1/3}(-z^{1/3}) = 0$

$\rightarrow x + y + z = 3(xyz)^{1/3}$

Now, cube again both side, we get $(x + y + z)^3 = 27\ xyz$.

Ex11. Select the correct interpretations

(A) $3^{2^3} = \left(3^2\right)^3$

(B) $3^{2^3} = 3^{\left(2^3\right)}$

(C) $3^{5^7} > 5^{7^3}$

(D) Real root of $\ln(x) = x \ln 2$ is $\left(\dfrac{1}{2}\right)^{(1/2)^{(1/2)^{(1/2)^{\cdots\cdots\infty}}}}$

Solution:

(A) $3^{2^3} = \left(3^2\right)^3 = 3^{2 \times 3} = 3^6$ Which is the wrong interpretation of power

(B) $3^{2^3} = 3^{\left(2^3\right)}$ is correct.

(C) $3^{5^7} > 5^{7^3} \rightarrow 3^{78125} > 5^{343}$ Correct

(D) $x = \left(\dfrac{1}{2}\right)^{(1/2)^{(1/2)^{(1/2)^{\cdots\cdots\infty}}}} \rightarrow x = \left(\dfrac{1}{2}\right)^x \rightarrow \ln(x) = -x \ln(2)$ Option (D) is wrong

So, **B** and **C** are only correct options.

Ex12. If $x^2 - y^2$ is a prime number, then show that $x^2 - y^2 = x + y$, where $x, y \in N$.

Solution:

$x^2 - y^2 = (x - y)(x + y)$ which is composite but it's given to be a prime, so its possible only if $x - y = 1$.

$\rightarrow x^2 - y^2 = x + y$.

Ex13. Select the corrects statements

(A) $2m$, $m^2 - 1$ and $m^2 + 1$ forms a Pythagorean triplet for $m > 1$ and $m \in N$.

(B) If N is a perfect square odd natural number then it can be written as sum of two consecutive natural numbers

(C) If $x \in R$, then $x^2 + (x+1)^2 + (x(x+1))^2 = (x(x+1)+1)^2$. i.e. $5^2 + 6^2 + 30^2 = (30+1)^2$.

(D) the number N = aaaaaa is divisible by 33

Solution:

All are correct options.

Ex14. The value of $\left(\sqrt[6]{\sqrt[3]{x^9}} \right)^4 \times \left(\sqrt[3]{\sqrt[6]{x^9}} \right)^4$ is

(A) x^{16} (B) x^{12} (C) x^8 (D) x^4

Solution:

$\left(\sqrt[6]{\sqrt[3]{x^9}} \right)^4 \times \left(\sqrt[3]{\sqrt[6]{x^9}} \right)^4 = \left(\sqrt[18]{x^9} \right)^4 \times \left(\sqrt[18]{x^9} \right)^4 = \left(\sqrt[18]{x^9} \right)^8 = \left(x^{9/18} \right)^8 = x^4$ so (D) is correct option.

Ex15. Select the Correct options

(A) $k \times 0 = 0$, where $k \in R$ (B) $\dfrac{0}{k} = 0, k \in R - \{0\}$

(C) $\dfrac{k}{0} = \text{Not Define (N.D.)}$ Where $k \in R$ (D) $xy = xz \rightarrow y = z$ only if $x \neq 0$.

Solution:

All are correct options

Ex16. If $a^x = b^y = c^z$ and $b^2 = ac$, then $\dfrac{1}{x} + \dfrac{1}{z}$ is

(A) $2/y$ (B) $1/y$ (C) $1/(2y)$ (D) $2y$

Solution:

Let $a^x = b^y = c^z = k \rightarrow a = k^{1/x}$, $b = k^{1/y}$, $c = k^{1/z}$

Now, as per question, $b^2 = ac \rightarrow k^{2/y} = k^{1/x} k^{1/z} = k^{\frac{1}{x}+\frac{1}{z}}$

So, $\dfrac{1}{x} + \dfrac{1}{z} = \dfrac{2}{y}$. Option (A) is correct

Ex17. The value of $\left(1-\dfrac{1}{3}\right)\left(1-\dfrac{1}{4}\right)\left(1-\dfrac{1}{5}\right)......\left(1-\dfrac{1}{n}\right)$ is equal to

    (A) $1/n$ (B) $2/n$ (C) $3/n$ (D) $4/n$

Solution:

$$\left(1-\frac{1}{3}\right)\left(1-\frac{1}{4}\right)\left(1-\frac{1}{5}\right)......\left(1-\frac{1}{n}\right)=\frac{2}{3}\times\frac{3}{4}\times\frac{4}{5}\times.....\times\frac{n-1}{n}=\frac{2}{n}.$$ Hence, option (B) is correct.

Ex18. If $2^{2x-y}=32$ and $2^{x+y}=16$ then $x^2 + y^2$ is equal to

    (A) 9   (B) 10   (C) 11   (D) 13

Solution:

As per question $2^{2x-y}=32=2^5 \to 2x - y = 5$ ....(1)

$2^{x+y}=16=2^4 \to x + y = 4$ ....(2)

From (1) and (2) we get, $x = 3$ and $y = 1$. Hence $x^2 + y^2 = 10$, so option (B) is correct.

Ex19. If $x > 1$, then the value of the expression $\dfrac{|x-1|+x}{2x-1}-\dfrac{|x-1|}{x-1}$ is

    (A) x (B) 1   (C) -1   (D) 0

Solution:

If $x > 1$ then $x - 1 > 0$ so $|x-1| = x - 1$.

Hence, $\dfrac{|x-1|+x}{2x-1}-\dfrac{|x-1|}{x-1}=\dfrac{x-1+x}{2x-1}-\dfrac{x-1}{x-1}=1-1=0 \ \forall \ x \in (1, \infty)$ so, Option (D) is correct

Ex20. The value of $3\cdot\overline{\overline{7}}-0\cdot\overline{9}$ is____

    (A) $3\cdot\overline{8}$ (B) $2\cdot\overline{7}$ (C) $2\cdot\overline{8}$ (D) None of these

Solution:

Let $x=3\cdot\overline{7}$ then $10x=37\cdot\overline{7} \to 9x=34 \to x=\dfrac{34}{9}$

$0\cdot\overline{9}=1$, so $3\cdot\overline{7}-0\cdot\overline{9}=\dfrac{34}{9}-1=\dfrac{25}{9}=2\cdot\overline{7}$. Hence, (B) is correct option

Ex21. Is the statement $\text{Irrational}^{\text{Irrational}} = \text{Irrational}$ true for any choice of irrational?

Solution: False

As you can see $\sqrt{2}^{\sqrt{2}} \in Q^C$ and $\left(\sqrt{2}^{\sqrt{2}}\right)^{\sqrt{2}}=\left(\sqrt{2}\right)^{\sqrt{2}\times\sqrt{2}}=\left(\sqrt{2}\right)^2=2$ which is a rational

Ex22. If $a+\dfrac{1}{b}=b+\dfrac{1}{c}=c+\dfrac{1}{a}$ where $(a \neq b \neq c \neq 0)$ then abc is equal to _____

Solution:

$a + \dfrac{1}{b} = b + \dfrac{1}{c} = c + \dfrac{1}{a} \rightarrow a^2bc + ac = ab^2c + ab = ab(c^2) + bc$ (got on multiplying each by 'abc')

Solve first two, we get $a^2bc + ac = ab^2c + ab \rightarrow abc\,(a - b) = a(b - c)$ ...... (1)

Similarly we get, $abc\,(b - c) = b(c - a)$ ...... (2) and $abc\,(a - c) = c(b - a)$ ...... (3)

Now multiply (1), (2) and (3)

We get $abc\,(a - b)\;abc\,(b - c)\;abc\,(a - c) = a(b - c)\;b(c - a)\;c(b - a) \rightarrow (abc)^3 = abc$ (as $a \neq b \neq c \neq 0$)

$(abc)^3 = abc \rightarrow abc[(abc)^2 - 1] = 0 \rightarrow abc = \pm 1$ since $abc \neq 0$.

So, $abc = \pm 1$ is only answer.

Ex23. Let x, y, z be positive reals. Which of the following implies x = y = z?

   (I) $x^3 + y^3 + z^3 = 3xyz$ (II) $x^3 + y^2z + yz^2 = 3xyz$

   (III) $x^3 + y^2z + z^2x = 3xyz$ (IV) $(x + y + z)^3 = 27\,xyz$

   (A) I, IV only   (B) I, II, IV only   (C) I, II & III only   (D) All of them

Solution:

(I) $x^3 + y^3 + z^3 = 3xyz = (x + y + z)(x^2 + y^2 + z^2 - xy - yz - zx) = \dfrac{1}{2}(x+y+z)\left((x-y)^2 + (y-z)^2 + (z-x)^2\right)$

     Since, x, y, z > 0 so, $x + y + z \neq 0 \rightarrow x^2 + y^2 + z^2 - xy - yz - zx = 0$

     $\rightarrow (x - y)^2 + (y - z)^2 + (z - x)^2 = 0 \rightarrow x = y = z$.

(II) As x, y, z > 0 so $x^3, y^2z, yz^2 > 0$

     $\rightarrow$ we can apply AM $\geq$ GM

     $\rightarrow \dfrac{x^3 + y^2z + yz^2}{3} \geq \left(x^3 y^3 z^3\right)^{1/3} \rightarrow x^3 + y^2z + yz^2 \geq 3xyz$

     But we are having equality so $x^3 = y^2z = yz^2 \rightarrow x = y = z$

(III) Here you can see x = 1, y = 2, z = 1 is satisfying the relation (Counter example)

(IV) $(x + y + z)^3 = 27\,xyz \rightarrow \dfrac{x + y + z}{3} = (xyz)^{1/3}$ i.e. AM = GM

     $\rightarrow x = y = z$

From the solutions, you can see I, II and IV are the correct statements, hence option (B) is correct.

Ex24. Let $n \geq 2$ and $n \in N$ then prove that

   (i) 2^n is the sum of two consecutive odd Naturals Numbers

   (ii) 3^n is the sum of three consecutive Natural Numbers

Solution:

(i)   Let $2^n = (2k - 1) + (2k + 1)$ i.e sum of two consecutive odd natural

     $\rightarrow 2^n = 4k \rightarrow k = 2^{n-2}$

     $\rightarrow$ Consecutive odd natural number = $2k - 1 = 2^{n-1} - 1$ and $2k + 1 = 2^{n-1} + 1$

     So, $2^n = (2^{n-1} - 1) \times (2^{n-1} + 1)$

(ii) Let $3^n = (k - 1) + k + (k + 1)$ where $k \in N$

$\rightarrow 3^n = 3k \rightarrow k = 3^{n-1}$. So, three consecutive numbers are $3^{n-1} - 1$, 3^{n-1} and $3^{n-1} + 1$

$\rightarrow 3^n = (3^{n-1} - 1) + 3^{n-1} + (3^{n-1} + 1)$

Ex25. Let k be an even natural number. Prove that '1' can't be express as the sum of reciprocals of k odd integers.

Solution:

Consider $n_1, n_2, n_3, \ldots\ldots, n_k$ are 'k' odd integers then

Let $1 = \dfrac{1}{n_1} + \dfrac{1}{n_2} + \dfrac{1}{n_3} + \ldots\ldots + \dfrac{1}{n_k}$ (If possible)

$$\dfrac{1}{n_1} + \dfrac{1}{n_2} + \dfrac{1}{n_3} + \ldots\ldots + \dfrac{1}{n_k}$$

$\rightarrow 1 = \dfrac{n_2 n_3 \ldots n_k + n_1 n_3 \ldots n_k + n_1 n_2 n_4 \ldots n_k + \ldots\ldots + n_1 n_2 n_3 \ldots n_{k-1}}{n_1 n_2 n_3 \ldots n_k}$

$\rightarrow n_2 n_3 \ldots n_k + n_1 n_3 \ldots n_k + n_1 n_2 n_4 \ldots n_k + \ldots\ldots + n_1 n_2 n_3 \ldots n_{k-1} = n_1 n_2 n_3 \ldots n_k$

$\rightarrow$ odd + odd + odd+ …..+ odd (k times) = odd

$\rightarrow$ Even = Odd (Not possible)

$\rightarrow$ '1' can't be express as the sum of reciprocals of k odd integers.

Ex26. Find the largest divisor of 1001001001 that does not exceed 10000.

Solution:

$1001001001 = 1001 \times 10^6 + 1001 = 1001(10^6 + 1) = 7 \times 11 \times 13 (10^6 + 1)$ ……..(1)

As we know $a^3 + 1 = (a + 1)(a^2 - a + 1) \rightarrow x^6 + 1 = (x^2)^3 + 1 = (x^2 + 1)(x^4 - x^2 + 1)$

$\rightarrow 10^6 + 1 = (10^2 + 1)(10^4 - 10^2 + 1) \rightarrow 10^6 + 1 = 101 \times 9901$ …..(2)

From (1) and (2) we can get

$1001001001 = 7 \times 11 \times 13 \times 101 \times 9901$

You can check that combinations of 7, 11, 13, 101 cannot generate the number between 9901 and 10000.

So, you the largest divisor does not exceed 10000 is 9901.

MATHSARC EDUCATION

A learning place to fulfill your dream of success!

MATHEMATICS   **PRACTICE PROBLEMS**

Mathsarc Education

A learning place to fulfill your dream of success!

SUBJECTIVE PROBLEMS

1. Find the square roots of following up to 4 places of decimals

    (i)   0.9   (ii) 2   (iii) 3   (iv) 123241

    (v) 213.9094   (vi) 42.0923   (vii) $0.\bar{6}$ (viii) 0.123

    (ix) $8 + 2\sqrt{15}$ (x) $\sqrt{50} + \sqrt{48}$ (xi) $35 - 12\sqrt{6} - 4\sqrt{15} + 6\sqrt{10}$

2. Solve: $\sqrt{9+2a} - \sqrt{2a} = \dfrac{5}{\sqrt{9+2a}}$

3. If $\sqrt{a} - \sqrt{12} = \sqrt{4} - \sqrt{a}$ then find the value of 'a'

4. If $x = \dfrac{5 - \sqrt{21}}{2}$ then find the value of $\left(x^3 + \dfrac{1}{x^3}\right) - 5\left(x^2 + \dfrac{1}{x^2}\right) + \left(x + \dfrac{1}{x}\right)$

5. Rationalize the denominator of $\dfrac{1}{\sqrt[3]{2} + 1}$

6. If $x = 9 + 4\sqrt{5}$, find the value of $\sqrt{x} - \dfrac{1}{\sqrt{x}}$

7. If $x \in R$, then solve the equation $\sqrt{x+1} + \sqrt{x-1} = 1$

8. Arrange the following in ascending order: $\sqrt[3]{9}, \sqrt[4]{11}, \sqrt[6]{17}$

9. Use a suitable identity to get each of the following products.

    (i) $(x + 3)(x + 3)$ (ii) $(2a - 7)(2a - 7)$ (iii) $\left(3a - \dfrac{1}{2}\right)\left(3a - \dfrac{1}{2}\right)$

    (iv) $(1.1m - 0.4)(1.1m + 0.4)$ (v) $(a^2 + b^2)(-a^2 + b^2)$ (vi) $(-a + b)(-a - b)$

    (vii) $(a - b)(a + b)(a^2 + b^2)$ (viii) $(1 - x)(1 + x)(1 + x^2)(1 + x^4)(1 + x^8)\ldots\left(1 + x^{2^n}\right)$

    (ix) $\left(a - \dfrac{b}{2} - 1\right)\left(a + \dfrac{b}{2} + 1\right)$ (x) $(2x - y)(2x + y)(4x^2 + y^2)$

10. Use the identity $(x + a)(x + b) = x^2 + (a + b)x + ab$ to find the following products

    (i) $(x + 3)(x + 7)$ (ii) $(4x + 5)(4x + 1)$ (iii) $(4x - 5)(4x - 1)$

    (iv) $(2x + 5y)(2x + 3y)$ (v) $(xyz - 4)(xyz - 2)$

11. Using $a^2 - b^2 = (a + b)(a - b)$, find

    (i) $51^2 - 49^2$ (ii) $(1.02)^2 - (0.98)^2$ (iii) $(12.1)^2 - (7.9)^2$

12. Find the True/ False statements

    (i) $2^{2^2} = 2^{\left(2^2\right)}$ (ii) $5^{2^3} = \left(5^2\right)^3 = 5^{2 \times 3}$

(iii) $\left(4^{3^2}\right)^2 = \left(4^{3^{2\times2}}\right) = 4^{3^4}$

(iv) $\left(4^{3^2}\right)^2 = \left(4^{3^2}\right) \times \left(4^{3^2}\right) = 4^{3^2+3^2} = 4^{3^2\times2}$

13. Simplify: $\dfrac{a-b}{a+b} + \dfrac{b-c}{b+c} + \dfrac{c-a}{c+a} + \dfrac{(a-b)(b-c)(c-a)}{(a+b)(b+c)(c+a)}$

14. (a) Factorize : $(x - y)^3 + (y - z)^3 + (z - x)^3$

(b) If $x^2 - x - 1 = 0$, then find the value of $x^3 - 2x + 1$

15. (a) Find all integers n such that $(n^2 - n - 1)^{n+2} = 1$

(b) If $x = \dfrac{4ab}{a+b}$, find the value of $\dfrac{x+2a}{x-2a} + \dfrac{x+2b}{x-2b}$

16. Find all the positive perfect cubes that divide 9^9.

17. (a) Find the largest prime factor of 203203.

(b) Find the last two (ten's and unit's) digits of $(2003)^{2003}$.

18. Find the number of perfect cubes between 1 and 1000009 which are exactly divisible by 9.

19. Factorize the following:

(i) $x^3 - 3x^2 + x - 3$

(ii) $4x^2 + \dfrac{1}{4x^2} - 9y^2 + 2$

(iii) $x^2 + 4y^2 - 9z^2 - 4xy$

(iv) $16 - x^4$

(v) $x^4 + x^2 + 1$

(vi) $5a^2 - 9b^2$

(vii) $9x^2 + y^2 + 4z^2 + 6xy - 4yz - 12xz$

(viii) $4x^2 + 9y^2 + 16z^2 + 12xy + 24yz + 16xz$

20. Simplify: $\dfrac{\left(a^2 - b^2\right)^3 + \left(b^2 - c^2\right)^3 + \left(c^2 - a^2\right)^3}{(a-b)^3 + (b-c)^3 + (c-a)^3}$

21. Prove that: $x^2 + y^2 + z^2 - xy - yz - zx = \dfrac{1}{2}\left[(x-y)^2 + (y-z)^2 + (z-x)^2\right]$

22. Show that $(a + 1)(a + 2)(a + 3)(a + 4) + 1$ is a perfect square

23. Find the square root of $(xy - yz - zx)^2 + 4xyz(x + y)$

24. If $x + y + z = 0$ then find the value of $\dfrac{x^2}{yz} + \dfrac{y^2}{xz} + \dfrac{z^2}{xy}$.

25. Solve the following for x

(i) $\sqrt{\left(\dfrac{4}{7}\right)^{x+4}} = \dfrac{7}{4}$

(ii) $\sqrt[3]{2x-3} - 4 = 0$

(iii) $\sqrt[5]{6x+2} = 2$

(iv) $\left(\dfrac{5}{3}\right)^{2x} \times \left(\dfrac{3}{5}\right)^{9x} = 1$

(v) $18^{4x-3} = \left(54\sqrt{2}\right)^{3x-4}$

26. If $9^{x+1} = (81)^{y+2}$ and $\left(\dfrac{1}{3}\right)^{3+x} = \left(\dfrac{1}{27}\right)^{3y}$ then find the value of x and y.

27. Express the following rational numbers in p/q form

(i) $0.\overline{6}$

(ii) $0.\overline{27}$

(iii) $3.1\overline{63}$

(iv) $12.23\overline{45}$

28. If $a + b + c \neq 0$ and $a^3 + b^3 + c^3 - 3abc = 0$ then prove that $a = b = c$.

29. Find the value of $\sqrt[3]{6+\sqrt[3]{6+\sqrt[3]{6+\sqrt[3]{6+.....\infty}}}}$

30. Simplify: $\left(1-\left(1-\left(1-x^2\right)^{-1}\right)\right)^{-1/2} \times \dfrac{1}{\sqrt{1-x^2}}$

31. Solve: $\displaystyle\sum_{r=1}^{n} \dfrac{4r}{4r^4+1}$

32. If $x + y = xy = 3$, find $x^3 + y^3$

33. Find all positive integers x, y, z satisfying the equation $3^x + y^2 = 5^z$.

34. Find all integer solutions (x, y) to the equation $x^2 + 3xy + 4006(x + y) + 2003^2 = 0$

35. Find all positive integers x, y such that $7^x - 3^y = 4$.

36. If a, b $\in$ R such that $a^2 + b^2 = 8$, $ab = 3$, Then value of $\dfrac{a^7 - a\cdot b^6 - b\cdot a^6 + b^7}{a+b}$?

37. If a, b, c $\in$ R such that $a + b + c \neq 0$, $a^3 + b^3 + c^3 = 12$, $abc = 4$ then value of $(a + b)(b + c)(c + a)$ is ___

38. If $\sqrt{2x+1} - \sqrt{x-3} = 2$, then value of $\dfrac{2x+8}{\sqrt{2x+1}+\sqrt{x-3}}$ is

39. Find 20 consecutive composite numbers.

OBJECTIVE PROBLEMS

1. If $a^x = b$, $b^y = c$ and $c^z = a$ then the value of xyz is
   (A) 0   (B) 1   (C) abc   (D) 1/abc

2. If $a + b + c = 1$, $ab + bc + ca = -1$ and $abc = -1$ then the value of $a^3 + b^3 + c^3$ is:
   (A) 1   (B) -1   (C) 2   (D) -2

3. The smallest natural number must be added to 23434563 to make it divisible by 11
   (A) 1   (B) 2   (C) 3   (D) 9

4. If $\dfrac{\sqrt{5}-1}{\sqrt{5}+1} + \dfrac{\sqrt{5}+1}{\sqrt{5}-1} = a + b\sqrt{5}$, where a, b $\in$ I, then
   (A) a = -3, b = 1   (B) a = 2, b = 1   (C) a = 1, b = -2   (D) a = 3, b = 0

5. Which one is not a transcendental number?
   (A) π (B) $\sqrt{3}$ (C) $\pi^{\sqrt{3}}$ (D) e

6. Last three digits of 2016×20015×2000017?
   (A) 160   (B) 080   (C) 180   (D) None of these

7. Three numbers are in ratio 2 : 3 : 4. The sum of their cube is 33957. Then constant of proportion, is
   (A) 3   (B) 9   (C) 7   (D) None of these

8. If $a^3 = 117 + b^3$ and $a = 3 + b$, then the value of a + b is (given that a > 0 and b > 0)
   (A) 7   (B) 9   (C) 11   (D) 13

9. If a and b are negative real numbers and c is a positive real number, then which of the following is/are correct?

1. $a - b < a - c$ 2. $a < b$, then $\dfrac{a}{c} < \dfrac{b}{c}$ 3. $\dfrac{1}{b} < \dfrac{1}{c}$

Select the correct answer using the code given below

(A) 1   (B) 2 only   (C) 3 only   (D) 2 and 3

10. If $\dfrac{61}{19} = 3 + \dfrac{1}{x + \dfrac{1}{y + \dfrac{1}{z}}}$ where x, y and z $\in$ N, then what is z equal to?

(A) 1   (B) 2   (C) 3   (D) 4

11. The maximum value of m $\in$ N, if the number $90 \times 42 \times 324 \times 55$ is divisible by 3^m.

(A) 8   (B) 7   (C) 6   (D) 5

12. Let $S = \{1, 2, 3, 4, 5, \ldots\ldots, 14\}$. The possible number of ordered pairs (a, b), where a, b $\in$ S and a $\neq$ b such that ab leaves remainder 1 when divided by 15, is

(A) 3   (B) 5   (C) 6   (D) None of these

13. Consider the following statements in respect of the expression $S_n = \dfrac{n(n+1)}{2}$, where n $\in$ I:

1. There are exactly two values of n for which $S_n = 861$.

2. $S_n = S_{(n+1)}$ and hence for any integer m, we have two values of n for which $S_n = m$.

Which of the above statements is/are correct?

(A) 1 only   (B) 2 only   (C) Both 1 and 2   (D) Neither 1 nor 2

14. Each of A, B, C and D has Rs. 100. A pays Rs 20 to B, who pays Rs 10 to C, who gets Rs 30 from D, In this context, which one of the following statements is *not* correct?

(A) C is the richest   (B) D is the poorest

(C) C has more than what A & D have together   (D) B is richer than D

15. A rectangle is divided into nine smaller rectangles as shown in the adjoining diagram. The areas of five of the smaller rectangles are shown in the diagram. The area of rectangle marked A, is

(The diagram is not drawn to scale)

(A) 1   (B) 4

(C) 12   (D) 6

1	2	
	3	4
A		16

16. Which one is a rational number?

(A) $(\pi)^{\frac{22}{7}} - \left(\dfrac{22}{7}\right)^{\pi}$ (B) $\pi - \dfrac{22}{7}$ (C) $\log_{\left(\frac{22}{7}\right)}(\pi)$ (D) $\log_{(27)^{2/7}} \sqrt{3}$

17. If $a + 10a + 100a + 1000a + 10000a = a + 1$, then a is

(A) 1   (B) $\dfrac{1}{11111}$ (C) $\dfrac{1}{11110}$ (D) None of these

18. If $A - B = A \div B$ Where A, B $\in$ Whole Numbers then A + B equals

(A) 0   (B) 6   (C) 8   (D) None of these

19. Select the correct statements:

    1. $\dfrac{\text{Even}}{\text{Even}} \times \text{Even} = \text{Odd (May be)}$

    2. $\dfrac{\text{Even}}{\text{Even}} \times \text{Even} = \text{Even (May be)}$

    (A) Only 1   (B) Only 2   (C) Both 1 and 2   (D) none of these

20. N is a positive integer such that 10N leaves a remainder 10 when divided by 100 i.e. $10\,N \equiv 70 \bmod 100$. What is the remainder when N is divided by 100?

    (A) 7   (B) 17   (C) 0   (D) Need more Information

21. Without actual division find out which of the following rational numbers has a terminating decimal expansion

    I: $17/50$ II: $11/30$ III: $\dfrac{121}{5^3 \times 2^{43}}$ IV: $\dfrac{1}{2^{32} \times 3^3}$

    (A) I and II only   (B) I and III only   (C) I, II and III   (D) All

22. If $\dfrac{1}{9} + \dfrac{1}{99} + \dfrac{1}{999} + \dfrac{1}{9999} + \ldots \ldots \infty = 0.abcde\ldots\ldots$, where a, b, c, d, e $\in \{0, 1, 2, 3, \ldots, 9\}$. Then the digit c is

    (A) 0   (B) 1   (C) 2   (D) Can't Determine

23. Mr. Laddu Lal often used to forget the notation $|\,.\,|$ (absolute value) . He writes $|a - b| = a - b$. In how many ordered pair (a, b), where a, b $\in \{0, 1, 2, 3, 4, \ldots\ldots, 10\}$, Mr. Laddu Lal be true?

    (A) 55   (B) 50   (C) 65   (D) 66

24. Select the correct statements

    (A) $4^{5000} \times 4^{5000} = 2^{19999} + 2^{19999}$ (B) $4^{5000} \times 4^{5000} > 2^{19999} + 2^{19999}$

    (C) $4^{5000} \times 4^{5000} < 2^{19999} + 2^{19999}$ (D) can't decide

25. If 'a' is a single digit natural number and $a^2 - a$ is divisible by 10, then numbers of values of a can take(s)?

    (A) 4   (B) 5   (C) 3   (D) 2

26. If $\left(x^2 + (x+1)^2\right)^{-1/2} = (x+2)^{-1}$, then x equals

    (A) -1   (B) 2   (C) 3   (D) 4

27. If $x^2 + y^2 - 4x + 4 = 0$, where x, y $\in$ R, then

    (A) $x^y = 1$ (B) $y^x = 0$ (C) $xy = 0$ (D) $|x| - |y| = 4$

28. Consider the following in respect of natural numbers a, b and c. Select the correct option(s)

    (A) LCM (ab, ac) = a LCM (b, c)   (B) HCF (ab, ac) = a HCF (b, c)

    (C) HCF (a, b) < LCM (a, b)   (D) HCF (a, b) divides LCM (a, b)

29. Select the correct options

    (A) $1000 \times (1000)^{2000} = 1000^{2001}$ (B) If $0 < x < 1$, then $x > x^2 > x^3 > x^{10}$

    (C) $\sqrt[5]{5^{5^5}} = 5^{5^4}$ (D) $2^{2^{2^2}} = 2^{16}$

30. Select the correct statement(s) (Rem ≡ Remainder)

(A) Euler's Theorem: Number of co-primes to $N = P_1^{\alpha_1} \times P_2^{\alpha_2} \times P_3^{\alpha_3} \times ...... \times P_n^{\alpha_n}$ where $P_1, P_2, P_3, ..., P_n$ are primes and $\alpha_1, \alpha_2, \alpha_3, ..., \alpha_n \in W$ are $\phi(N) = N\left(1 - \dfrac{1}{P_1}\right)\left(1 - \dfrac{1}{P_2}\right)\left(1 - \dfrac{1}{P_3}\right).....\left(1 - \dfrac{1}{P_n}\right)$.

(B) Fermat's Theorem: If P is a prime number, P and N are co-primes such that N < P then $\text{Rem}\left(\dfrac{N^P}{P}\right) = N$ and $\text{Rem}\left(\dfrac{N^{P-1}}{P}\right) = 1$ or $N^{P-1} - 1$ is divisible by P.

(C) Fermat's Little Theorem: $a^P \equiv a \pmod{p}$ or $a^P - a$ is an integral multiple of prime p for any integer 'a'.

(D) Wilson's Theorem: If P is a prime number then $\text{Rem}\left(\dfrac{(P-1)!}{P}\right) = P - 1$ & $\text{Rem}\left(\dfrac{(P-2)!}{P}\right) = 1$.

31. If $x^{p^q} = \left(x^p\right)^q$ then p = ?

(A) $q^{1/q}$ (B) q^q (C) $q^{\frac{1}{q-1}}$ (D) $q^{\frac{1}{q+1}}$

32. If $\dfrac{b}{a} = 2$ and $\dfrac{c}{b} = 3$ then value of $\dfrac{a+b}{b+c}$ is

(A) 1/3   (B) 3/8   (C) 3/5   (D) 2/3

33. Let $a_0 = 0$ and $a_n = 3a_{n-1} + 1$ for $n \geq 1$. Then the remainder obtained dividing a_{2010} by 11 is

(A) 0   (B) 7   (C) 3   (D) 4

34. If a, b are natural numbers such that $2013 + a^2 = b^2$, then the minimum possible value of ab is

(A) 671   (B) 668   (C) 658   (D) 645

35. In the figure given, a rectangle of perimeter 76 units is divided into 7 congruent rectangles.

What is the perimeter of each of the smaller rectangles?

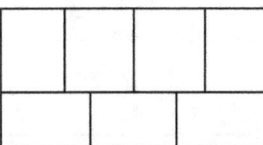

(A) 38   (B) 32

(C) 28   (D) 19

EXERCISE · HINT AND SOLUTIONS

PUZZLE # 5 : 185

IQ Test # 3 35°

IQ #131 & 27 Rings.

PUZZLE # 6: Gold is in the first box

IQ Test #2

	2	
6	8	5
4	1	3
	7	

ANSWER KEY – EXERCISE 1

Q. NO.	ANSWER	Q. NO.	ANSWER	Q. NO.	ANSWER
1	A	5	C	9	A
2	C	6	B	10	C
3	C	7	B	11	D
4	D	8	B		

ANSWER KEY – EXERCISE 2

Q. NO.	ANSWER	Q. NO.	ANSWER	Q. NO.	ANSWER
1	C	5	D	9	D
2	B	6	C	10	D
3	B	7	C	11	C
4	C	8	A		

ANSWER KEY – EXERCISE 3

Q. NO.	ANSWER	Q. NO.	ANSWER	Q. NO.	ANSWER
1	C	8	A	15	C
2	D	9	A	16	A
3	B	10	C	17	D
4	A	11	B	18	D
5	C	12	C	19	C
6	A	13	D	20	B
7	C	14	A		

ANSWER KEY – SAMPLE TEST PAPER

Q. NO.	ANSWER	Q. NO.	ANSWER	Q. NO.	ANSWER
1	C	8	B	15	B, C, D
2	B	9	C	16	7
3	D	10	D	17	1
4	B	11	A,B, C	18	0
5	B	12	A, B, D	19	2
6	B	13	C, D	20	1
7	B	14	B, D		

SOLUTIONS – EXERCISE - 1

1. **A**

2. **C.**

$$R = \frac{30^{65} - 29^{65}}{30^{64} + 29^{64}} \rightarrow (30^{64} + 29^{64})R = 30^{65} - 29^{65} \rightarrow (R + 29)\,29^{64} = (30 - R)\,30^{64}$$

$$\rightarrow \frac{R + 29}{30 - R} = \left(\frac{30}{29}\right)^{64} \rightarrow \frac{R + 29}{30 - R} > 1 \rightarrow \frac{2R - 1}{R - 30} < 0 \rightarrow R \in (0.5, 30)\,.\text{So more appropriate answer is (C).}$$

If we calculate the value of R from calculator, it's come out to be ≈ 23.95

3. **C**

Since $[\sqrt{2046}] = [\sqrt{2047}] = [\sqrt{2048}] = [\sqrt{2049}] = 45$, where [.] = GIF. So, answer is 2003 + 45 = 2048. Therefore, (C) option is correct.

4. **D**

You can verify the results by assuming some numbers. You will get (D) as correct option.

5. **C**

Let $y = \sqrt{\frac{1}{2}\sqrt{\frac{1}{2}\sqrt{\frac{1}{2}\sqrt{\frac{1}{2}}}}.......\infty} \rightarrow y = \sqrt{(y/2)}$ on squaring both side we get $y^2 = y/2$

$y(y - 1/2) = 0$. Since $y \neq 0$ so $y = \frac{1}{2}$ is the answer. Hence option (C) is correct.

6. **B**

Let $y = \sqrt{1 + 2\sqrt{1 + 2\sqrt{1 + 2\sqrt{1 +}}}..........} \rightarrow y = \sqrt{1 + 2y}$. On squaring both sides we get $y^2 = 1 + 2y$

$\rightarrow y^2 - 2y - 1 = 0 \rightarrow y = \frac{2 \pm \sqrt{8}}{2} \rightarrow 1 \pm \sqrt{2}$ but $y > 0$ so $y \neq 1 - \sqrt{2}$. Hence $y = 1 + \sqrt{2}$ is the answer.

7. **B**

LCM of {2, 3, 8, 9} = 72.

$$2^{1/2} = \left(2^{36}\right)^{1/72} ....(1) \qquad\qquad 3^{1/3} = \left(3^{24}\right)^{1/72} .... (2)$$

$$8^{1/8} = \left(8^{9}\right)^{1/72} = \left(2^{27}\right)^{1/72} ....(3) \qquad\qquad 9^{1/9} = \left(9^{8}\right)^{1/72} = \left(3^{16}\right)^{1/72} ....(4)$$

From (1), (2), (3) & (4) we can say $2^{36} > 2^{27}$ and $3^{24} > 3^{16}$, now comparison is in between 2^{36} and 3^{24}. Here, we can bring them at same ground as $2^{36} = (2^6)^6 = (64)^6$ and $3^{24} = (3^4)^6 = 81^6$ so here you can see 81 > 64, hence $3^{24} > 2^{36}$. So, option (B) is correct.

8. **B**

Let the numbers are x and y, then as per question $x^2 - y^2 = 60$.

$\rightarrow (x - y)(x + y) = 1 \times 2 \times 2 \times 3 \times 5$. So possible pairs (x – y, x + y) are

5×12, 2× 30, 4×15, 1×60, 3×20, 6×10.

Here you can observe x – y + x + y = 2x i.e. Even natural numbers.

Here pairs, 2×30 and 6× 10 are helpful for us.

If x – y = 2 and x + y = 30 → (x, y) = (16, 14). For another pair 6×10, we have

If x – y = 6 and x + y = 10 → (x, y) = (8, 2). Hence only two pairs are possible.

9. A

Let the consecutive odd natural numbers are x – 1, x + 1, x + 3. Then as per question

200 < x – 1+ x + 1+ x + 3 < 400 → 200 < 3 (x + 1) < 400. Here perfect square is a multiple of 3.

225, 256, 289, 324, 361 are the perfect squares between 200 and 400 and 225 is only multiple of 3, so √225 = 15. (A) Option is correct.

10. C

$$\sqrt[3]{\left(7^{a+b-c}\right)\left(7^{b+c-a}\right)\left(7^{c+a-b}\right)} = \sqrt[3]{7^{(a+b+c)}} = 7^{\frac{a+b+c}{3}} = 7^3 .$$

11. D

You can justify the result by taking suitable examples.

SOLUTIONS – EXERCISE · 2

1. C

2. B

$$\underbrace{111....111}_{2010 \text{ times}}\underbrace{222......222}_{2011 \text{ times}}5 = \underbrace{111....111}_{2010 \text{ time}} \times 10^{2012} + \underbrace{222....222}_{2011 \text{ times}} \times 10 + 5 = \frac{10^{2010}-1}{9} \times 10^{2012} + 2 \times \frac{10^{2011}-1}{9} \times 10 + 5$$

$$= \frac{1}{9}\left(10^{4022} - 10^{2012} + 2 \times 10^{2012} - 20 + 45\right) = \frac{1}{9}\left(\left(10^{2011}\right)^2 + 2 \times 5 \times 10^{2011} + 5^2\right) = \left(\frac{10^{2011}+5}{3}\right)^2 .$$

Hence, the square root is $\dfrac{10^{2011}+5}{3}$. So, option (B) is correct.

3. B

125 + 8 = 133.

4. C

7! = 1×2×3×4×5×6×7 = 5040 so the given equation is $x! = \dfrac{(5040)!}{5040} = 5039!$, hence x = 5039.

5. D

Let 1000 = R so given equation reduced to $\dfrac{1}{a} + \dfrac{1}{b} = \dfrac{1}{R}$.

For positive integer R we can say a > R and b > R. Let a = R + p and b = R + q where p, q ∈ N.

$\dfrac{1}{a} + \dfrac{1}{b} = \dfrac{1}{R}$ → R(a + b) = ab → R(R + p + R + q) = (R + p)(R + q) → R² = pq. Hence number of ordered pair (a, b) = Number of divisors of R² i.e. 1000².

1000² = 2⁶×5⁶. So total number of divisors = (1 + 6)(1 + 6) = 49.

6. C

7⁴ = 2401 so, 7²⁰⁰⁸ = (7⁴)⁵⁰² = (2400 + 1)⁵⁰² = 2400 I + 1. So last two digits = 01. (C) Option is correct.

7. C

x + y = xy → x + y – xy – 1 = – 1 → y(1 – x) – (1 – x) = – 1.

(y – 1) (x – 1) = 1, i.e. product of two integers = 1, it's possible only if both are either 1 or – 1 simultaneously. So, possible ordered pairs (x, y) are (2, 2) or (0, 0). Hence, (C) is correct option.

8.   A

Sum of digit's 2 can be in two ways: 2 + 0 or 1 + 1.

Let N be an 11 digit number between 10^{10} and 10^{11}. For 11 digit number, 11th place = 1, then other digit '1' has 10 choices. There is one more number i.e. 2×10^{10}. Hence, (A) is correct option.

9.   D

Number of elements in set T = n (T) = $\dfrac{467-3}{8}+1=59$.

The possible pairs when their sum is divisible by 470 are (3, 467), (11, 459), (19, 451),......., (231, 239) and there will remain one unpaired number 235. Now, only one number can be member of set S at a time from a mentioned pair. So maximum possible number of elements in S is = 29 + 1 = 30.

Therefore, (D) option is correct.

10.   D

Firstly let x = n + 7 to make the equation easier to deal with...

New expression $\dfrac{(x-6)^2}{x} \in I$, $(x^2 - 36x + 36)/x = x - 36 + (36/x)$. So all we need to do is find the positive divisors of 36 which are x = 1, 2, 3, 4, 6, 9, 12, 18, 36 → n = 2, 5, 11, 29. Hence sum = 47.

11.   C

$12112211122211112222_3 = 548458448 8_9$

ANSWER KEY – PRACTICE PROBLEMS

1. (i) 0.9486   (ii) 1.4142   (iii) 1.7320   (iv) 351.0569

    (v) 14.6256   (vi) 6.4878   (vii) 0.8164   (viii) 0.3507

    (ix) $\sqrt5 + \sqrt3$ (x) $2^{1/4}(\sqrt3 + \sqrt2)$ (xi) $3\sqrt2 - 2\sqrt3 + \sqrt5$

2. a = 8   3. a = $4 + 2\sqrt3$ 4. 0   5. $\dfrac{2^{2/3} - 2^{1/3} + 1}{3}$

6. 4   7. x = ∅   8. $\sqrt[6]{17} < \sqrt[4]{11} < \sqrt[3]{9}$

9. (i) $x^2 + 6x + 9$ (ii) $4a^2 - 28a + 49$ (iii) $9a^2 - 3a + (1/4)$ (iv) $1.21\ m^2 - 0.16$

    (v) $b^4 - a^4$ (vi) $a^2 - b^2$ (vii) $a^4 - b^4$ (viii) $1 - x^{2^{n+1}}$

    (ix) $a^2 - \left(\dfrac{b}{2}+1\right)^2 = a^2 - \dfrac{b^2}{4} - b - 1$ (x) $16x^4 - y^4$

10. (i) $x^2 + 10x + 21$ (ii) $16x^2 + 24x + 5$ (iii) $16x^2 - 24x + 5$ (iv) $4x^2 + 16xy + 15y^2$

    (v) $(xyz)^2 - 6\,xyz + 8$

11. (i) 200   (ii) 0.08   (iii) 84

12. (i) True   (ii) False   (iii) False   (iv) True

13. 0

14. (a) $3(x - y)(y - z)(z - x)$ (b) 2   15. (a) n = 2 or – 1   (b) 2

16. The positive perfect cubes that divide 9^9 are:

    $1^3, 3^3, (3^2)^3, (3^3)^3, (3^4)^3, (3^5)^3, (3^6)^3, (3^7)^3, (3^9)^3$ i.e. 10 numbers

17. (a) 29   (b) Ten's place 2 and unit place 7   18. 33

19.   (i) $(x - 3)(x^2 + 1)$ (ii) $\left(2x + \dfrac{1}{2x} + 3y\right)\left(2x + \dfrac{1}{2x} - 3y\right)$

      (iii) $(x - 2y - 3z)(x - 2y + 3z)$ (iv) $(2 - x)(2 + x)(2^2 + x^2)$

      (v) $(x^2 + x + 1)(x^2 - x + 1)$ (vi) $(\sqrt{5}a - 3b)(\sqrt{5}a + 3b)$

      (vii) $(3x + y - 2z)^2$ (viii) $(2x + 3y + 4z)^2$

20. $3(a + b)(b + c)(c + a)$ 23. $xy + yz + zx$ 24. 3

25.   (i)   - 6   (ii) 67/2   (iii)   5   (iv) 0

      (v)   6

26. $x = 33/7$ and $y = 6/7$

27.   (i)   2/3   (ii)   3/11   (iii) 174/55   (iv) 6729/550

29.   2   30. 1   31. $1 - \dfrac{1}{2n^2 + 2n + 1}$ 32. 0

33. $x = 2$, $y = 4$ and $z = 2$ 34. $(x, y) = (-1334, -446224)$, $(-2003, 0)$, and $(1336001, -446224)$.

35. $x = y = 1$ 36. 110   37. 32   38. 4

39.   20! + 2, 20! + 3, 20! + 4, ..., 20! + 20, 20! + 21 (There are infinite sets of solutions)

OBJECTIVE PROBLEMS

1. B	2. A	3. B	4. D
5. B	6. B	7. C	8. A
9. D	10. C	11. B	12. D
13. A	14. C	15. D	16. D
17. C	18. B	19. C	20. D
21. B	22. C	23. D	24. A
25. A	26. A, C	27. A, B, C	28. A, B, C, D
29. A, B, C, D	30. A, B, C, D	31. C	32. B
33. A	34. C	35. C	

	NOTES MAKING
Que / Page No.	**NOTES**

POINTS TO REMEMBER:

❖ .

❖ .

❖ .

❖ .

❖ .

OTHERS: ---

MODULUS AND INTERVAL METHOD (WAVY CURVE)

"If you just work on stuff that you like and you're passionate about, you don't have to have a master plan with how things will play out"

Mark Zuckerberg (Co-Founder & CEO - Facebook)

ABOUT MODULUS AND INTERVAL METHOD

The Modulus (Absolute Value Function) and Interval Method are the building blocks of the competitive exam preparation and learning mathematics by heart. These require in solving various questions, equations and in-equations. Although the topics are simple but students face problems in start.

In 1806, Jean – Robert Argand introduce the term *modulus, meaning unit of measure in French, specifically for the complex absolute value.* The notation $|x|$, x within vertical bar was introduce by Karl Weierstrass in 1841.

The absolute value is closely related to the idea of distance. The absolute value of a real number or complex number is the shortest distance of that number from the origin. The absolute value of the difference of two real or complex numbers is the distance between them.

Even we will study modulus function in chapter 'Relation and Function' again.

The interval method is to know the sign of a function or expression in different region. The method is *not* to know the exact value of the given expression but to get an idea of sign of expression.

"The journey of thousand miles begins with a first step"

MODULUS OR ABSOLUTE VALUE FUNCTION ($|.|$)

$|x|$ is defined as the absolute value of x or the shortest distance of point x on real number line from origin.

Ex1:   (i) $|-2| = 2$ (ii) $|-2.3| = 2.3$

       (iii) $|7| = 7$ (iv) $|3.8| = 3.8$

       (v) $|0| = 0$ (vi) $|4+3i| = \sqrt{4^2 + 3^2} = 5$

Observation

       (i) $|-2| = -(-2) = 2$ as $-2 < 0$ (ii) $|-2.3| = -(-2.3) = 2.3$ as $-2.3 < 0$

       (iii) $|7| = 7$ as $7 \geq 0$ (iv) $|3.8| = 3.8$ as $3.8 \geq 0$

       (v) $|0| = 0$ as $0 \geq 0$

(vi) $|4+3i| = \sqrt{4^2 + 3^2} = 5$, ie. distance of point Z = 4 + 3i from origin in Argand plane. Please have a look in figure shown below.

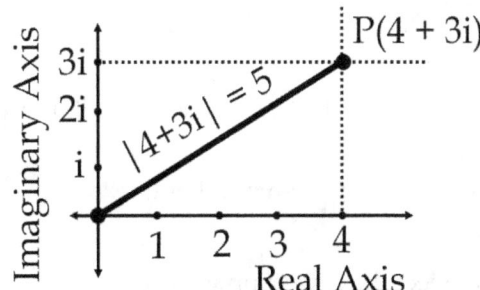

So, in general we can say $|x| = x$ if $x \geq 0$ and $|x| = -x$ if $x < 0$. In piecewise function we can write

$$|x| = \begin{cases} x & \text{if } x \geq 0 \\ -x & \text{if } x < 0 \end{cases}.$$

ABSOLUTE VALUE Vs NUMBER LINE

Please observe the figure and get best out of it.

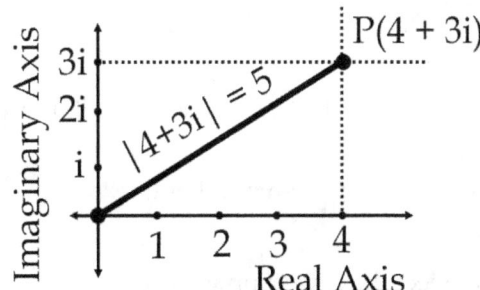

From number line, we can say that $|x|$ is the distance of real number x from origin.

Ex2:   If Mr Laddu Lal is standing at x = -10 on the number line then how far is he away from origin?

Answer:

      Obviously, the answer will be 10 i.e. $|x| = |-10| = 10$

Ex3: If $|x| = 3$ then select the correct options

(A) $x = -3$ 　　　(B) $x = 3$ 　　　(C) $\sqrt{x^2} = |x|$ 　　　(D) $\sqrt{(-2)^2} = -2$

Solution:

$\sqrt{x^2} \neq x \, \forall \, x \in R$ and $\sqrt{x}$ is always a non negative real number. i.e. $\sqrt{x} \geq 0 \, \forall \, x \geq 0$. So, options A, B and C are correct.

Ex4: If $|x - 3| = 4$ then find the value of 'x'.

Solution:

METHOD - I

$|x - 3| = 4 \rightarrow x - 3 = \pm 4 \rightarrow x = 3 \pm 4 \rightarrow x = 7$ or -1.

METHOD – II

The distance of real number x from $3 = |x - 3| = 4$. So, the points on the number line which are 4 units away from 3 are 7 and -1.

Note: $|x - a|$ represent the distance of x from 'a' on number line.

Ex5: If $|x - 2| = 5$ & $|y| = 4$ then select the correct options

(A) $x = -1$ or 7 　　　(B) $y = \pm 4$ 　　　(C) $x + y \in \{-5, 3, 11\}$

(D) Sum of all the different possible values of $|x| + |y| = 16$.

Solution:

$|x - 2| = 5 \rightarrow x - 2 = \pm 5$, so $x = 7$ or -3.

$|y| = 4 \rightarrow y = \pm 4$. So, all are correct options. i.e. A, B, C and D.

Ex6: If $|xy| = 6$ & $x, y \in I$ then minimum value of $|x|^{|y|} + y + x$, is

(A) 4 　　　(B) 2 　　　(C) 3 　　　(D) 1

Solution:

$|xy| = 6$ and $x, y \in I \rightarrow (x, y) \in \{(1, 6), (-1, -6), (2, 3), (-2, -3)\}$

To get minimum value of $|x|^{|y|} + y + x$, both x and y must be negative and $|x|^{|y|}$ must be Lowest.

So, $\left(|x|^{|y|} + y + x \right)\Big|_{MIN} = |-2|^{|-3|} - 2 - 3 = 3$. Therefore option (C) is correct.

Ex7: The number of real roots of the equation $|x|^2 - 3|x| + 2 = 0$, is

(A) 1 　　　(B) 2 　　　(C) 3 　　　(D) 4

Solution:

$|x|^2 - 3|x| + 2 = 0 \rightarrow (|x| - 1)(|x| - 2) = 0 \rightarrow |x| = 1$ or 2, hence x = 1, - 1, 2 or - 2. So, option (D) is correct.

Ex8: Find the real roots of the equation $|x|^2 - 2|x| - 3 = 0$.

Solution:

$|x|^2 - 2|x| - 3 = 0 \rightarrow |x|^2 - 3|x| + |x| - 3 = 0 \rightarrow |x|(|x| - 3) + 1(|x| - 3) = 0$

$\rightarrow (|x| + 1)(|x| - 3) = 0 \rightarrow |x| = -1$ or 3 but $|x| \neq -1$ as $|x| \geq 0 \forall x \in R$, so $|x| = 3 \rightarrow x = \pm 3$ are the roots.

GRAPH OF Y = |X|

To plot the curve of y = |x|, we need to make a table as shown

x	0	1	- 1	2	- 2	3	- 3
y	0	1	1	2	2	3	3

Even we can have piece wise function as $y = |x| = \begin{cases} x & \text{if } x \geq 0 \\ -x & \text{if } x < 0 \end{cases}$ to plot the curve

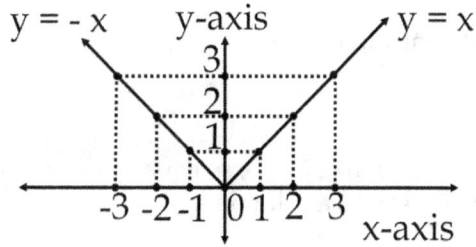

PROPERTIES OF |X|

(i) $\sqrt{x^2} = |x| \rightarrow |x|^2 = x^2 \ \forall \ x \in R$

(ii) $||x|| = |x| = |-x| \ \forall \ x \in C$

(iii) $|xy| = |x| \cdot |y| \ \forall \ x, y \in C$. The property extend to $|xyzw| = |x||y||z||w|$ & $|x^n| = |x|^n$

(iv) If $|x| = 0$ then x = 0

(v) $\left|\dfrac{x}{y}\right| = \dfrac{|x|}{|y|}$, where $x, y \in C$ & $|y| \neq 0$

(vi) Triangular Inequality $\left||x| - |y|\right| \leq |x \pm y| \leq |x| + |y| \ \forall \ x, y \in C$

(vii) If $x, y \in R$ and $|x + y| = |x| + |y|$ then both x and y are of same sign, i.e. $xy \geq 0$.

(viii) If $x, y \in R$ and $|x - y| = |x| + |-y|$ then both x and - y are of same sign, i.e. $xy \leq 0$.

(ix) $|x| = x \rightarrow x \geq 0$, the property can be extended to function as well. $|f(x)| = f(x) \rightarrow f(x) \geq 0$.

(x) $|x| + x = 0$ or $|x| = -x \rightarrow x \leq 0$. Similarly, $|f(x)| = -f(x) \rightarrow f(x) \leq 0$.

(xi) If $|x| + |y| = 0 \rightarrow x = 0$ and $y = 0 \ \forall \ x, y \in C$

(x) $|x| \geq 0 \ \forall \ x \in R$

(xi) $x^2 + y^2 = 0 \rightarrow x = 0, y = 0$ only in real numbers not in complex numbers.

(xii) $1 \leq |\sin(x)| + |\cos(x)| \leq \sqrt{2} \ \forall \ x \in R$

Ex9: Solve the equation $|2x - 3| = 8$.

Solution:

The definition of $y = |x| = \begin{cases} x & \text{if } x \geq 0 \\ -x & \text{if } x < 0 \end{cases}$ but you needs to think that the definition can be applied to

any function f(x) and it will look like $y = |f(x)| = \begin{cases} f(x) & \text{if } f(x) \geq 0 \\ -f(x) & \text{if } f(x) < 0 \end{cases}$.

METHOD · I

$|2x - 3| = 8 \rightarrow 2x - 3 = \pm 8 \rightarrow 2x = 3 \pm 8 \rightarrow x = 11/2 \text{ or } -5/2$.

METHOD – II

$|2x - 3| = 8 \rightarrow 2\left|x - \dfrac{3}{2}\right| = 8 \rightarrow \left|x - \dfrac{3}{2}\right| = 4$, x which are 4 unit apart from $x = 3/2$ are $\dfrac{11}{2}$ & $-\dfrac{5}{2}$ as

$\dfrac{3}{2} + 4 = \dfrac{11}{2}$ & $\dfrac{3}{2} - 4 = -\dfrac{5}{2}$.

METHOD – III

$y = |2x - 3| = \begin{cases} 2x - 3 & \text{if } 2x - 3 \geq 0 \rightarrow x \geq \dfrac{3}{2} \\ -(2x - 3) & \text{if } 2x - 3 < 0 \rightarrow x < \dfrac{3}{2} \end{cases}$. Here we need to make cases, since $|2x - 3|$ is

behaving differently for different values of x.

Case (I) $x \geq \dfrac{3}{2}$

In this case mode $|2x - 3|$ will open like $2x - 3$, so equation $|2x - 3| = 8 \rightarrow 2x - 3 = 8 \rightarrow x = 11/2$ which lies in the region $x \geq 3/2$, so, it's a solution.

Case (II) $x < \dfrac{3}{2}$

In this case $|2x - 3|$ will open like $-(2x - 3)$ or $3 - 2x$, so $|2x - 3| = 8 \rightarrow -(2x - 3) = 8$ $\rightarrow 2x = -5 \rightarrow x = -5/2$ which lies in the region $x < 3/2$ so, it's also a solution.

Note: Please don't make mindset about methods (good or bad). Each one is equally important.

Ex10: The exhaustive solution set of x if absolute value of $x \leq 3$, is

(A) $0 \leq x \leq 3$ (B) $-3 \leq x \leq 0$ (C) $-3 < x < 3$ (D) $-3 \leq x \leq 3$

Solution:

The set of real numbers whose distance from origin is less than equal to 3 are $-3 \leq x \leq 3$, hence (D) is correct option.

Ex11: Find the number of solutions of equation $|xy| = |y|$, hence plot the solution curve on Cartesian Co-ordinate system.

Solution:

$|xy| = |y| \rightarrow |x||y| - |y| = 0 \rightarrow |y|(|x| - 1) = 0 \rightarrow$ either $|y| = 0$ or $|x| - 1 = 0$ or both are zero simultaneously.

(i) If $|y| = 0 \rightarrow y = 0$ and $x \in R$, x – axis is solution set

(ii) If $|x| - 1 = 0 \rightarrow x = \pm 1$ and $y \in R$

∴ there are infinite solution of given equation. Look at the solution set shown below.

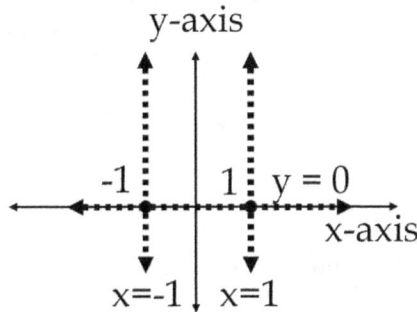

Ex12: Solve the following Equations

(i) $|x + 2| = 2(3 - x)$

(ii) $|3x - 2| + x = 11$

(iii) $||x - 1| + 2| = 1$

(iv) $|x - 2|^2 + |x - 2| - 2 = 0$

(v) $x^2 - |x| - 2 = 0$

Solution:

(i) $|x + 2| = 2(3 - x)$. The main problem with the question is to open the $|.|$ part or how to deal with modulus.

As per definition of modulus $|x + 2| = x + 2$ if $x + 2 \geq 0 \rightarrow x \geq -2$ and $|x + 2| = -(x + 2)$ if $x + 2 < 0$ i.e. $x < -2$. Now we can write the same in piecewise function as

$|x + 2| = \begin{cases} x + 2 & \text{if } x \geq -2 \\ -(x + 2) & \text{if } x < -2 \end{cases}$. You can see, $|x + 2|$ is behaving differently for different value of x.

can we deal it simultaneously? No, so we need to make cases and divide a real number line in two regions as shown in figure.

$$|x+2| = ?$$

$$x < -2 \quad x = -2 \quad x \geq -2$$

x-axis

$$-(x + 2) \qquad (x + 2)$$

Case (I) x < - 2

For $x < -2$, we can write $|x + 2| = -(x + 2)$. Hence the equation $|x + 2| = 2(3 - x)$ becomes

$-(x + 2) = 2(3 - x) \rightarrow -x - 2 = 6 - 2x \rightarrow 2x - x = 6 + 2 \rightarrow x = 8$ but 8 is $\not< -2$ so we have no solution in the region $x < -2$.

Case (II) x ≥ - 2

For $x \geq -2$, we can write $|x + 2| = x + 2$. Hence the equation $|x + 2| = 2(3 - x)$ becomes

$x + 2 = 2(3 - x) \rightarrow x + 2 = 6 - 2x \rightarrow 3x = 6 - 2 \rightarrow x = 4/3$

Which is in the region $x \geq -2$, so $x = 4/3$ is the only solution.

(ii) $|3x - 2| + x = 11$

Observe: $|3x - 2| = 0 \rightarrow x = 2/3$

As per definition of modulus, $|3x-2| = \begin{cases} 3x-2 & \text{if } x \geq \frac{2}{3} \\ -(3x-2) & \text{if } x < \frac{2}{3} \end{cases}$. So make 2 cases as shown

Case (I) $x < \frac{2}{3}$

$|3x - 2| + x = 11 \rightarrow -(3x - 2) + x = 11 \rightarrow -3x + 2 + x = 11$

$\rightarrow -2x = 9$ so, $x = -9/2$ which is within the working region $x < 2/3$, hence $x = -9/2$ is a solution.

Case (II) $x \geq \frac{2}{3}$

$|3x - 2| + x = 11 = 3x - 2 + x = 11 \rightarrow 4x = 13 \rightarrow x = 13/4$ which is in required region hence its also a solution.

So, equation $|3x - 2| + x = 11$ have 2 roots, $x = -\frac{9}{2}$ and $\frac{13}{4}$.

(iii) $||x - 1| + 2| = 1 \rightarrow |x - 1| + 2 = \pm 1$. So, we need to have two cases

$|x - 1| + 2 = 1$ and another $|x - 1| + 2 = -1$

Case (I) $|x - 1| + 2 = 1$

$|x - 1| + 2 = 1 \rightarrow |x - 1| = -1$ but $|x - 1| \geq 0 \ \forall \ x \in R$, so $|x - 1| \neq -1$ hence no solution in the case.

Case (II) $|x - 1| + 2 = -1$

$|x - 1| + 2 = -1 \rightarrow |x - 1| = -3$ this is also similar to case (I).

$|x - 1| \neq -3 \ \forall \ x \in R$. So, no solution in the case

∴ the equation has no real roots. i.e. no solution.

(iv) $|x-2|^2 + |x-2| - 2 = 0$

Let $|x - 2| = t$, so equation become $t^2 + t - 2 = 0 \rightarrow t = 1$ or $- 2$.

i.e. $|x - 2| = 1$ or $- 2$ but $|x - 2| \neq -2$ as $|x - 2| \geq 0 \ \forall \ x \in R$, so $|x - 2| = 1$

$\rightarrow x - 2 = \pm 1 \rightarrow x = 3$ or 1

Roots of the equation $|x-2|^2 + |x-2| - 2 = 0$ are $x = 3$ and 1.

(v) $x^2 - |x| - 2 = 0$. As $x^2 = |x|^2$ so the equation can be rewrite as $|x|^2 - |x| - 2 = 0$

$\rightarrow (|x| - 2)(|x| + 1) = 0$. $|x| + 1 \neq 0$ for any real x so $|x| - 2 = 0 \rightarrow x = \pm 2$.

Ex13: The number of real roots of the equation $x^2 - 4x - 3|x - 2| + 6 = 0$ is/are?

(A) 0   (B) 2   (C) 3   (D) 4

Solution

To solve the equation $x^2 - 4x - 3|x - 2| + 6 = 0$ we have two methods

METHOD – I $|x-2| = \begin{cases} x-2 & \text{if } x \geq 2 \\ 2-x & \text{if } x < 2 \end{cases}$

Case (I) x ≥ 2

The equation $x^2 - 4x - 3|x - 2| + 6 = 0 \rightarrow x^2 - 4x - 3(x - 2) + 6 = 0 \rightarrow x^2 - 7x + 12 = 0$

$\rightarrow (x - 4)(x - 3) = 0 \rightarrow x = 3, 4$ both belongs to the region $x \geq 2$ so $x = 3$ and 4 are solutions.

Case (II) x < 2

The equation $x^2 - 4x - 3|x - 2| + 6 = 0 \rightarrow x^2 - 4x - 3(2 - x) + 6 = 0 \rightarrow x^2 - x = 0 \rightarrow x = 0, 1$ both are less than 2 so these are also the solutions.

Hence, $x = 3, 4, 0$ and 1 are the real roots of equation $x^2 - 4x - 3|x - 2| + 6 = 0$.

METHOD – II

Observe the equation $x^2 - 4x - 3|x - 2| + 6 = 0$.

We can rewrite the equation as $x^2 - 4x + 4 - 3|x - 2| + 2 = 0 \rightarrow (x - 2)^2 - 3|x - 2| + 2 = 0$.

Now, $(x - 2)^2 = |x - 2|^2$ so, $(x - 2)^2 - 3|x - 2| + 2 = 0 \rightarrow |x - 2|^2 - 3|x - 2| + 2 = 0$

$\rightarrow |x - 2| = 1$ or 2, so $|x - 2| = 1 \rightarrow x = 1$ and 3 and $|x - 2| = 2 \rightarrow x = 0$ and 4.

∴ Equation $x^2 - 4x - 3|x - 2| + 6 = 0$ have 0, 1, 2 and 4 as 4 real roots. Option (D) is correct.

Let's have an exercise to check our learning, so solve each and every problem in the exercise. ☺

EXERCISE - 1

1. $|x + 1| + |x - 1| + |x + 2| + |x - 2| = 0$ then x has

   (A) The values {1, –1, 2, –1}

   (B) the values $\left\{\dfrac{1}{2}, -\dfrac{1}{2}, \dfrac{3}{2}, -\dfrac{3}{3}\right\}$

   (C) No solution

   (D) none of these

2. If $\dfrac{1}{|x|} + \dfrac{1}{x} = 1$ then the value of x is/are

   (A) 1   (B) 2   (C) 3   (D) None of these

3. If ordered pairs (x, y) satisfies $|3x - 1| = 2$ and $|y - 1| = 1$ then the correct option(s) is/are

   (A) Only 3 pairs (x, y) are possible

   (B) 4 pairs (x, y) are possible

   (C) The maximum area formed by line joining pairs = 8/3 sq. unit

   (D) No of possible matrices $\begin{vmatrix} -x & y \\ -y & x \end{vmatrix}$ are 4.

4. The minimum value of $|x - 1| + |x - 3| + |x - 4|$ is

   (A) 5   (B) 4   (C) 3   (D) None of these

5. The number of integral values of x satisfying $|(x - 1)(x - 7)| > x^2 - 8x + 7$, are

   (A) 5   (B) 4   (C) 3   (D) None of these

Ex14. Solve the following equations

(i) $|x-1| + |x-2| = 2$ (ii) $|x-1| + |x-2| = 3$

(iii) $|x| - |x-1| = 1$ (iv) $|2x-1| + |x| = 5$

Solution:

(i) $|x-1| + |x-2| = 2$, here you can observe that we have modulus at two terms and their behavior will be as shown $|x-1| = \begin{cases} x-1 & \text{if } x \geq 1 \\ 1-x & \text{if } x < 1 \end{cases}$ & $|x-2| = \begin{cases} x-2 & \text{if } x \geq 2 \\ 2-x & \text{if } x < 2 \end{cases}$. Here you can see both are having different conditions on x. so, let's have a map of $|x-1|$ and $|x-2|$ on real number line.

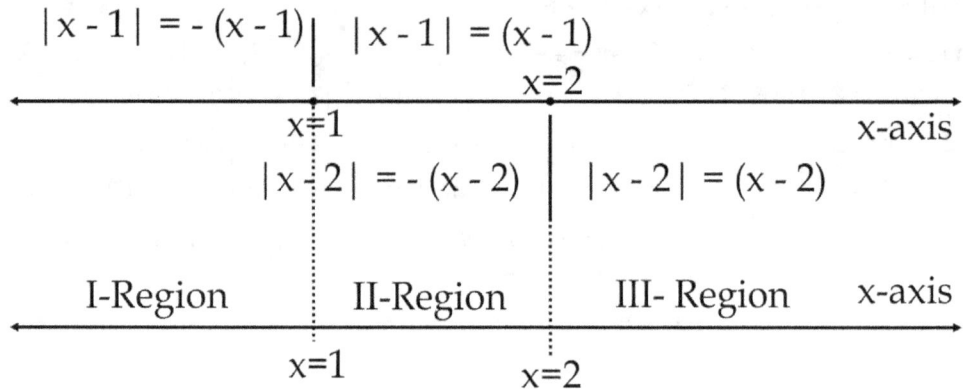

To solve the problem, we have to make 3 cases (For each region) and search solution in each region. Think! Why we need to take three cases?

CASE (I) I - Region i.e. $x \leq 1$

In this case, the equation $|x-1| + |x-2| = 2$ become $-(x-1) - (x-2) = 2 \rightarrow -2x + 3 = 2 \rightarrow x = 1/2$ which is in the region $x \leq 1$, so it's a solution.

CASE (II) II - Region i.e. $1 < x < 2$ (*observe the inequality sign we may also take like $1 < x \leq 2$*)

In this case, the equation $|x-1| + |x-2| = 2$ become $(x-1) - (x-2) = 2 \rightarrow x - x + 1 = 2 \rightarrow 1 = 2$ which is not possible, so there is no solution in the region $1 < x < 2$.

CASE (III) III - Region i.e. $x \geq 2$

In this case, the equation $|x-1| + |x-2| = 2$ become $(x-1) + (x-2) = 2 \rightarrow 2x = 5 \rightarrow x = 5/2$ which is in the region $x \geq 2$, so it's also a solution.

Hence, $x = 1/2$ and $5/2$ are the only solution.

Note: (i)   What we did here is, we search solution in different region on real number line.

(ii)   if we take union of all three region then we get full real number line. $(-\infty, 1] \cup (1, 2) \cup [2, \infty) = R$

(iii) Equality sign is as per our choice.

(iv) $x = 1$ and $x = 2$ are the points where $|x-2| = 0$, $|x-1| = 0$ or they are changing their expression about these points.

(ii) $|x - 1| + |x - 2| = 3$

Observe, $|x - 1| = 0$ at $x = 1$ and $|x - 2| = 0$ at $x = 2$

So, as per definition $|x - 1| = \begin{cases} x - 1 & \text{if } x \geq 1 \\ 1 - x & \text{if } x < 1 \end{cases}$ and $|x - 2| = \begin{cases} x - 2 & \text{if } x \geq 2 \\ 2 - x & \text{if } x < 2 \end{cases}$

Now plot number line as shown below (Observe it.)

Case (I) $x \leq 1$

$|x - 1| + |x - 2| = 3 \rightarrow -(x - 1) - (x - 2) = 3 \rightarrow -2x + 3 = 3 \rightarrow x = 0.$

Which is in the region $x \leq 1$, hence it's a solution

Case (II) $1 < x \leq 2$

$|x - 1| + |x - 2| = 3 \rightarrow (x - 1) - (x - 2) = 3 \rightarrow 1 = 3$ which is not possible hence no solution in the region $1 < x \leq 2$.

Case (III) $x > 2$

$|x - 1| + |x - 2| = 3 \rightarrow (x - 1) + (x - 2) = 3 \rightarrow 2x = 6 \rightarrow x = 3$ which lies in the region $x > 2$ so it's also a solution. Hence, $x = 0$ and 3 are the only solution.

(iii) $|x| - |x - 1| = 1$

Observe $|x| = 0$ at $x = 0$ and $|x - 1| = 0$ at $x = 1$

Case (I) $x \leq 0$

$|x| - |x - 1| = 1 \rightarrow -x - (-(x - 1)) = 1 \rightarrow -x + x - 1 = 1 \rightarrow -1 = 1$ which is not possible, hence no solution in the region

Case (II) $0 < x < 1$

$|x| - |x - 1| = 1 \rightarrow x - (-(x - 1)) = 1 \rightarrow x + x - 1 = 1 \rightarrow x = 1$ which is not in the region $0 < x < 1$, so no solution.

Observe: $x = 1$ *will be the solution if we take the case as $0 < x \leq 1$.*

Case (III) $x \geq 1$

$|x| - |x - 1| = 1 \rightarrow x - (x - 1) = 1 \rightarrow 1 = 1$ which is true for all values of $x \geq 1$.

Hence $x \in [1, \infty)$ is the solution.

(iv) $|2x - 1| + |x| = 5$

Case (I) $x < 0$

$|2x - 1| + |x| = 5 \rightarrow -(2x - 1) - x = 5 \rightarrow -3x = 4 \rightarrow x = -4/3$ which is a solution.

Case (II) $0 \leq x \leq 1/2$

$|2x - 1| + |x| = 5 \rightarrow -(2x - 1) + x = 5 \rightarrow -x = 4 \rightarrow x = -4.$

Case (III) x > 1/2

$|2x - 1| + |x| = 5 \to (2x - 1) + x = 5 \to 3x = 6 \to x = 2$ which is also a solution.

So, x = - 4/3 and 2 are the only roots of the equation $|2x - 1| + |x| = 5$.

I hope, you may get the concept of opening the modulus. Ask questions and find the solution your own. Do some experiment and verify your result by putting it into original equation.

Ex15. Solve the equation $|2x - 1| = |x + 2|$.

Solution:

Concept: $|x| = |y| \to x = \pm y \, \forall \, x, y \in R$

$|2x - 1| = |x + 2| \to 2x - 1 = \pm (x + 2)$

Case (I) $2x - 1 = (x + 2) \to x = 3$

Case (II) $2x - 1 = - (x + 2) \to x = - 1/3$

So, x = 3 and – 1/ 3 are the roots of equation $|2x - 1| = |x + 2|$.

Ex16. Find the maximum and minimum value of $|x + y|$ if $x \in [-1, 2]$ & $y \in [-1, 1]$

Solution:

Concept: As per Triangular Inequality $\left||x| - |y|\right| \le |x \pm y| \le |x| + |y| \, \forall \, x, y \in C$.

So, $||x| - |y|| \le |x + y| \le |x| + |y|$

Hence, $|x + y|_{MAX} = |x|_{MAX} + |y|_{MAX} = 2 + 1 = 3$ so $|x + y| \le 3 \, \forall \, x \in [-1, 2]$ & $y \in [-1, 1]$.

$|x + y|_{MIN} = ||x| - |y||_{MIN} = 0$. Think!

Ex17. Find the minimum value of the expression $F(x) = |x - 1| + |x - 2| + |x - 3| + \ldots + |x - 50|$

Where $x \in R$

Solution:

The minimum value of F(x) will meet at mid i.e. at x = 25

So, $F(x)\big|_{MIN} = 24 + 23 + 22 + 21 + \ldots + 2 + 1 + 0 + 1 + 2 + 3 + \ldots + 25 = 625$

Method - II

Concept: $|x - y| \le |x| + |y|$

$|x - k| + |x - (51 - k)| \ge |(x - k) - (x - (51 - k))| = 51 - 2k \, \forall \, k = 1, 2, 3, \ldots 25$

i.e. $|x - 1| + |x - 50| \ge 49$

$|x - 2| + |x - 49| \ge 47$, $\quad |x - 3| + |x - 48| \ge 49$ and so on $|x - 25| + |x - 26| \ge 1$

So, $|x - 1| + |x - 2| + |x - 3| + \ldots + |x - 50| \ge 1 + 3 + 5 + \ldots + 47 + 49$

Hence, $F(x)|_{MIN} = 1 + 3 + 5 + \ldots + 47 + 49 = 25 \times 25 = 625$.

Ex18. Solve the equation $|2x - 1| - |3 - 4x| = x$

Solution:

Case (I) $x \le 1/2$

$|2x - 1| - |3 - 4x| = x \rightarrow |2x - 1| - |4x - 3| = x \rightarrow - (2x - 1) - (-(4x - 3)) = x \rightarrow - 2x + 1 + 4x - 3 = x$

$\rightarrow x = 2$ which is not in the region $x \le 1/2$ so, no solution.

Case (II) 1/2 < x < 3/4

$|2x - 1| - |3 - 4x| = x \rightarrow |2x - 1| - |4x - 3| = x \rightarrow 2x - 1 + 4x - 3 = x \rightarrow 5x = 4 \rightarrow x = 4/5$

Which is not in the range $1/2 < x < 3/4$, so no solution in this region

Case (III) x ≥ 3/4

$|2x - 1| - |3 - 4x| = x \rightarrow |2x - 1| - |4x - 3| = x \rightarrow 2x - 1 - 4x + 3 = x \rightarrow 3x = 2 \rightarrow x = 2/3$ which is not in the range $x \ge 3/4$ so there is no solution.

Hence, the equation $|2x - 1| - |3 - 4x| = x$ has no real roots

Ex19. Plot the region $|x| + |y| = 1$ and hence find its area enclosed

Solution:

The equation $|x| + |y| = 1$ in first quadrant will looks like

'x + y = 1' and 'x - y = 1' in fourth quadrant.

Observe the equation in different quadrant as shown.

The points of intersections are

A(1, 0), B(0, 1), C(- 1, 0) and D(0, - 1)

So are of enclosed region

$= \sqrt{2} \times \sqrt{2} = 2$ sq. units

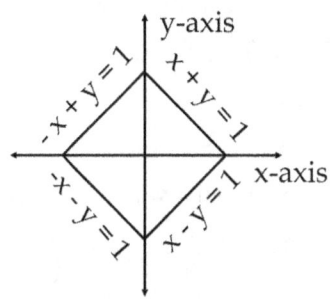

Ex20. How many integral values of x satisfy the inequality $14 < |x - 5| + |x + 3| + |x + 7| < 25$?

    (A) 9   (B) 11   (C) 13   (D) 10

Solution:

x = 5, - 3 and - 7 break the number line in 4 region as shown below

x= - 7   x = - 3   x = 5   x-axis

Case (I) x ≤ - 7 or x ∈ (- ∞, - 7]

$14 < - (x - 5 + x + 3 + x + 7) < 25 \rightarrow 14 < - (3x + 5) < 25$

$\rightarrow 3x + 5 + 14 < 0$ and $3x + 5 + 25 > 0$

$\rightarrow x < - 19/3$ and $x > - 10 \rightarrow - 10 < x < - 19/3$.

We need to take intersection of $x \le - 7$ and $- 10 < x < - 19/3$ which is $- 10 < x \le - 7$. So integral values of x in $- 10 < x \le - 7$ are $- 9, - 8$ and $- 7$.

Case (II) – 7 < x ≤ - 3 or x∈ (- 7, - 3]

The In-equation $14 < |x - 5| + |x + 3| + |x + 7| < 25$ becomes $14 < 5 - x - x - 3 + x + 7 < 25$.

$\rightarrow 14 < - x + 9 < 25 \rightarrow 5 < - x < 16 \rightarrow - 5 > x > - 16 \rightarrow x\in(- 16, - 5)$ but $(- 16, - 5) \cap(- 7, - 3] = (- 7, - 5)$

So, $x = - 6$ is an integral solution in the region.

Case (III) – $3 \leq x \leq 5$ or $x \in [-3, 5]$

The In-equation $14 < |x - 5| + |x + 3| + |x + 7| < 25$ becomes $14 < 5 - x + x + 3 + x + 7 < 25$.

$\rightarrow 14 < x + 15 < 25 \rightarrow -1 < x < 10 \rightarrow x \in (-1, 10)$

And $(-1, 10) \cap [-3, 5] = (-1, 5]$ hence, integral solutions are $x = 0, 1, 2, 3, 4$ and 5.

Case (IV) $x > 5$ or $x \in (5, \infty)$

The In-equation $14 < |x - 5| + |x + 3| + |x + 7| < 25$ becomes $14 < x - 5 + x + 3 + x + 7 < 25$

$\rightarrow 14 < 3x + 5 < 25 \rightarrow 9 < 3x < 20 \rightarrow 3 < x < 20/3 \rightarrow x \in (3, 20/3)$

And $(3, 20/3) \cap (5, \infty) = (5, 20/3)$, hence integral solution is $x = 6$.

So, solution set of inequality is $\{-9, -8, -7, -6, 0, 1, 2, 3, 4, 5, 6\}$. So, total numbers of integral solution are 11. Option (B) is correct.

Ex21. The number of integral values of x satisfying the equation $\dfrac{|x^2 - 1|}{x - 2} = x$, is/are

Solution:

$x \neq 2$ as equation will not defined for $x = 2$.

Case (I) $x \in (-\infty, -1] \cup [1, \infty)$

As $x^2 - 1 \geq 0 \ \forall \ x \in (-\infty, -1] \cup [1, \infty)$, hence $\dfrac{|x^2 - 1|}{x - 2} = x \rightarrow \dfrac{x^2 - 1}{x - 2} = x \rightarrow x^2 - 1 = x(x - 2)$

$\rightarrow x^2 - 1 = x^2 - 2x \rightarrow x = 1/2$ but it doesn't lies in the region $x \in (-\infty, -1] \cup [1, \infty)$, so no solution

Case (II) $x \in (-1, 1)$

As $x^2 - 1 < 0 \ \forall \ x \in (-1, 1)$ so $\dfrac{|x^2 - 1|}{x - 2} = x \rightarrow \dfrac{1 - x^2}{x - 2} = x \rightarrow 1 - x^2 = x(x - 2) \rightarrow 1 - x^2 = x^2 - 2x$

$\rightarrow 2x^2 - 2x - 1 = 0 \rightarrow x = \dfrac{1 \pm \sqrt{3}}{2}$ but $x = \dfrac{1 + \sqrt{3}}{2} \notin (-1, 1)$ $\& x = \dfrac{1 - \sqrt{3}}{2} \in (-1, 1)$ so equation have only

one root $x = \dfrac{1 - \sqrt{3}}{2}$ but zero integral roots.

Ex22. Solve the Equation $\dfrac{|2x - 1|}{x - 2} + \dfrac{x}{|x - 1|} = 1$.

Solution:

$|2x - 1|$ and $|x - 1|$ change expression about $x = 1/2$ and 1 respectively and $x \neq 1$ and 2 for equation to be defined.

Case (I) $x \leq 1/2$

$\dfrac{|2x - 1|}{x - 2} + \dfrac{x}{|x - 1|} = 1 \rightarrow \dfrac{1 - 2x}{x - 2} + \dfrac{x}{1 - x} = 1 \rightarrow (1 - 2x)(1 - x) + x(x - 2) = (x - 2)(1 - x)$

$\rightarrow 1 - 3x + 2x^2 + x^2 - 2x = -x^2 + 3x - 2 \rightarrow 4x^2 - 8x + 3 = 0 \rightarrow 4x^2 - 6x - 2x + 3 = 0$

$\rightarrow 2x(2x - 3) - 1(2x - 3) = 0 \rightarrow (2x - 1)(2x - 3) = 0 \rightarrow x = 1/2$ and $3/2$ but $x \neq 3/2$

So, in this region $x \leq 1/2$ we have only one root $x = 1/2$.

Case (II) $1/2 < x \leq 1$

$\dfrac{|2x-1|}{x-2} + \dfrac{x}{|x-1|} = 1 \rightarrow \dfrac{2x-1}{x-2} + \dfrac{x}{1-x} = 1 \rightarrow (2x-1)(1-x) + x(x-2) = (x-2)(1-x)$

$\rightarrow -2x^2 + 3x - 1 + x^2 - 2x = -x^2 + 3x - 2 \rightarrow 2x = 1 \rightarrow x = 1/2$

But $x = 1/2$ is not in the region $1/2 < x \leq 1$, hence no solution in this region.

Case (III) $x > 1$

$\dfrac{|2x-1|}{x-2} + \dfrac{x}{|x-1|} = 1 \rightarrow \dfrac{2x-1}{x-2} + \dfrac{x}{x-1} = 1 \rightarrow (2x-1)(x-1) + x(x-2) = (x-2)(x-1)$

$\rightarrow 2x^2 - 3x + 1 + x^2 - 2x = x^2 - 3x + 2 \rightarrow 2x^2 - 2x - 1 = 0 \rightarrow x = \dfrac{1 \pm \sqrt{3}}{2}$ but $x \neq \dfrac{1-\sqrt{3}}{2} < 1$ so $x = \dfrac{1+\sqrt{3}}{2}$

is the only solution.

So, the equation $\dfrac{|2x-1|}{x-2} + \dfrac{x}{|x-1|} = 1$ have two real roots $x = \dfrac{1}{2}$ & $\dfrac{1+\sqrt{3}}{2}$.

Ex23. Plot the curve of following expressions

    (i) $y = |x - 1|$ (ii) $y = |2x - 1|$

Solution:

(i) $y = |x-1| = \begin{cases} x-1 & \text{if } x \geq 1 \\ 1-x & \text{if } x < 1 \end{cases} \rightarrow y = |x-1| = \begin{cases} y = x-1 & \text{if } x \geq 1 \\ y = 1-x & \text{if } x < 1 \end{cases}$

Observe the curve of $y = |x-1|$ as shown in figure.

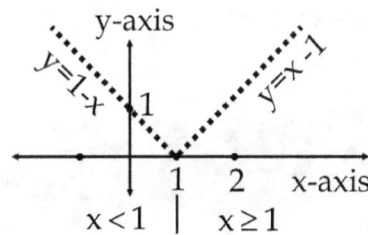

(ii) $y = |2x-1| = \begin{cases} 2x-1 & \text{if } x \geq \dfrac{1}{2} \\ 1-2x & \text{if } x < \dfrac{1}{2} \end{cases} \rightarrow y = |2x-1| = \begin{cases} y = 2x-1 & \text{if } x \geq \dfrac{1}{2} \\ y = 1-2x & \text{if } x < \dfrac{1}{2} \end{cases}$

Observe the curve of $y = |2x-1|$ as shown in figure.

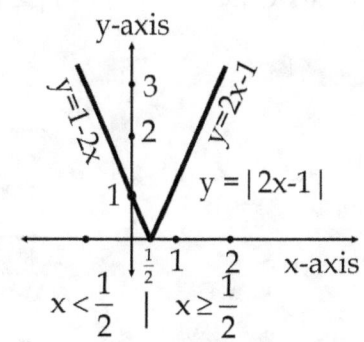

Ex24. Transform any general curve $y = f(x)$ to $y = |f(x)|$.

Solution:

As per definition $y = |f(x)| = \begin{cases} f(x) & \text{if } f(x) \geq 0 \\ -f(x) & \text{if } f(x) < 0 \end{cases}$

Let the curve of $y = f(x)$ is as shown in figure, and then looks at its transformation

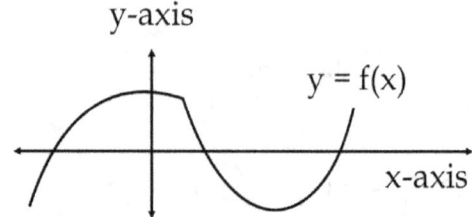

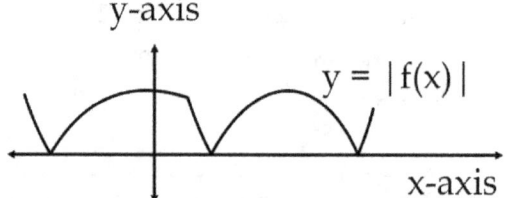

Taken mirror image of negative part of $y = f(x)$ w.r.t. x – axis and keeping positive part as it's.

Ex25. If point $P(x, y)$ lies on the curve min $\{|x|, |y - 1|\} = 2$, where $P(x, y)$ is in X-Y plane (2D Cartesian Co-ordinate system) then

(A) Minimum value of OP is 1 where O being origin

(B) Minimum value of $(|a| + |b - 1|) = 4$ where $(a, b) = P(x, y)$

(C) Four distinct points $P(x, y)$ will satisfy the equation $x^2 + (y - 1)^2 = 8$

(D) If min $\{|x|, |y - 1|\} \leq 2$, then region is symmetric about the lines $x = 0$ and $y = 1$

Solution:

First we need to sketch the curve min $\{|x|, |y - 1|\} = 2$ in Cartesian co-ordinate system.

(i) $|y - 1| = |x| \rightarrow y - 1 = \pm x \rightarrow y = 1 \pm x$

So, $y = 1 + x$ and $y = 1 - x$ are two lines where $|y - 1| = |x|$.

Case (i) If $|x| \leq |y - 1|$ then $|x| = $ min $\{|x|, |y - 1|\} = 2 \rightarrow |x| = 2 \rightarrow x = \pm 2$

Case (ii) If $|y - 1| \leq |x|$ then $|y - 1| = $ min $\{|x|, |y - 1|\} = 2 \rightarrow |y - 1| = 2 \rightarrow y = 3$ or -1

Observe the figure and looks at the shaded region as shown

Light Gray Colour: $|x| \leq |y - 1|$

Dark Gray Colour: $|y - 1| \geq |x|$

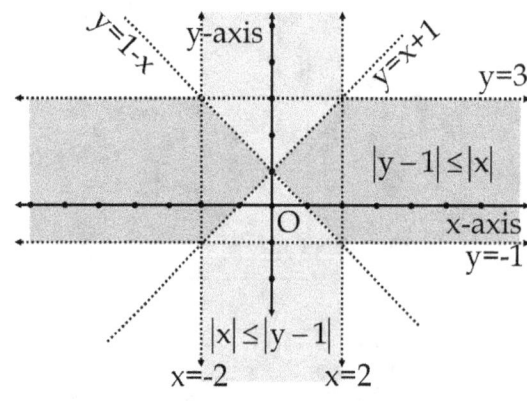

Now look at the curve of min $\{|x|, |y - 1|\} = 2$ as shown below.

(A)   OP will be minimum if it is at either (2, -1) or at (- 2, - 1). $OP_{MIN} = \sqrt{5}$. So at two points it's possible.

Hence option (A) is wrong

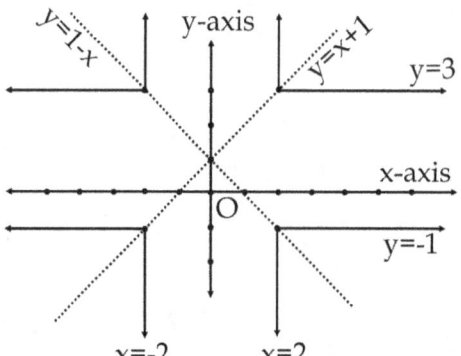

(B)   Minimum value of $(|a| + |b - 1|) = 4$ and it will meet at four points (2, 3), (2, -1), (- 2, 3) and (- 2, -1). Hence, (B) is correct option.

(C) $x^2 + (y - 1)^2 = 8$ is a circle having centre (0, 1) and it will have four intersection point with the curve

min $\{|x|, |y - 1|\} = 2$ at (2, 3), (2, -1), (- 2, 3) and (- 2, -1). So, (C) is correct option.

(D)   min $\{|x|, |y - 1|\} \leq 2$ is the shaded region including the boundaries as shown in first figure.

Hence, (B), (C) and (D) are correct options.

Let's have a mind refreshing problem of my book available in online market.

Mind Refreshment Problem # 1

An admission panel is taking interview of three children and their moms one-by-one in such a way that mother interviews before her child. In how many ways can admission panel take interview?

Author: Ramesh Chandra   **Flipkart Book :** Permutation and Combinations

Mind Refreshment Problem # 2

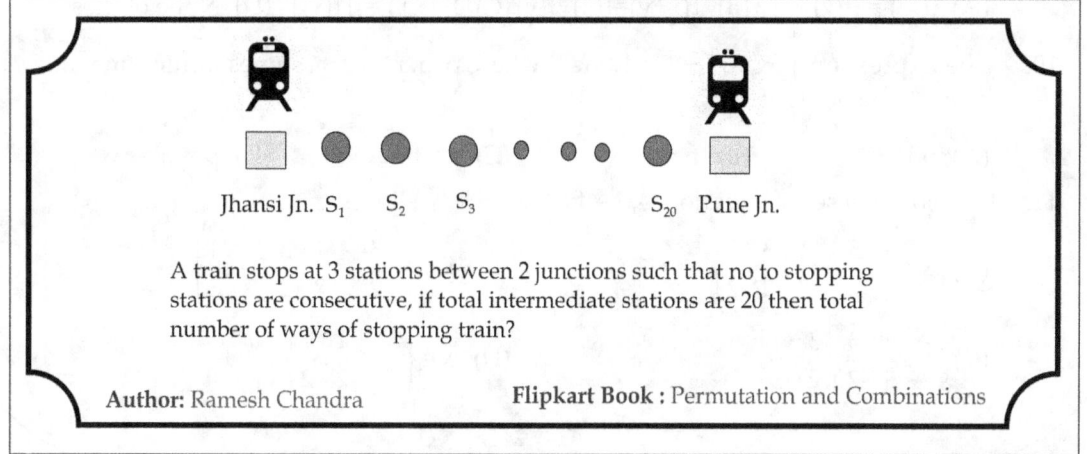

A train stops at 3 stations between 2 junctions such that no to stopping stations are consecutive, if total intermediate stations are 20 then total number of ways of stopping train?

Author: Ramesh Chandra   **Flipkart Book :** Permutation and Combinations

EXERCISE - 2

1. The area enclosed by $2|x| + 3|y| \leq 6$, is

   (A) 18 sq. units   (B) 45 sq. units   (C) 12 sq. units   (D) 10 sq. units

2. Solve for x: $\dfrac{|x-3|}{x-1} + \dfrac{x}{|x|} = 2$

   (A) $x = 3/2, 2$ (B) $x = 2, -1$ (C) $x = 3/2, -1$ (D) $x = 2$ only

3. The set of values of a for which the equation $x^2 = |x - a|$ have exactly 3 real roots

   (A) $\{0, 1, -1\}$ (B) $\{0, \frac{1}{2}, -\frac{1}{2}\}$ (C) $\{\frac{1}{2}, -\frac{1}{2}\}$ (D) $\{0, \frac{1}{4}, -\frac{1}{4}\}$

4. The real solution set of equation: $\sqrt{x - 2\sqrt{x-1}} + \sqrt{x + 2\sqrt{x-1}} = 2$, is

   (A) $x \leq 2$

   (C) $1 \leq x \leq 2$

   (B) $1, 3/2$ and 2 only

   (D) None of these

5. The area bounded by three curves $|x + y| = 1$, $|x| = 1$ and $|y| = 1$, is equal to

   (A) 4 sq. unit   (B) 3 sq. unit   (C) 2 sq. unit   (D) 1 sq. unit

6. The number of solutions (x, y) of the equation $1! + 2! + 3! + ...+ (|x|)! = y^2$ where $x \neq 0$, $|x| \leq 6$ and $y \in I$ is/are

   (A) 4   (B) 8   (C) 6   (D) None of these

7. How many integral values of x satisfying $2 < |x| - 2|x - 1| \leq 5$

   (A) 3 in $x \in (-\infty, 0)$ (B) 1 in $x \in [0, 1]$ (C) 0 in $x \in (1, \infty)$ (D) None of these

8. If x satisfies $|x - 1| + |x - 2| + |x - 3| \geq 6$, then

   (A) $0 \leq x \leq 4$ (B) $x \leq -2$ or $x \geq 4$ (C) $x \leq 0$ or $x \geq 4$ (D) None of these

9. Let $f(x) = |x - 1|$. Then

   (A) $f(x^2) = (f(x))^2$ (B) $f(x+y) = f(x) + f(y)$ (C) $f(|x|) = |f(x)|$ (D) None of these

10. The Expression $|px - q| + r|x|$, $x \in R$ where $p, q, r \in R^+$, assumes minimum value only at one point, if

    (A) $p \neq q$ (B) $r \neq q$ (C) $r \neq p$ (D) $p = q = r$

11. The solution set of equation: $4x^2 - 4x + 9 = 6|2x - 1|$, is

    (A) $x \in \left\{-\dfrac{3}{2}, -\dfrac{1}{2}, \dfrac{3}{2}, \dfrac{5}{2}\right\}$

    (B) $x \in \left\{-\dfrac{3}{2}, -\dfrac{1}{2}, \dfrac{3}{2}, \dfrac{5}{2}, \dfrac{1}{3}\right\}$

    (C) $x \in \left\{-\dfrac{3}{2}, \dfrac{3}{2}, \dfrac{1}{3}\right\}$

    (D) $x \in \left\{\dfrac{3}{2}, -\dfrac{1}{2}, \dfrac{1}{2}\right\}$

We have only answer key for this exercise and let readers to think and find solution their own. ☺☺☺

INTERVAL METHOD (WAVY CURVE)

Before actually move on to wavy curve we have to have some basic questionnaires on inequalities

Q1: If a, b ∈ R such that ab ≥ 0 then what can be said about a and b?

(A) Both a and b are of same sign

(B) If a = 0 then b ∈ R or If b = 0 then a ∈ R

(C) Both a and b can be zero simultaneously

(D) If cd < 0 then c and d are of opposite sign.

Solution:

You can see, A, B and C are correct options. Even option D is correct if c, d ∈ R

For option (D), we have an counter example $c = i = \sqrt{(-1)}$ and d = 2 i then cd = - 2 < 0. So, you need to take care in complex numbers.

Q2: If a > 0 then what can be said about 'a'?

(A) 'a' is any positive real number

(B) 'a' can take any value in the set (0, ∞)

(C) 'a' can be 0.0001, 1.2, 3.1, π, e etc

(D) 'a' cannot be zero, - 2, - 3, - 2.3 etc

Solution:

All answers are correct.

Q3: If a > b, where a, b ∈ R then select the correct statements

(A) $\frac{1}{a} < \frac{1}{b}$ Where a, b ≠ 0

(B) ac > bc ∀ c ∈ R

(C) ab > 0 and $\frac{1}{ab} > 0$ have same solutions

(D) a + c > b + c ∀ c ∈ R

Solution:

(A) you can take different values of a and b to justify the result.

If $a > b \to \frac{1}{a} < \frac{1}{b}$ only if both a and b are of same sign.

If $a > b \to \frac{1}{a} > \frac{1}{b}$ only if both a and b are of opposite sign or specifically a > 0 > b.

Option (A) is incorrect.

(B) If a > b → ac > bc only if c > 0

If a > b → ac < bc only if c < 0

Sign of inequality will change if we multiply both side by a negative no.

Ex. 4 > 3 → 4×(- 2) < 3 × (- 2) → - 8 < - 6 True

Ex. 4 > 3 → 4 × 2 > 3 × 2 → 8 > 6 True. So, option (B) is incorrect.

Option (C) and (D) are correct.

Q4: If ab = 0, then

(A) a = 0 and b ∈ R

(B) b = 0 and a ∈ R

(C) Either a = 0 or b = 0

(D) Either a = 0 or b = 0 or a = b =0

Solution:

Option (D) is correct.

Q5: If $a^2 \leq 0$, where $a \in R$, then

(A) No real value of a

(B) $a = 0$ only

(C) if $a^2 + b \leq 0$ then $b \leq -a^2 \to b \in (-\infty, 0]$

(D) If $\dfrac{a^2}{b} \leq 0 \to b < 0$ where $a^2 \leq 0$

Solution:

Options (B) and (C) are correct.

If $a^2 \leq 0$ and $\dfrac{a^2}{b} \leq 0 \to b \in R - \{0\}$.

Q6. Select the correct statements for $a, b, c, d \in R$

(A) If $a^2 b \geq 0$ & $a = 0$ then $b \in R$

(B) $\dfrac{a}{b} > 0 \to ab > 0$ but $\dfrac{a}{b} \geq 0 \not\to ab \geq 0$

(C) If $\dfrac{a^2 b}{c} > 0$ then $\dfrac{b}{c} > 0$ & $a \in R - \{0\}$

(D) $\dfrac{a}{b} \geq \dfrac{c}{d} \to ad \geq bc$ where $b \neq 0, d \neq 0$

Solution:

Take different values of a, b, c and d to analyze each and every statement.

Option (A), (B) & (C) are correct.

Q7: Select the correct option(s)

(A) $(x - 1)^2 \geq 0$ and $(x - k)^n \geq 0$ have same solution $x \in R$, where n is even natural number & $k \in R$.

(B) $(x-1)^{1/3} > 0$ and $(x-1)^{1/2} > 0$ have same solutions for x

(C) $\dfrac{1}{x} \leq 0 \to x \in (-\infty, 0)$ and $x \leq 0 \to x \in (-\infty, 0]$

(D) $\dfrac{a}{b} \geq \dfrac{c}{d} \to ad \geq bc$ Where a, b, c and d $\in R$ and b, d $\neq 0$

Solution:

$(x-1)^{1/3} > 0 \to x > 0 \to x \in (0, \infty)$

$(x-1)^{1/2} > 0 \to \sqrt{x-1} > 0$ true only for those real value of x satisfying $(x - 1) > 0 \to x > 1 \to x \in (1, \infty)$

But $(0, \infty) \neq (1, \infty)$ so, (B) is wrong option.

$\dfrac{a}{b} \geq \dfrac{c}{d} \to ad \geq bc$ only if both b and d are positive numbers.

Hence, options (A) and (C) are correct.

Now it's a time to explain interval method or wavy curve in detail. I will discuss it by some theory, examples and will set rules of working in different situations.

Ex1. Find the set of values of x for which the factor x – a is positive (i.e x – a > 0), zero (i.e. x – a = 0) and negative (i.e. x – a < 0), where a is any real constant.

Solution:

The factor x – a is zero at x = a i.e. x – a = 0 at x = a.

The factor x – a is positive $\forall$ x $\in$(a, ∞) i.e. x – a > 0 $\forall$ x $\in$(a, ∞)

The factor x – a is negative $\forall$ x $\in$(- ∞, a) i.e. x – a < 0 $\forall$ x $\in$(- ∞, a)

Look at the sign curve of (x - a) and (a - x) on number line and observe its sign

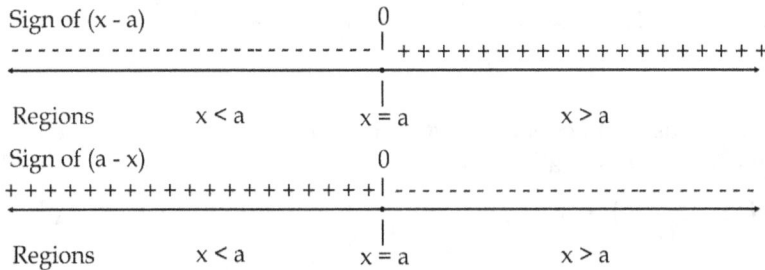

Let's have sign curve of following factors (x - 1), (x - 1)³, (x - 1)⁵, (x - 1)$^{1/3}$ and (x - 1)$^{-1/3}$.

Sign of (x - 1)

Regions   x < 1   x > 1

Sign of (x - 1)³

Sign of (x - 1)⁵

Sign of (x - 1)$^{1/3}$

Sign of (x - 1)$^{-1/3}$

x = 1

Here you can observe (x - 1), (x - 1)³, (x - 1)⁵, (x - 1)$^{1/3}$ and (x - 1)$^{-1/3}$ have same sign curve.

So (x - 1) > 0, (x - 1)³ > 0, (x - 1)⁵ > 0, (x - 1)$^{1/3}$ > 0 and (x - 1)$^{-1/3}$ > 0

have same solution x > 0 i.e. x $\in$ (1, ∞)

(x - 1) $\leq$ 0, (x - 1)³ $\leq$ 0, (x - 1)⁵ $\leq$ 0 and (x - 1)$^{1/3}$ $\leq$ 0

have same solution x $\leq$ 0 i.e.   x $\in$(- ∞, 1] but (x - 1)$^{-1/3}$ $\leq$ 0 have solution x $\in$ (-∞, 1). Why?

Since, In-equation $(x-1)^{-1/3} = \dfrac{1}{(x-1)^{1/3}} \leq 0$ will not define for x = 1 as denominator is exactly zero.

Let's move on another example having multiple of two factors

Ex2. Find the solution set of following inequalities

(i) $x(x-1) > 0$ (ii) $x(x-1) < 0$ (iii) $\dfrac{x}{x-1} < 0$ (iv) $\dfrac{x-1}{x} < 0$

(v) $x(x-1) \geq 0$ (vi) $x(x-1) \leq 0$ (vii) $\dfrac{x}{x-1} \leq 0$ (viii) $\dfrac{x-1}{x} \leq 0$

Solution:

The question is very basic and concept developing so please observe correctly and look it's solution. Please take care of equality sign, dark circle point and hollow circle (ND: Not Define)

Note: (a) $(+) \times (+) = (+)$ (b) $(+) \times (-) = (-)$ (c) $(-) \times (+) = (-)$ (d) $(-) \times (-) = (+)$

Here is the sign curve of x, (x - 1), x(x - 1), $\dfrac{x}{x-1}$ and $\dfrac{x-1}{x}$

Please take reference of sign curve of corresponding product.

(i) $x(x-1) > 0 \rightarrow x(x-1)$ is positive hence it's in the region $x < 0$ or $x > 1 \rightarrow x \in (-\infty, 0) \cup (1, \infty)$

So, solution set of $x(x-1) > 0$ is $x \in (-\infty, 0) \cup (1, \infty)$.

(ii) $x(x-1) < 0 \rightarrow x(x-1)$ is negative hence it's in the region $0 < x < 1 \rightarrow x \in (0, 1)$

So, solution set of $x(x-1) < 0$ is $x \in (0, 1)$.

(iii) $\dfrac{x}{x-1} < 0 \rightarrow \dfrac{x}{x-1}$ is negative which is in the region $0 < x < 1 \rightarrow x \in (0, 1)$. So solution set is $x \in (0, 1)$.

(iv) $\dfrac{x-1}{x} < 0 \rightarrow \dfrac{x-1}{x}$ is negative which is in the region $0 < x < 1 \rightarrow x \in (0, 1)$. So solution set is $x \in (0, 1)$.

Observe: (ii), (iii) & (iv) have same solution set. Can you answer the question, why? I hope you got answer

(v) $x(x-1) \geq 0 \rightarrow x(x-1)$ is non-negative or it can be zero or positive

$x(x-1) \geq 0 \rightarrow x(x-1) > 0$ or $x(x-1) = 0$

$x(x-1) > 0 \rightarrow x \in (-\infty, 0) \cup (1, \infty)$ and $x(x-1) = 0 \rightarrow x = 0$ or 1

So, combined solution set of x(x - 1) ≥ 0 is x∈(- ∞, 0) ∪ (1, ∞) ∪{0, 1} → x∈(- ∞, 0] ∪ [1, ∞)

(vi) x (x - 1) ≤ 0 → x (x - 1) < 0 or x (x - 1) = 0 → x∈(0, 1) or x = 0 or 1

So, combined solution set of x(x - 1) ≤ 0 is x∈(0, 1)∪{0, 1} → x∈[0, 1]

(vii) $\dfrac{x}{x-1} \le 0$ is equivalent to $\dfrac{x}{x-1} < 0$ or $\dfrac{x}{x-1} = 0$

$\dfrac{x}{x-1} < 0 \rightarrow x\in(0, 1)$ and $\dfrac{x}{x-1} = 0 \rightarrow x = 0, x \neq 1$ (since, at x = 1, in-equation is not define)

So, combined solution set of $\dfrac{x}{x-1} \le 0$ is x∈(0, 1) or {0} → x∈ [0, 1)

(viii) $\dfrac{x-1}{x} \le 0$ is equivalent to $\dfrac{x-1}{x} < 0$ or $\dfrac{x-1}{x} = 0$

$\dfrac{x-1}{x} < 0 \rightarrow x\in(0,1)$ and $\dfrac{x-1}{x} = 0 \rightarrow x = 1, x \neq 0$

So, combined solution set of $\dfrac{x-1}{x} \le 0$ is x∈(0, 1]

Observe the wave (figure) and compare the solution set of following

(A) (i) & (v);   (ii) & (vi);   (iii) & (vii);   (iv) & (viii)

Now, let's have questions with solution sets to compare and develop thinking.

(i)   x(x - 1) > 0 → x∈(- ∞, 0) ∪ (1, ∞)   (ii)   x (x - 1) < 0 → x∈(0, 1)

(iii) $\dfrac{x}{x-1} < 0 \rightarrow x\in(0, 1)$ (iv) $\dfrac{x-1}{x} < 0 \rightarrow x\in(0, 1)$

(v)   x(x - 1) ≥ 0 → x∈(- ∞, 0] ∪ [1, ∞)   (vi)   x (x - 1) ≤ 0 → x∈[0, 1]

(vii) $\dfrac{x}{x-1} \le 0 \rightarrow x\in$ [0, 1)   (viii) $\dfrac{x-1}{x} \le 0 \rightarrow x\in(0, 1]$

Ex3. Solve the following inequalities

(i)   x(x -1)³ (x - 2)⁵ ≥ 0   (ii) $\dfrac{x(x-1)^3}{x-2} \le 0$ (iii) $\dfrac{x(x-2)}{(x-1)^3} \le 0$ (iv) $\dfrac{x(x-1)^3}{x-2} > 0$

Solution:

(i)   Look at the sign curve for x(x -1)³ (x - 2)⁵ ≥ 0

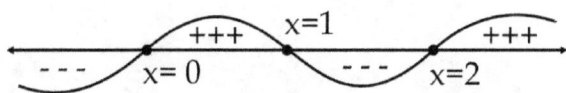

Note: If leading coefficients of 'x' are positive then extreme right part will always be positive.

You can observe that in the region 0 ≤ x ≤ 1 or x ≥ 2, inequality x(x -1)³ (x - 2)⁵ ≥ 0 satisfies.

So, solution set is: x ∈ [0, 1] ∪ [2, ∞).

(ii) Look at the sign curve for $\dfrac{x(x-1)^3}{x-2} \le 0$ and observe x = 2.

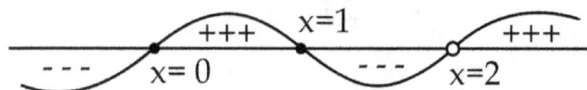

As x ≠ 2, so we will get hole at x = 2.

In the region x ≤ 0 or 1 ≤ x < 2 the inequality $\dfrac{x(x-1)^3}{x-2} \le 0$ is true.

So, solution set is: x ∈ (-∞, 0] ∪ [1, 2).

(iii) Look at the sign curve for $\dfrac{x(x-2)}{(x-1)^3} \le 0$ and observe x = 1.

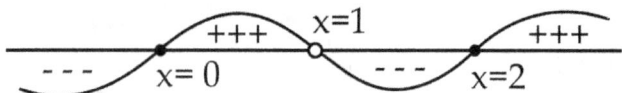

Here, x ≠ 1, for inequality to be defined.

Solution set: x ∈ (- ∞, 0] ∪ (1, 2].

(iv) Look at the sign curve for $\dfrac{x(x-1)^3}{x-2} > 0$ and observe x = 0, 1 & 2.

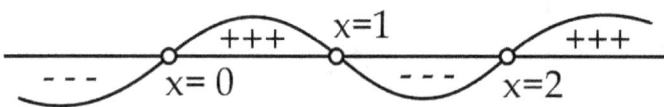

At x = 0 & 1, we have 0 > 0 which is not true and at x = 2 we have ND > 0 which is also wrong.

So, x ≠ 0, 1 and 2 as solutions. Hence, solution set: x ∈ (0, 1) ∪ (2, ∞).

Ex4. Plot the sign curve of following (*Must see solution*)

(i) $(x - 2)^2$ or $(2 - x)^2$ (ii) $(x - 1)^4 (x - 2)^6 x^{4/3}$ (iii) $\dfrac{x^2(x-3)^6}{(x-1)^{2016}(x-2)^4}$ (iv) $x^2 + x + 1$

Solution:

(i) $(x - 2)^2$ or $(2 - x)^2$ will have same sign curve as $(x - 2)^2 = (2 - x)^2$

Look at the sign curve of $(x - 2)^2$ or $(2 - x)^2$ and observe

Sign is not changing at x = 2.

If we have inequality like $(x - 2)^2 \ge 0$ then its solution must be x ∈ R

If we have inequality like $(x - 2)^2 > 0$ then solution must be x ∈ R – {2}

If we have inequality like $(x - 2)^2 < 0$ then solution must be $x \in \emptyset$

If we have inequality like $(x - 2)^2 \leq 0$ then solution must be $x \in \{2\}$

(ii) The sign curve of $(x - 1)^4 (2 - x)^6 x^{4/3}$ is as shown

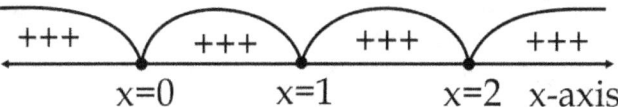

Know the answer of various related inequality.

(a) $(x - 1)^4 (2 - x)^6 x^{4/3} \geq 0 \rightarrow x \in R$

(b) $(x - 1)^4 (2 - x)^6 x^{4/3} \leq 0 \rightarrow x \in \{0, 1, 2\}$

(c) $(x - 1)^4 (2 - x)^6 x^{4/3} > 0 \rightarrow x \in R - \{0, 1, 2\}$

(d) $(x - 1)^4 (2 - x)^6 x^{4/3} < 0 \rightarrow x \in \emptyset$

(iii) Observe the sign plot of $\dfrac{x^2 (x-3)^6}{(x-1)^{2016} (x-2)^4}$ as shown

At $x = 1, 2$ we have holes and at $x = 0, 3$ we have inclusions.

Know the answer of various related inequality.

(a) $\dfrac{x^2 (x-3)^6}{(x-1)^{2016} (x-2)^4} \geq 0 \rightarrow x \in R - \{1, 2\}$

(b) $\dfrac{x^2 (x-3)^6}{(x-1)^{2016} (x-2)^4} \leq 0 \rightarrow x \in \{0, 3\}$

(c) $\dfrac{x^2 (x-3)^6}{(x-1)^{2016} (x-2)^4} > 0 \rightarrow x \in R - \{0, 1, 2, 3\}$

(c) $\dfrac{x^2 (x-3)^6}{(x-1)^{2016} (x-2)^4} < 0 \rightarrow x \in \emptyset$

(iv) **First we needs to know about quadratic expressions ($y = ax^2 + bx + c$) behavior. ($D = b^2 - 4ac$)**

Statement # 1: if $a > 0$ and Discriminant $D \leq 0$ then $ax^2 + bx + c \geq 0\ \forall\ x \in R$.

Statement # 2: if $a > 0$ and Discriminant $D < 0$ then $ax^2 + bx + c > 0\ \forall\ x \in R$.

Statement # 3: if $a < 0$ and Discriminant $D \leq 0$ then $ax^2 + bx + c \leq 0\ \forall\ x \in R$.

Statement # 4: if $a < 0$ and Discriminant $D < 0$ then $ax^2 + bx + c < 0\ \forall\ x \in R$.

For $x^2 + x + 1$: $a = 1 > 0$ and $D = 1^2 - 4 \times 1 \times 1 = -3 < 0 \rightarrow x^2 + x + 1 > 0\ \forall\ x \in R$ as per statement # 2.

Observe the sign plot of $x^2 + x + 1$ as shown

Know the answer of various related inequality.

(a) $x^2 + x + 1 < 0 \rightarrow x \in \emptyset$

(b) $x^2 + x + 1 \leq 0 \rightarrow x \in \emptyset$

(c) $x^2 + x + 1 > 0 \rightarrow x \in R$

(d) $x^2 + x + 1 \geq 0 \rightarrow x \in R$

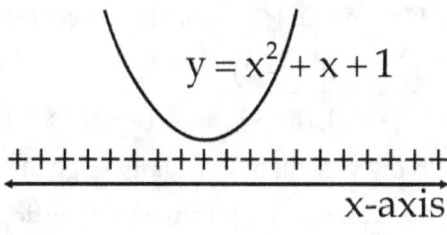

Note: sign is not changing at $x = a$ for factor $(x - a)^{\text{even}}$

Ex5. Plot the sign curve of $\dfrac{(x-1)^2(2x-1)^5}{x(x-3)^3(x-2)^4}$.

Solution:

Observe the sign plot of $\dfrac{(x-1)^2(2x-1)^5}{x(x-3)^3(x-2)^4}$ as shown

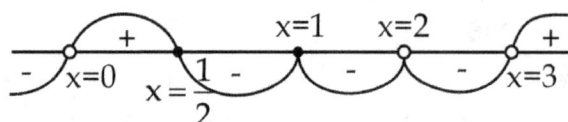

Note: Observe, sign is not changing at x = 1 and 2, as factors $(x - 1)^2$ and $(x - 2)^4$ have even powers.

Observe, sign is changing at x = 0, 1/2 and 3 as factors x, $(2x - 1)^5$ and $(x - 3)^3$ have odd powers.

Know the answer of various related inequality.

(a) $\dfrac{(x-1)^2(2x-1)^5}{x(x-3)^3(x-2)^4} \geq 0 \rightarrow x \in (0, 1/2] \cup (3, \infty) \cup \{1\}$

(b) $\dfrac{(x-1)^2(2x-1)^5}{x(x-3)^3(x-2)^4} \leq 0 \rightarrow x \in (-\infty, 0) \cup [1/2, 3) - \{2\}$

(c) $\dfrac{(x-1)^2(2x-1)^5}{x(x-3)^3(x-2)^4} > 0 \rightarrow x \in (0, 1/2) \cup (3, \infty)$

(d) $\dfrac{(x-1)^2(2x-1)^5}{x(x-3)^3(x-2)^4} < 0 \rightarrow x \in (-\infty, 0) \cup (1/2, 3) - \{1, 2\}$ or $x \in (-\infty, 0) \cup (1/2, 1) \cup (1, 2) \cup (2, 3)$

Ex6. Plot the sign curve of $\dfrac{x^3(1-x)^5(x-2)^4}{(x-3)^4(x-4)}$ and solve the following inequalities

(i) $\dfrac{x^3(1-x)^5(x-2)^4}{(x-3)^4(x-4)} \geq 0$

(ii) $\dfrac{x^3(1-x)^5(x-2)^4}{(x-3)^4(x-4)} \leq 0$

(iii) $\dfrac{x^3(1-x)^5(x-2)^4}{(x-3)^4(x-4)} > 0$

(iv) $\dfrac{x^3(1-x)^5(x-2)^4}{(x-3)^4(x-4)} < 0$

Solution:

First make each leading coefficients of x, positive. Why?

If we do this then we always get (+) positive sign at right extreme of number line.

$$\dfrac{x^3(1-x)^5(x-2)^4}{(x-3)^4(x-4)} = -\dfrac{x^3(x-1)^5(x-2)^4}{(x-3)^4(x-4)}$$

Now we will have negative sign at right extreme for the case. Think!

We can make it positive by reverting inequality. Look at the sign curve

$$\frac{x^3(1-x)^5(x-2)^4}{(x-3)^4(x-4)} \quad \text{Sign}$$

$$\frac{x^3(x-1)^5(x-2)^4}{(x-3)^4(x-4)} \quad \text{Sign}$$

To avoid confusion, always have (+) ve coefficients of x and have (+) sign at extreme right of number line.

(i) $\dfrac{x^3(1-x)^5(x-2)^4}{(x-3)^4(x-4)} \geq 0 \rightarrow -\dfrac{x^3(x-1)^5(x-2)^4}{(x-3)^4(x-4)} \geq 0 \rightarrow \dfrac{x^3(x-1)^5(x-2)^4}{(x-3)^4(x-4)} \leq 0$

$\rightarrow x \in (-\infty, 0] \cup [1, 4) - \{3\}$ or $x \in (-\infty, 0] \cup [1, 2] \cup (2, 3) \cup (3, 4)$

As $x \neq 3, 4$ but $x = 0, 1$ & 2 are solution of inequality.

(ii) $\dfrac{x^3(1-x)^5(x-2)^4}{(x-3)^4(x-4)} \leq 0 \rightarrow -\dfrac{x^3(x-1)^5(x-2)^4}{(x-3)^4(x-4)} \leq 0 \rightarrow \dfrac{x^3(x-1)^5(x-2)^4}{(x-3)^4(x-4)} \geq 0$

$\rightarrow x \in [0, 1] \cup (4, \infty) \cup \{2\}$

As $x = 2$ is solution so we need to include it in the solution.

(iii) $\dfrac{x^3(1-x)^5(x-2)^4}{(x-3)^4(x-4)} > 0 \rightarrow -\dfrac{x^3(x-1)^5(x-2)^4}{(x-3)^4(x-4)} > 0 \rightarrow \dfrac{x^3(x-1)^5(x-2)^4}{(x-3)^4(x-4)} < 0$

$\rightarrow x \in (-\infty, 0) \cup (1, 4) - \{2, 3\}$

As, $x \neq 0, 1, 2, 3$ and 4 due to inequality. So, we have to remove these points if included in solution.

(iv) $\dfrac{x^3(1-x)^5(x-2)^4}{(x-3)^4(x-4)} < 0 \rightarrow -\dfrac{x^3(x-1)^5(x-2)^4}{(x-3)^4(x-4)} < 0 \rightarrow \dfrac{x^3(x-1)^5(x-2)^4}{(x-3)^4(x-4)} > 0$

$\rightarrow x \in (0, 1) \cup (4, \infty)$. As, $x \neq 0, 1, 2, 3$ and 4 due to inequality

Ex7. Plot the sign curve of $\dfrac{x}{x}$ and hence answer the following inequalities

(i) $\dfrac{x}{x} \geq 0$ (ii) $\dfrac{x}{x} \leq 0$ (iii) $\dfrac{x}{x} > 0$

Solution:

$\dfrac{x}{x}$ is not define at $x = 0$. So, $\dfrac{x}{x} = \begin{cases} \text{Not Define (ND)} & \text{if } x = 0 \\ 1 & \text{if } x \neq 0 \end{cases}$. Look at the sign curve

(i) $\dfrac{x}{x} \geq 0 \rightarrow x \in R - \{0\}$ (ii) $\dfrac{x}{x} \leq 0 \rightarrow x \in \emptyset$

(iii) $\dfrac{x}{x} > 0 \rightarrow x \in R - \{0\}$

Ex8. Solve the following inequalities

(i) $\dfrac{x(x-1)^3(x-2)^2}{(x-1)} \geq 0$

(ii) $\dfrac{x(x-1)^3(x-2)^2}{(x-1)} \leq 0$

(iii) $\dfrac{x^2(x-1)^{301}(2x-3)^7(x-3)^{111}}{(x-1)^{2016}(4-3x)^{24}(x-5)^{2/3}} \geq 0$

(iv) $\dfrac{x^2(x-1)^{301}(2x-3)^7(x-3)^{111}}{(x-1)^{2016}(4-3x)^{24}(x-5)^{2/3}} < 0$

Solution:

(i) Observe the sign curve of $\dfrac{x(x-1)^3(x-2)^2}{(x-1)}$

As , $x \neq 1$ so $\dfrac{x(x-1)^3(x-2)^2}{(x-1)} \rightarrow \left[(x-1)^2(x-2)^2\right]x$.

I have separated even power factors. We will not have sign change at x = 1 & 2 due to even power.

So, $\dfrac{x(x-1)^3(x-2)^2}{(x-1)} \geq 0 \rightarrow \left[(x-1)^2(x-2)^2\right]x \geq 0 \rightarrow x \in [0, \infty) \cup \{1, 2\} = [0, \infty)$ but $x \neq 1$ so solution

set for the inequality is $x \in [0, \infty) - \{1\}$.

(ii) $\dfrac{x(x-1)^3(x-2)^2}{(x-1)} \leq 0 \rightarrow \left[(x-1)^2(x-2)^2\right]x \leq 0 \rightarrow x \in (-\infty, 0] \cup \{2\}$.

(iii) As, (x - 1) factor is repeating in Numerator and Denominator. First cancel them by saying $x \neq 1$. So

$\dfrac{x^2(x-1)^{301}(2x-3)^7(x-3)^{111}}{(x-1)^{2016}(4-3x)^{24}(x-5)^{2/3}} \rightarrow \dfrac{x^2(2x-3)^7(x-3)^{111}}{(x-1)^{1715}(4-3x)^{24}(x-5)^{2/3}}$

$= \left[\dfrac{x^2}{(4-3x)^{24}(x-5)^{2/3}}\right]\left|\dfrac{(2x-3)^7(x-3)^{111}}{(x-1)^{1715}}\right|$

Observe that I have separated even power factors which will not play any rolls in sign change.

Only, odd power factor product changes sign. So,

$\dfrac{x^2(x-1)^{301}(2x-3)^7(x-3)^{111}}{(x-1)^{2016}(4-3x)^{24}(x-5)^{2/3}} \geq 0 \rightarrow \left[\dfrac{x^2}{(4-3x)^{24}(x-5)^{2/3}}\right]\left|\dfrac{(2x-3)^7(x-3)^{111}}{(x-1)^{1715}}\right| \geq 0$

Observe the sign curve of $\dfrac{x^2(x-1)^{301}(2x-3)^7(x-3)^{111}}{(x-1)^{2016}(4-3x)^{24}(x-5)^{2/3}} \equiv \left[\dfrac{x^2}{(4-3x)^{24}(x-5)^{2/3}}\right]\dfrac{(2x-3)^7(x-3)^{111}}{(x-1)^{1715}}$

$\left|\dfrac{x^2}{(4-3x)^{24}(x-5)^{2/3}}\right|\dfrac{(2x-3)^7(x-3)^{111}}{(x-1)^{1715}} \geq 0 \rightarrow x \in (1, 3/2] \cup [3, \infty) \cup \{0\} - \{4/3, 5\}.$

METHOD - II

In this method we are going to plot only those points on number line, having odd powers and will consider the even powers roll at the time of writing solution.

As, $\left|\dfrac{x^2}{(4-3x)^{24}(x-5)^{2/3}}\right|\dfrac{(2x-3)^7(x-3)^{111}}{(x-1)^{1715}} \geq 0$

For equality, $x = 0, 3/2, 3$ are solution and $x \neq 4/3, 5$ and 1. Now I am plotting the curve for

$\dfrac{(2x-3)^7(x-3)^{111}}{(x-1)^{1715}} \geq 0$. Look at the wavy curve shown below.

Hence, $\dfrac{(2x-3)^7(x-3)^{111}}{(x-1)^{1715}} \geq 0 \rightarrow x \in (1, 3/2] \cup [3, \infty)$ but we need to consider even power factors so

$\left|\dfrac{x^2}{(4-3x)^{24}(x-5)^{2/3}}\right|\dfrac{(2x-3)^7(x-3)^{111}}{(x-1)^{1715}} \geq 0 \rightarrow x \in (1, 3/2] \cup [3, \infty) \cup \{0\} - \{4/3, 5\}$

(iv) $\dfrac{x^2(x-1)^{301}(2x-3)^7(x-3)^{111}}{(x-1)^{2016}(4-3x)^{24}(x-5)^{2/3}} < 0 \rightarrow \left|\dfrac{x^2}{(4-3x)^{24}(x-5)^{2/3}}\right|\dfrac{(2x-3)^7(x-3)^{111}}{(x-1)^{1715}} < 0$

So, $\dfrac{(2x-3)^7(x-3)^{111}}{(x-1)^{1715}} < 0 \rightarrow x \in (-\infty, 1) \cup (3/2, 3).$

Hence, $\left|\dfrac{x^2}{(4-3x)^{24}(x-5)^{2/3}}\right|\dfrac{(2x-3)^7(x-3)^{111}}{(x-1)^{1715}} < 0 \rightarrow x \in (-\infty, 1) \cup (3/2, 3) - \{0\}.$

Ex9. Solve the following Inequalities

(i) $\dfrac{1}{x} < 2$

(ii) $x \geq \dfrac{1}{x}$

(iii) $\dfrac{x-3}{5} \geq \dfrac{3-x}{8x}$

(iv) $\dfrac{2x}{2x^2+5x+2} > \dfrac{1}{x+1}$

Solution:

(i)   The problem seems very simple but good for concept learning.

Wrong Way: $\frac{1}{x} < 2 \to 1 < 2x \to x > \frac{1}{2} \to x \in \left(\frac{1}{2}, \infty\right)$. Can you identify the wrongness in the method?

Think! $x = -3$ satisfy $\frac{1}{x} < 2$ but not arriving in the solution set $x \in \left(\frac{1}{2}, \infty\right)$.

Concept:

(a) $\frac{a}{b} > c \to a > bc$ only if $b > 0$

(b) $\frac{a}{b} > c \to a < bc$ only if $b < 0$

So, avoid cross multiplication as far as possible.

Correct way: $\frac{1}{x} < 2 \to \frac{1}{x} - 2 < 0 \to \frac{1-2x}{x} < 0 \to \frac{2x-1}{x} > 0$

$\frac{2x-1}{x} > 0 \to x \in (-\infty, 0) \cup (1/2, \infty)$

(ii) $x \geq \frac{1}{x} \not\to x^2 - 1 \geq 0$. Think!

Correct way: $x \geq \frac{1}{x} \to x - \frac{1}{x} \geq 0 \to \frac{x^2 - 1}{x} \geq 0 \to \frac{(x-1)(x+1)}{x} \geq 0$

$\frac{(x-1)(x+1)}{x} \geq 0 \to x \in [-1, 0) \cup [1, \infty)$

(iii) $\frac{x-3}{5} \geq \frac{3-x}{8x} \not\to \frac{1}{5} \geq -\frac{1}{8x}$. Why? (No cancellation in In-equations, It causes loss of genuine roots)

Correct way: $\frac{x-3}{5} \geq \frac{3-x}{8x} \to \frac{x-3}{5} - \frac{3-x}{8x} \geq 0 \to \frac{x-3}{5} + \frac{x-3}{8x} \geq 0$

$\to (x-3)\left(\frac{1}{5} + \frac{1}{8x}\right) \geq 0 \to \frac{(x-3)(8x+5)}{40x} \geq 0$

$\frac{(x-3)(8x+5)}{40x} \geq 0 \to x \in \left[-\frac{5}{8}, 0\right) \cup [3, \infty)$

(iv) $\dfrac{2x}{2x^2+5x+2} > \dfrac{1}{x+1} \to \dfrac{2x}{2x^2+5x+2} - \dfrac{1}{x+1} > 0 \to \dfrac{2x(x+1)-(2x^2+5x+2)}{(2x^2+5x+2)(x+1)} > 0$

$\to \dfrac{-(3x+2)}{(2x+1)(x+2)(x+1)} > 0 \to \dfrac{(3x+2)}{(2x+1)(x+2)(x+1)} < 0$

$$x=-2 \qquad x=-1 \qquad x=-\dfrac{2}{3} \qquad x=-\dfrac{1}{2}$$

$\to x \in (-2,-1) \cup \left(-\dfrac{2}{3}, -\dfrac{1}{2}\right)$ is solution set.

Let's have an exercise to know and check our learning. ☺ ☺ ☺

EXERCISE - 3

1. Solve the following Inequalities

   (i) $2x^2 - 5x + 2 \leq 0$

   (ii) $x^2 - 5x + 6 \geq 0$

   (iii) $x^2 - 4x + 6 \leq \dfrac{6}{x+1}$

   (iv) $\dfrac{1}{x+1} \geq \dfrac{x}{3}$

   (v) $\dfrac{x^2+x+1}{x^2-x+1} \leq 0$

   (vi) $\dfrac{x^2+x+1}{x^2-x+1} > 0$

   (vii) $\dfrac{x^2-4x-21}{(x-1)(x-3)^2} \geq 0$

   (viii) $\dfrac{x^2-4x-21}{(x-1)(x-3)^2} \leq 0$

   (ix) $\dfrac{(x^2+1)(3-x)}{x^5(x-1)^7(x-5)^{2016}} \leq 0$

   (x) $\dfrac{(x^2-1)(x+1)^2}{(x^3+1)(x+3)^4} < 0$

   (xi) $\dfrac{8x^2+16x-51}{(2x-3)(x+4)} > 3$

   (xii) $\dfrac{2x}{2x^2+5x+2} > \dfrac{1}{x+1}$

   (xiii) $\dfrac{2x-1}{2x^3+3x^2+x} > 0$

   (xiv) $\dfrac{x^2-2x+3}{x^2-4x+3} > -3$

2. The greatest integer x for which the inequality $\dfrac{x-3}{x^2+9x-22} < 0$ is satisfied, is

   (A) -12   (B) -11   (C) 2   (D) 3

3. The set of real numbers of x for which $x^2 - |x+2| + x > 0$, is

   (A) $(-\infty, -2) \cup (2, \infty)$
   (B) $(-\infty, -\sqrt{2}) \cup (\sqrt{2}, \infty)$
   (C) $(-\infty, -1) \cup (1, \infty)$
   (D) $(\sqrt{2}, \infty)$

4. Solve the following inequalities

   (i) $2 \leq 4 - 2x < 4$

   (ii) $-2 \leq \dfrac{1}{2} - \dfrac{2x}{3} \leq \dfrac{11}{6}, x \in N$

   (iii) $-75 < 3(x-2) \leq 0, x \in W$

   (iv) $7x - 2 \leq 3x - 4 < 3 - x$

5. Solve the following system of inequalities

   (i) $\begin{cases} 11-2x < 6-x \\ x-6 < \dfrac{3x-7}{2} \end{cases}$

   (ii) $\begin{cases} x-3 > \dfrac{3-x}{4} \\ \dfrac{1+x}{3} \geq -2 - \dfrac{x}{4} \end{cases}$

6. The values of k for which the inequality $x^2 - (k-3)x - k + 6 > 0$ valid $\forall x \in R$, is_____?

We have answer key only for the exercise and let readers to think and find solution their own. ☺☺☺

INEQUALITY EQUIVALENCE

There is no as such definition in mathematics but we are creating it to our understanding.

Definition:

If two factors have same sign scheme over real number line then these factors are said to be in 'Inequality Equivalence'.

Ex1. $|x| - 1$ and $(x - 1)(x + 1)$ have same sign scheme over number line as shown in figure. So $|x| - 1$ and $(x - 1)(x + 1)$ are in inequality equivalence.

$$|x| - 1 \qquad \underset{x = -1 \qquad x = 1}{\overset{+++ \qquad --- \qquad +++}{\rule{6cm}{0.4pt}}}$$

$$(x - 1)(x + 1) \qquad \underset{x = -1 \qquad x = 1}{\overset{+++ \qquad --- \qquad +++}{\rule{6cm}{0.4pt}}}$$

Ex2. $e^x - 1$ and x have same sign scheme so they are in 'Inequality equivalence'.

$$e^x - 1 \qquad \underset{x = 0}{\overset{------- \qquad +++++}{\rule{6cm}{0.4pt}}}$$

$$x \qquad \underset{x = 0}{\overset{------- \qquad +++++}{\rule{6cm}{0.4pt}}}$$

Now, question is, why are we studying the topic?

Many times, we need to manipulate our mathematical model to find the solution in easy way and it helps us in achieving the same.

Ex3. Find the solution set of $\dfrac{x\left(3^x - 1\right)(x-4)^2}{(x-3)(x-5)^4} > 0$.

Solution:

Here you can observe that $3^x - 1$ and x are inequality equivalence so

$\dfrac{x\left(3^x - 1\right)(x-4)^2}{(x-3)(x-5)^4} > 0$ and $\dfrac{x^2(x-4)^2}{(x-3)(x-5)^4} > 0$ have same solution set.

Hence, $\dfrac{x^2(x-4)^2}{(x-3)(x-5)^4} > 0 \rightarrow x \in (3,\infty) - \{4, 5\}$ So, $\dfrac{x\left(3^x - 1\right)(x-4)^2}{(x-3)(x-5)^4} > 0 \rightarrow x \in (3,\infty) - \{4, 5\}$

Ex4. Find the set of values of x satisfying $\left(e^x - 1\right)x(x-1) \geq 0$

Solution:

$e^x - 1$ & x are inequality equivalent so solution set of $\left(e^x - 1\right)x(x-1) \geq 0$ & $x^2(x-1) \geq 0$ are same.

So, $x^2(x - 1) \geq 0 \rightarrow x \in [1, \infty) \cup \{0\}$. Hence, $\left(e^x - 1\right)x(x-1) \geq 0 \rightarrow x \in [1,\infty) \cup \{0\}$.

Question: Is inequality equivalence work in the inequality $\left(e^x - 1\right) \geq 1$?

Answer: NO. Think!

Note: inequality equivalence work only in the inequality of type (f(x) > 0, f(x) < 0, f(x) ≥ 0, f(x) ≤ 0)

DOMAIN RELATED INEQUALITY

Ex5. Find the solution set of $\dfrac{\sqrt{x}\,(4x-1)^2\,(x+2)^3\,(x-2)}{(2x-1)} \le 0$

Solution:

Here you can see $\sqrt{x}$ is defined only for $x \ge 0$ and $\sqrt{x} \ge 0 \;\forall\; x \in [0, \infty)$.

So, given inequality is defined only for $x \ge 0$ hence, we have to find our solution only in $x \in [0, \infty)$

Look at the sign curve for $\dfrac{\sqrt{x}\,(x-2)}{(2x-1)}$

$\dfrac{\sqrt{x}\,(4x-1)^2\,(x+2)^3\,(x-2)}{(2x-1)} \le 0 \rightarrow x \in \left(\dfrac{1}{2}, 2\right] \cup \left\{\dfrac{1}{4}\right\}$

Ex6. Find the solution set of $\dfrac{(1-2x)^{3/4}\,(x-1)^3\,(x-2)^2}{x^3\,(x^2+1)(x+1)^5\,(x+2)^4} \le 0$

Solution:

Here $(1-2x)^{3/4}$ is defined only if $1-2x \ge 0 \rightarrow x \le 1/2$. So we have to find the solution only in the set $x \in (-\infty, 1/2]$ and $(1-2x)^{3/4} \ge 0 \;\forall\; x \in (-\infty, 1/2]$

Look at the sign curve of $\left[\dfrac{(1-2x)^{3/4}}{(x^2+1)(x+2)^4}\right]\left[(x-1)^3(x-2)^2\right]\dfrac{1}{x^3\,(x+1)^5} \le 0$

$\dfrac{(1-2x)^{3/4}\,(x-1)^3\,(x-2)^2}{x^3\,(x^2+1)(x+1)^5\,(x+2)^4} \le 0 \rightarrow \left[\dfrac{(1-2x)^{3/4}}{(x^2+1)(x+2)^4}\right]\left[(x-1)^3(x-2)^2\right]\dfrac{1}{x^3\,(x+1)^5} \le 0$

$\rightarrow x \in (-\infty, -1) \cup (0, \text{½}] - \{-2\}$

Ex7. Solution of inequation $\dfrac{(x-9)(x+7)\,|x-10|}{|x-1|} \le 0$, is

(A) $(-7, 9) \cup \{10\} - \{1\}$ (B) $[-7, 9] - \{1\}$ (C) $[-7, 9] \cup \{10\} - \{1\}$ (D) None of these

Solution:

$|x - 10|$ & $|x - 1|$ will not play any role in sign change scheme as $|x - 10| \ge 0$ & $|x - 1| \ge 0 \;\forall\; x \in R$.

$\dfrac{(x-9)(x+7)\,|x-10|}{|x-1|} \le 0 \rightarrow \left[\dfrac{|x-10|}{|x-1|}\right](x-9)(x+7) \le 0 \rightarrow x \in [-7, 9] \cup \{10\} - \{1\}$.

Hence, (C) is correct option.

Ex8. If $\dfrac{(x-5)|x+2|}{(x^2+x+5)(x^2-4x-5)}>0$, then x satisfies

(A) $(-1, 5)$ (B) $(-1, \infty) - \{5\}$ (C) $[-1, \infty)$ (D) $(5, \infty)$

Solution:

First observe $x^2 + x + 5 > 0 \ \forall \ x \in R$ as $a = 1 > 0$ and $D = b^2 - 4ac = -19 < 0$. So, it will not play any role in sign changing scheme. Similarly, $|x + 2| \geq 0 \ \forall \ x \in R$ so, observe the new inequation format

$$\frac{(x-5)|x+2|}{(x^2+x+5)(x^2-4x-5)}>0 \rightarrow \left\lfloor\frac{|x+2|}{x^2+x+5}\right\rfloor\frac{(x-5)}{(x^2-4x-5)}>0 \rightarrow \left\lfloor\frac{|x+2|}{x^2+x+5}\right\rfloor\frac{(x-5)}{(x-5)(x+1)}>0$$

As $x \neq 5$ so $\left\lfloor\dfrac{|x+2|}{x^2+x+5}\right\rfloor\dfrac{(x-5)}{(x-5)(x+1)}>0 \rightarrow \left\lfloor\dfrac{|x+2|}{x^2+x+5}\right\rfloor\dfrac{1}{(x+1)}>0$

$\rightarrow x \in (-1, \infty) - \{5\}$. So, option (B) is correct.

Ex9. Let $A = \left\{x : \dfrac{|x(x-1)|(x+1)^{\frac{3}{2}}\ln(x+2)}{(x^2-2x)^2(x-1)(e^x-2)} \geq 0\right\}$ and $B = \left\{a : a > 0, 2 + \sin\theta = x^2 + \dfrac{a}{x^2} \ \forall x \in R - \{0\}, \ \theta \in R\right\}$

are two sets, then .... (given $\log_{10} 2 = 0.3010, |.|$ represent modulus function)

(A) $A \subseteq B$ (B) $B \subseteq A$ (C) $A \cap B = \phi$ (D) $A \cap B = (0, \ln 2)$

Solution:

First solve $\dfrac{|x(x-1)|(x+1)^{\frac{3}{2}}\ln(x+2)}{(x(x-2))^2(x-1)(e^x-2)} \geq 0$ to find set A.

$x \neq 0, 2, 1, \ln 2$ but $x = -1$ can be. so now inequation reduced to $\dfrac{(x+1)^{\frac{3}{2}}\ln(x+2)}{(x-1)(e^x-2)} \geq 0$

but $\ln(x+2)$ & $(x+1)^{\frac{3}{2}}$ is defined simultaneously for $x \geq -1$

so above inequation is equivalent to $\dfrac{(x+1)}{(x-1)(x-\ln 2)} \geq 0 \ \forall x \geq -1$

$\Rightarrow x \in [-1, \ln 2) \cup (1, \infty) - \{0, 2\}$ so $A = [-1, \ln 2) \cup (1, \infty) - \{0, 2\}$ ....(1)

$2 + \sin\theta = x^2 + \dfrac{a}{x^2} \ \forall x \in R - \{0\}, \theta \in R, a > 0$

$\dfrac{a}{x^2} + x^2 \geq 2\sqrt{a} \Rightarrow 2 + \sin\theta \geq 2\sqrt{a}$

Now, if smallest value of $2 + \sin\theta$ satisfies above Inequality then this hold true $\forall \theta \in R$.

$1 \geq 2\sqrt{a} \Rightarrow a \leq \dfrac{1}{4}$ but $a > 0$, so $a \in \left(0, \dfrac{1}{4}\right]$ ....(2)

Since $\log_{10} 2 = 0.3$ then $\ln 2 > \log_{10} 2 = 0.3$

So $\ln 2 > \dfrac{1}{4}$, hence $B \subseteq A$. Option (B) is correct.

MODULUS INEQUALITIES

Let's have some preliminary examples to understand inequalities

Ex1. $|x| < 2 \rightarrow -2 < x < 2$ or $x \in (-2, 2)$

Ex2. $|x| \leq 2 \rightarrow -2 \leq x \leq 2$ or $x \in [-2, 2]$

Ex3. $|x| < -2 \rightarrow x \in \emptyset$

Ex4. $|x| > 3 \rightarrow x < -3$ or $x > 3 \rightarrow x \in (-\infty, -3) \cup (3, \infty)$

Ex5. $|x| \geq 3 \rightarrow x \leq -3$ or $x \geq 3 \rightarrow x \in (-\infty, -3] \cup [3, \infty)$

Ex6. $|x| > -3 \rightarrow x \in R$

Ex7. $2 < |x| \leq 3 \rightarrow x \in [-3, -2) \cup (2, 3]$

Understand each and every above basic inequality which helps us in understanding following general inequalities.

If $k \in R^+$ constant then

(i) $\quad |f(x)| < k \rightarrow -k < f(x) < k$

Ex1. Solve: $|2x^2 - x - 1| < 2$

Solution:

$\quad\quad |2x^2 - x - 1| < 2 \rightarrow -2 < 2x^2 - x - 1 < 2$

Case (i) $-2 < 2x^2 - x - 1$

$\rightarrow 2x^2 - x + 1 > 0$ true $\forall x \in R$ as $a = 2 > 0$ and $D = b^2 - 4ac = -7 < 0$.

So, solution set is $x \in R$ ....(1)

Case (ii) $2x^2 - x - 1 < 2$

$\rightarrow 2x^2 - x - 3 < 0 \rightarrow (2x - 3)(x + 1) < 0 \rightarrow x \in (-1, 3/2)$ ....(2)

From (1) and (2) we get the common values of x as $x \in (-1, 3/2)$

Think! Why are we selecting common values of x as answer?

(ii) $\quad |f(x)| \leq k \rightarrow -k \leq f(x) \leq k$

Ex2. Solve: $\left|\dfrac{x}{x-1}\right| \leq 1$

Solution:

$\quad\quad \left|\dfrac{x}{x-1}\right| \leq 1 \rightarrow -1 \leq \dfrac{x}{x-1} \leq 1$ so we have two cases to solve this inequality

Case (i) $-1 \leq \dfrac{x}{x-1}$

Wrong Way: $-1 \leq \dfrac{x}{x-1} \not\rightarrow -(x-1) \leq x$ as we don't know the nature of $(x - 1)$ (+ ve or - ve)

Now, if we did it then it will give us only those solution at which $(x - 1) > 0$. Think about!

Correct Way: $-1 \leq \dfrac{x}{x-1} \rightarrow \dfrac{x}{x-1} + 1 \geq 0 \rightarrow \dfrac{2x-1}{x-1} \geq 0 \rightarrow x \in \left(-\infty, \dfrac{1}{2}\right] \cup (1, \infty)$ .....(1)

Case (ii) $\dfrac{x}{x-1} \leq 1$

$$\frac{x}{x-1} \le 1 \to \frac{x}{x-1} - 1 \le 0 \to \frac{1}{x-1} \le 0 \to x \in (-\infty, 1) \dots\dots (2)$$

From (1) and (2) we get solution set as $x \in \left(-\infty, \frac{1}{2}\right] \cup (1, \infty)$ and $x \in (-\infty, 1) \to x \in \left(-\infty, \frac{1}{2}\right]$

So, answer is $x \in \left(-\infty, \frac{1}{2}\right]$

(iii) $|f(x)| < -k \to x \in \emptyset$ as $|f(x)| \ge 0 \, \forall \, x \in D_f$ (where D_f = Domain of f(x))

Ex3. Solve: $|2x - 1| < -3$

Solution:

There is no real values of x which satisfy $|2x - 1| < -3$ as $|2x - 1| \ge 0 \, \forall \, x \in R$.

(iv) $|f(x)| > k \to f(x) < -k$ or $f(x) > k \to f(x) \in (-\infty, -k) \cup (k, \infty)$

Ex4. Solve: $\left|1 - \frac{1}{2x}\right| > 3$

Solution:

$$\left|1 - \frac{1}{2x}\right| > 3 \to 1 - \frac{1}{2x} < -3 \text{ or } 1 - \frac{1}{2x} > 3 \quad \text{(Observe the meaning of 'or')}$$

Case (i) $1 - \frac{1}{2x} < -3$

$$1 - \frac{1}{2x} < -3 \to 4 - \frac{1}{2x} < 0 \to \frac{8x-1}{2x} < 0 \to x \in \left(0, \frac{1}{8}\right) \quad \dots\dots(1)$$

Case (ii) $1 - \frac{1}{2x} > 3$

$$1 - \frac{1}{2x} > 3 \to 2 + \frac{1}{2x} < 0 \to \frac{4x+1}{2x} < 0 \to x \in \left(-\frac{1}{4}, 0\right) \quad \dots\dots (2)$$

From (1) and (2) solution set is $x \in \left(0, \frac{1}{8}\right) \cup \left(-\frac{1}{4}, 0\right) \to x \in \left(-\frac{1}{4}, \frac{1}{8}\right) - \{0\}$

(v) $|f(x)| \ge k \to f(x) \le -k$ or $f(x) \ge k \to f(x) \in (-\infty, -k] \cup [k, \infty)$

Ex5. Solve: $|x^2 - 2x + 2| \ge 4$

Solution:

$|x^2 - 2x + 2| \ge 4 \to x^2 - 2x + 2 \le -4$ or $x^2 - 2x + 2 \ge 4$

Case (i) $x^2 - 2x + 2 \le -4$

$\to x^2 - 2x + 6 \le 0 \to (x - 1)^2 + 5 \le 0$ not possible so, $x \in \phi$ ..... (1)

Case (ii) $x^2 - 2x + 2 \ge 4$

$\to x^2 - 2x - 2 \ge 0 \to \left(x - \left(1 + \sqrt{3}\right)\right)\left(x - \left(1 - \sqrt{3}\right)\right) \ge 0 \to x \in (-\infty, 1 - \sqrt{3}] \cup [1 + \sqrt{3}, \infty)$. ..... (2)

From (1) and (2) we get $x \in (-\infty, 1 - \sqrt{3}] \cup [1 + \sqrt{3}, \infty)$ as answer.

(vi) $|f(x)| > -k \rightarrow x \in D_f$ (where D_f = Domain of function $f(x)$)

Ex6. Solve for real values of x: $\left|\sqrt{2-x}\right| > -3$

Solution:

$\sqrt{2-x}$ is real only if $2 - x \geq 0 \rightarrow x \leq 2$.

So, for $x \in (-\infty, 2]$, $\left|\sqrt{2-x}\right| > -3$ is always true. Hence, $x \in (-\infty, 2]$ is the solution for the inequation.

Ex7. Solve for real values of x: $\left|2 - \sqrt{2-x}\right| \leq 0$

Solution:

The inequation is true only if $\left|2 - \sqrt{2-x}\right| = 0$ as $\left|2 - \sqrt{2-x}\right| \geq 0 \; \forall x \leq 2$

$\rightarrow 2 - \sqrt{2-x} = 0 \rightarrow x = -2$

(vii) If $k_1, k_2 \in R^+$ such that $k_1 < |f(x)| \leq k_2 \rightarrow -k_2 \leq f(x) < -k_1$ or $k_1 < f(x) \leq k_2$

Ex8. Solve: $3 < |x^2 - x| \leq 4$

Solution:

The inequality $3 < |x^2 - x| \leq 4$ is combinations of two inequation

(i) $3 < |x^2 - x|$ and (ii) $|x^2 - x| \leq 4$

If we solve (i) $3 < |x^2 - x| \rightarrow x^2 - x < -3$ or $x^2 - x > 3$

$x^2 - x < -3 \rightarrow x^2 - x + 3 < 0 \rightarrow x \in \phi$ ....(1)

$x^2 - x > 3 \rightarrow x^2 - x - 3 > 0 \rightarrow \left(x - \dfrac{1+\sqrt{13}}{2}\right)\left(x - \dfrac{1-\sqrt{13}}{2}\right) > 0 \rightarrow x \in \left(-\infty, \dfrac{1-\sqrt{13}}{2}\right) \cup \left(\dfrac{1+\sqrt{13}}{2}, \infty\right)$ ..(2)

From (1) and (2) we get $3 < |x^2 - x| \rightarrow x \in \left(-\infty, \dfrac{1-\sqrt{13}}{2}\right) \cup \left(\dfrac{1+\sqrt{13}}{2}, \infty\right)$ .... (3)

If we solve (ii) $|x^2 - x| \leq 4 \rightarrow -4 \leq x^2 - x \leq 4$

$-4 \leq x^2 - x \rightarrow x^2 - x + 4 \geq 0$ true $\forall x \in R$. .... (4)

$x^2 - x \leq 4 \rightarrow x^2 - x - 4 \leq 0 \rightarrow \left(x - \dfrac{1-\sqrt{17}}{2}\right)\left(x - \dfrac{1+\sqrt{17}}{2}\right) \leq 0 \rightarrow x \in \left[\dfrac{1-\sqrt{17}}{2}, \dfrac{1+\sqrt{17}}{2}\right]$ ..... (5)

From (4) and (5) we get $|x^2 - x| \leq 4 \rightarrow x \in \left[\dfrac{1-\sqrt{17}}{2}, \dfrac{1+\sqrt{17}}{2}\right]$ ......(6)

From (3) and (6) we get $3 < |x^2 - x| \leq 4 \rightarrow x \in \left[\dfrac{1-\sqrt{17}}{2}, \dfrac{1-\sqrt{13}}{2}\right) \cup \left(\dfrac{1+\sqrt{13}}{2}, \dfrac{1+\sqrt{17}}{2}\right]$

METHOD · II

$3 < |x^2 - x| \leq 4 \rightarrow -4 \leq x^2 - x < -3$ or $3 < x^2 - x \leq 4$

Case (i) $-4 \leq x^2 - x < -3$

$-4 \leq x^2 - x \rightarrow x^2 - x + 4 \geq 0 \rightarrow 0$ true $\forall x \in R$ ......(1)

$x^2 - x < - 3 \rightarrow x^2 - x + 3 < 0 \rightarrow x \in \phi$ ....(2)

from (1) and (2) we have $x \in \phi$ .... (3)

Case (ii) $3 < x^2 - x \le 4$

$3 < x^2 - x \rightarrow x^2 - x - 3 > 0 \rightarrow \left(x - \dfrac{1+\sqrt{13}}{2}\right)\left(x - \dfrac{1-\sqrt{13}}{2}\right) > 0 \rightarrow x \in \left(-\infty, \dfrac{1-\sqrt{13}}{2}\right) \cup \left(\dfrac{1+\sqrt{13}}{2}, \infty\right)$ ....(4)

$x^2 - x \le 4 \rightarrow x^2 - x - 4 \le 0 \rightarrow \left(x - \dfrac{1-\sqrt{17}}{2}\right)\left(x - \dfrac{1+\sqrt{17}}{2}\right) \le 0 \rightarrow x \in \left[\dfrac{1-\sqrt{17}}{2}, \dfrac{1+\sqrt{17}}{2}\right]$ ......(5)

from (4) and (5) we get, $3 < x^2 - x \le 4 \rightarrow x \in \left[\dfrac{1-\sqrt{17}}{2}, \dfrac{1-\sqrt{13}}{2}\right) \cup \left(\dfrac{1+\sqrt{13}}{2}, \dfrac{1+\sqrt{17}}{2}\right]$ ......(6)

From (3) and (6) we get $3 < |x^2 - x| \le 4 \rightarrow x \in \left[\dfrac{1-\sqrt{17}}{2}, \dfrac{1-\sqrt{13}}{2}\right) \cup \left(\dfrac{1+\sqrt{13}}{2}, \dfrac{1+\sqrt{17}}{2}\right]$

Ex9. The solution set of the inequality $|x|^3 - 2x^2 - 4|x| + 3 < 0$ is

Solution:

Notice that $|x| = 3$ is a root of the equation $|x|^3 - 2x^2 - 4|x| + 3 = 0$. Then original inequality

rewritten as $(|x| - 3)(|x|^2 + |x| - 1) < 0$, that is $\left(|x|-3\right)\left(|x|-\dfrac{-1+\sqrt{5}}{2}\right)\left(|x|-\dfrac{-1-\sqrt{5}}{2}\right) < 0$.

$\rightarrow \left(|x|-3\right)\left(|x|-\dfrac{-1+\sqrt{5}}{2}\right) < 0$ since $\left(|x|-\dfrac{-1-\sqrt{5}}{2}\right) > 0 \; \forall \; x \in R$.

Hence, using wavy curve we get, $x \in \left(-3, -\dfrac{\sqrt{5}-1}{2}\right) \cup \left(\dfrac{\sqrt{5}-1}{2}, 3\right)$.

SOME SPECIAL CONCEPTS:

Ex10. Solve for $x \in R$: $|x^2 + 6x + 7| = |x^2 + 4x + 4| + |2x + 3|$

Solution:

Concept: $|a + b| = |a| + |b| \rightarrow ab \ge 0$

$|x^2 + 6x + 7| = |x^2 + 4x + 4| + |2x + 3| \rightarrow (x^2 + 4x + 4)(2x + 3) \ge 0 \rightarrow (x + 2)^2(2x + 3) \ge 0$

$\rightarrow x \in \left[-\dfrac{3}{2}, \infty\right) \cup \{-2\}$

Ex11. Solve the following Inequalities

(i) $|x^2 - 7x + 12| > x^2 - 7x + 12$ (ii) $|3x - 5| - |2x + 3| > 0$

(iii) $|x^2 - 2x - 3| < 3x - 3$ (iv) $|x^2 - 5x| > |x|^2 - 5|x|$

Solution:

(i) $|x^2 - 7x + 12| > x^2 - 7x + 12$

Concept: $|a| > a$ only if $a < 0$

$|x^2 - 7x + 12| > x^2 - 7x + 12 \rightarrow x^2 - 7x + 12 < 0 \rightarrow (x - 4)(x - 3) < 0 \rightarrow x \in (3, 4)$

(ii) $|3x - 5| - |2x + 3| > 0$

For solving above inequality we have two way to solve

(a) Either split in 3 cases or

(b) $|3x - 5| > |2x + 3|$ and square both side

I prefer (b) as $|3x - 5| > |2x + 3| > 0$ so inequality will not change on squaring.

Generally we avoid squaring as it causes increase in roots

(Ex. $x = 2 \rightarrow x = 2$, on squaring both side we get $x^2 = 4 \rightarrow x = \pm 2$)

$\rightarrow |3x - 5|^2 > |2x + 3|^2 \rightarrow 9x^2 - 30x + 25 > 4x^2 + 12x + 9$

$\rightarrow 5x^2 - 42x + 16 > 0 \rightarrow (5x - 2)(x - 8) > 0 \rightarrow x \in \left(-\infty, \dfrac{2}{5}\right) \cup (8, \infty)$

(iii) $|x^2 - 2x - 3| < 3x - 3$

$x^2 - 2x - 3 = 0 \rightarrow (x - 3)(x + 1) = 0 \rightarrow x = -1, 3$

Look at the curve of $y = x^2 - 2x - 3$

$y = x^2 - 2x - 3$ $y = \left|x^2 - 2x - 3\right|$

To solve above inequality we have to make two cases

Case (i) $-1 < x < 3$

Here you can observe $x^2 - 2x - 3 < 0 \; \forall \; x \in (-1, 3)$

So $|x^2 - 2x - 3| < 3x - 3 \rightarrow -(x^2 - 2x - 3) < 3x - 3 \rightarrow x^2 + x - 6 > 0$

$\rightarrow (x + 3)(x - 2) > 0 \rightarrow x \in (-\infty, -3) \cup (2, \infty)$.

So, solution in the region $x \in (-1, 3)$ is $x \in (2, 3)$........(1)

Case (ii) $x \in (-\infty, -1] \cup [3, \infty)$

Here you can observe $x^2 - 2x - 3 \geq 0 \; \forall \; x \in (-\infty, -1] \cup [3, \infty)$

So, inequation $|x^2 - 2x - 3| < 3x - 3 \rightarrow x^2 - 2x - 3 < 3x - 3 \rightarrow x^2 - 5x \leq 0$

$\rightarrow x(x - 5) \leq 0 \rightarrow x \in [0, 5]$

So, solution set in the region $x \in (-\infty, -1] \cup [3, \infty)$ is $x \in [3, 5]$.......(2)

Hence, from (1) and (2), $|x^2 - 2x - 3| < 3x - 3 \rightarrow x \in (2, 5]$

(iv) $|x^2 - 5x| > |x|^2 - 5|x| \rightarrow |x(x - 5)| > |x|^2 - 5|x|$

Case (i) $0 \leq x \leq 5$

$|x(x - 5)| > |x|^2 - 5|x| \rightarrow -(x^2 - 5x) > x^2 - 5x$

$\rightarrow 2x^2 - 10x < 0 \rightarrow 2x(x - 5) < 0 \rightarrow x \in (0, 5)$

So, solution set in $x \in [0, 5]$ is $x \in (0, 5)$ ....(1)

Case (ii) x∈ (- ∞, 0)

$|x(x-5)| > |x|^2 - 5|x| \rightarrow x^2 - 5x > x^2 + 5x \rightarrow 10x < 0 \rightarrow x < 0$

So, solution set in the region x∈ (- ∞, 0) is x∈ (- ∞, 0)..........(2)

Case (iii) x∈ (5, ∞)

$|x(x-5)| > |x|^2 - 5|x| \rightarrow x^2 - 5x > x^2 - 5x \rightarrow 0 > 0$

This is not possible. So, no solution in the region. i.e. x∈ φ...........(3)

From (1), (2) & (3) we get final solution as x ∈ (- ∞, 5) – {0}.

Ex12. Solve the following equations

(i) $\left|\dfrac{x}{x-1}\right| + |x| = \dfrac{x^2}{|x-1|}$　　　　　　(ii) $|x^2 + 4x + 3| + 2x + 5 = 0$

(iii) $|x^2 - 3x - 4| = 9 - |x^2 - 1|$　　　　(iv) $|2x^2 - 3x| = |2x - 1|$

Solution:

(i) First observe the equation $\left|\dfrac{x}{x-1}\right| + |x| = \dfrac{x^2}{|x-1|}$　　　$\left(\text{Observation}: \dfrac{x^2}{|x-1|} = \left|\dfrac{x^2}{x-1}\right| = \left|\dfrac{x}{x-1} + x\right|\right)$

So, equation $\left|\dfrac{x}{x-1}\right| + |x| = \dfrac{x^2}{|x-1|} \rightarrow \left|\dfrac{x}{x-1}\right| + |x| = \left|\dfrac{x}{x-1} + x\right|$

Concept: $|a + b| = |a| + |b| \rightarrow ab \geq 0$ where a, b ∈ R

$\left|\dfrac{x}{x-1}\right| + |x| = \left|\dfrac{x}{x-1} + x\right| \rightarrow \left(\dfrac{x}{x-1}\right)x \geq 0 \rightarrow x\in (1, \infty) \cup \{0\}$

(ii) $|x^2 + 4x + 3| + 2x + 5 = 0$

Observe $x^2 + 4x + 3$: $x^2 + 4x + 3 = (x + 1)(x + 3)$

$x^2 + 4x + 3 \geq 0 \ \forall \ x \in (-\infty, -3] \cup [-1, \infty)$ and $x^2 + 4x + 3 < 0 \ \forall \ x\in (-3, -1)$

Case (i) x ∈ (- ∞, - 3] ∪ [- 1, ∞)

$|x^2 + 4x + 3| + 2x + 5 = 0 \rightarrow x^2 + 4x + 3 + 2x + 5 = 0 \rightarrow x^2 + 6x + 8 = 0$

$\rightarrow x = - 2, - 4$ but $x \neq - 2$ as $-2 \notin (-\infty, -3] \cup [-1, \infty)$

So, x = - 4 is the solution we met in this region

Case (ii) x∈ (- 3, -1)

$|x^2 + 4x + 3| + 2x + 5 = 0 \rightarrow - (x^2 + 4x + 3) + 2x + 5 = 0 \rightarrow x^2 + 2x - 2 = 0$

$\rightarrow x = - 1 - \sqrt{3}, - 1 + \sqrt{3}$. But $x \neq - 1 + \sqrt{3}$ in the region

Hence, equation $|x^2 + 4x + 3| + 2x + 5 = 0$ has two roots x = - 4 & - 1 - $\sqrt{3}$

(iii) $|x^2 - 3x - 4| = 9 - |x^2 - 1| \rightarrow |(x + 1)(x - 4)| = 9 - |(x - 1)(x + 1)|$

$|x^2 - 3x - 4| = 0$ at x = - 1 and 4; $x^2 - 1 = 0$ at x = - 1 and 1.

$$\left|x^2-3x-4\right|=\begin{cases} x^2-3x-4 & \text{If } x\in(-\infty,-1]\cup[4,\infty) \\ -(x^2-3x-4) & \text{If } x\in(-1,4)\end{cases} \text{ and } \left|x^2-1\right|=\begin{cases} x^2-1 & \text{if } x\in(-\infty,-1]\cup[1,\infty) \\ 1-x^2 & \text{if } x\in(-1,1)\end{cases}$$

So we need to make 4 cases

x = -1   x = 1   x = 4

Case (i) x ≤ -1

$|x^2 - 3x - 4| = 9 - |x^2 - 1| \rightarrow x^2 - 3x - 4 = 9 - (x^2 - 1) \rightarrow 2x^2 - 3x - 14 = 0 \rightarrow (x + 2)(2x - 7) = 0$
$\rightarrow x = -2$ or $x = 7/2$ but $x \neq 7/2$ in the region

Case (ii) - 1 < x < 1

$|x^2 - 3x - 4| = 9 - |x^2 - 1| \rightarrow -(x^2 - 3x - 4) = 9 + x^2 - 1 \rightarrow 2x^2 - 3x + 4 = 0$ has no real roots as D < 0.

Case (iii) 1 ≤ x < 4

$|x^2 - 3x - 4| = 9 - |x^2 - 1| \rightarrow -(x^2 - 3x - 4) = 9 - (x^2 - 1) \rightarrow 3x = 6 \rightarrow x = 2$ is the solution in the region

Case (iv) x ≥ 4

$|x^2 - 3x - 4| = 9 - |x^2 - 1| \rightarrow x^2 - 3x - 4 = 9 - (x^2 - 1) \rightarrow 2x^2 - 3x - 14 = 0 \rightarrow (x + 2)(2x - 7) = 0$
$\rightarrow x = -2$ or $x = 7/2$ but $x \neq -2$ & $7/2$ in the region

Hence, x = 2 and - 2 are the only roots of the equation $|x^2 - 3x - 4| = 9 - |x^2 - 1|$

(iv) $|2x^2 - 3x| = |2x - 1|$

On square both side, we get equation as:

$4x^4 + 9x^2 - 12x^3 = 4x^2 + 1 - 4x \rightarrow 4x^4 - 12x^3 + 5x^2 + 4x - 1 = 0$

Let $P(x) = 4x^4 - 12x^3 + 5x^2 + 4x - 1$ and $P(1) = 0 \rightarrow (x - 1)$ is a factor of $P(x)$ (Using Factor Theorem)

$P(x) = (x - 1)(4x^3 - 8x^2 - 3x + 1)$

Similarly, we can factorize $4x^3 - 8x^2 - 3x + 1 = (2x + 1)(2x^2 - 5x + 1)$.

So, $P(x) = 4x^4 - 12x^3 + 5x^2 + 4x - 1 = (x - 1)(2x + 1)(2x^2 - 5x + 1) = 0$.

(we were lucky enough in factorization)

$\rightarrow x = 1, -\dfrac{1}{2}, \dfrac{5-\sqrt{17}}{4}$ & $\dfrac{5+\sqrt{17}}{4}$

METHOD · II

$|2x^2 - 3x| = |2x - 1|$

$2x^2 - 3x = \pm (2x - 1)$

Case (i) 2x² - 3x = 2x - 1

$2x^2 - 3x = 2x - 1 \rightarrow 2x^2 - 5x + 1 \rightarrow x = \dfrac{5 \pm \sqrt{17}}{4}$

Case (ii) 2x² - 3x = - (2x - 1)

$2x^2 - x - 1 = 0 \rightarrow x = 1$ and $- 1/2$

Here making cases will leads us to long and time taking solution.

EXERCISE - 4

1. If $\dfrac{4}{x+2} > 3-x$ then $x \in$

   (A) $(-1, 2)$

   (B) $(-2, -1) \cup (2, \infty)$

   (C) $(2, 5)$

   (D) $(-\infty, -2) \cup (-1, 2)$

2. Solution to $|x-1| + |x-7| = 6$ is given by $x \in$

   (A) $[1, 7]$

   (B) $(1, 7)$

   (C) $(-\infty, -1-\sqrt{3}] \cup (\sqrt{3}-1, \infty)$

   (D) $\{1, 2, 3, 4, 5, 6, 7\}$

3. If $1 \le \dfrac{2x}{x^2+1} \le 2$ then $x \in$

   (A) R (B) $(2, \infty)$ (C) $\{1\}$ (D) none of these

4. If $\dfrac{x^2-4x+5}{x^2+5x+6} \ge 0$, then $x \in$

   (A) $R - \{-2, -3\}$

   (B) $(-\infty, -3) \cup (-2, \infty)$

   (C) $(-3, -2)$

   (D) R

5. If $x \le \dfrac{6}{x-5}$, then $x \in$

   (A) $(-\infty, -1] \cup (5, 6]$ (B) $(-1, 5)$ (C) $(0, 5)$ (D) $[-1, 6] - \{5\}$

6. If $\dfrac{x^2-|x|-12}{x-3} \ge 0$ then $x \in$

   (A) $[-4, 3) \cup [4, \infty)$ (B) $[0, 5] - \{3\}$ (C) $(1, 5] - \{3\}$ (D) none of these

7. If $\left|\dfrac{3x}{x^2-4}\right| \le 1$, then $x \in$

   (A) $R - \{-2, 2\}$

   (B) $[-1, 1]$

   (C) $(-\infty, -4] \cup [-1, 1] \cup [4, \infty)$

   (D) $(-\infty, -4] \cup (-2, 1] \cup (2, \infty)$

8. If $\dfrac{(x^2-1)(x+2)(x+1)^2}{(x-2)(x+1)} \ge 0$, then x lies in the interval

   (A) $(-\infty, -2] \cup [1, 2)$

   (B) $[-2, 1] \cup (2, \infty) - \{-1\}$

   (C) $(-2, -1) \cup (2, \infty) - \{1\}$

   (D) $(-\infty, -2] \cup (-1, 1] \cup (2, \infty)$

9. If $\dfrac{1}{|x|-3} < \dfrac{1}{2}$ then $x \in$

   (A) $(-\infty, -5) \cup (5, \infty)$

   (B) $(-2, 2) - \{0\}$

   (C) $(-\infty, -5) \cup (-3, 3) \cup (5, \infty)$

   (D) $(-5, -3) \cup (3, 5)$

SOME SOLVED EXAMPLES

Ex1. The minimum value of $f(x) = ||x + 2| - 2|x - 2|| + |x|$, is

    (A) 1/2   (B) 1/3   (C) 2/3   (D) 1

Solution:

The critical points are $x = 0$, $x = -2$, $x = 2$ and when $x + 2 = \pm 2(x - 2)$

i.e. $x = 2/3$ and $x = 6$. The minimum value occurs at $x = 2/3$.

So, option (C) is correct.

Ex2. For $a \in R$ if $|x + a - 3| + |x - 2a| = |2x - a - 3|$ is true $\forall x \in R$ then exhaustive set of value of a is

    (A) {- 4, 4}   (B) {- 3, 2}   (C) {- 2, 2}   (D) {1}

Solution:

$(x + a - 3)(x - 2a) \geq 0 \ \forall x \in R$

$\rightarrow x^2 - x(a + 3) + 2a (3 - a) \geq 0 \ \forall x \in R$

$\rightarrow (a + 3)^2 - 8a (3 - a) \leq 0$

$\rightarrow 9 a^2 - 18a + 9 \leq 0 \rightarrow a = 1$. So, option (D) is correct.

Ex3. The value of x for which $\dfrac{1}{x-1} > \dfrac{1}{x+1}$ is

    (A) $(-\infty, -1) \cup (1, \infty)$ (B) $(-1, 1)$ (C) $(0, \infty)$ (D) $x \in \phi$

Solution:

$$\frac{1}{x-1} > \frac{1}{x+1} \rightarrow \frac{1}{x-1} - \frac{1}{x+1} > 0 \rightarrow \frac{x+1-x+1}{(x+1)(x-1)} > 0 \rightarrow \frac{2}{(x-1)(x+1)} > 0$$

    +   -   +

    $x = -1$ $x = 1$

$\rightarrow x \in (-\infty, -1) \cup (1, \infty)$. Hence, option (A) is correct.

Ex4. If $|e^x - 4| < 3$ then x belongs to

    (A) $(0, \ln 7)$ (B) $(\ln 7, \infty)$ (C) $(-\infty, 0)$ (D) $(-1, 0)$

Solution:

$|e^x - 4| < 3 \rightarrow -3 < e^x - 4 < 3 \rightarrow 1 < e^x < 7$

$\rightarrow \ln (1) < x < \ln (7) \rightarrow 0 < x < \ln (7) \rightarrow x \in (0, \ln 7)$. Hence, option (A) is correct.

Ex5. Let $P_1(x) = |x| - 2$ and $P_{i+1}(x) = P_1(P_i(x)) \ \forall i = 1, 2, 3, ...,$ then for some $n \in N$

    (A) $P_n(x) = 0$ has atleast 2n real and distinct roots $\forall n \in N$

    (B) $P_n(x) = 0$ has atmost 2n real and distinct roots $\forall n \in N$

    (C) $P_n(x) = 0$ has exactly 2n real and distinct roots $\forall n \in N$

    (D) $P_n(x) = 0$ has exactly (n + 1) real and distinct roots $\forall n \in N$

Solution:

$P_1(x) = |x| - 2$

$P_{i+1}(x) = P_1(P_i(x)) \; \forall \; i = 1, 2, 3,\ldots \rightarrow P_2(x) = P_1(P_1(x)) = ||x| - 2| - 2$

$P_3(x) = P_1(P_2(x)) = |P_2(x)| - 2 = |||x| - 2| - 2| - 2$ and so on…

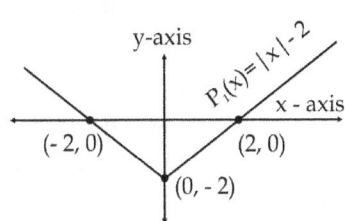 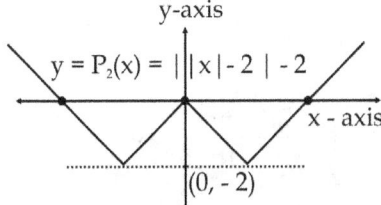

And this series will continue.

Hence, $P_n(x) = 0$ will have exactly $n + 1$ distinct real roots. Option (D) is correct.

Ex6. Let the area enclosed by the curve $|x - 60| + |y| = |x/4|$ be $4(n!)$, then the value of n is _____

Solution:

$|x - 60| + |y| = |x/4| \rightarrow |y| = |x/4| - |x - 60|$

Let $x > 60$ then $|y| = 60 - (3/4)x$.

To get roots: $|y| = 0 \rightarrow 60 - (3/4)x = 0 \rightarrow x = 80$

For $0 < x < 60$, $|y| = (5/4)x - 60$

To get roots: $|y| = 0 \rightarrow (5/4)x - 60 = 0 \rightarrow x = 48$

Look at the figure of curve $|y| = |x/4| - |x - 60|$

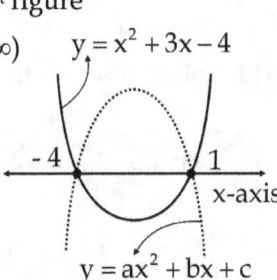

Where (60, 15), (48, 0), (80, 0) and (60, -15) are
points of intersections of curves $|y| = (5/4)x - 60$ and $|y| = 60 - (3/4)x$.

So, Area $= 32 \times 15 = 480 = 4 \times 5!$. Hence, $n = 5$

Ex7. If $f: R - \{-4, 1\} \rightarrow R$ is defined as $f(x) = \dfrac{ax^2 + bx + c}{x^2 + 3x - 4}$, where $f(x) < 0 \; \forall \; x \in R - \{-4, 1\}$. If $|f(2)| = 2$,

then the root(s) of the equation $|ax^2 + bx + c| = |x^2 + 3x - 4|$ is/are _____

Solution:

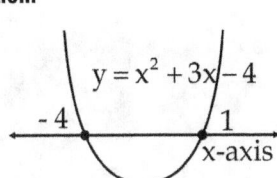

$x^2 + 3x - 4 = 0 \rightarrow (x + 4)(x - 1) = 0 \rightarrow x = -4$ or 1

The curve of $y = x^2 + 3x - 4$ is as shown in 1st figure

Observe, $x^2 + 3x - 4 > 0 \; \forall \; x \in (-\infty, -4) \cup (1, \infty)$

and $x^2 + 3x - 4 < 0 \; \forall \; x \in (-4, 1)$.

As $f(x) < 0 \; \forall \; x \in R - \{-4, 1\}$, will be true only in one situation as
shown in 2nd figure.

$|f(2)| = 2 \rightarrow |ax^2 + bx + c|_{AT \, x=2} = 12$ and $|x^2 + 3x - 4|_{at \, x=2} = 6$.

So, the curves $y = |ax^2 + bx + c|$ and $y = |x^2 + 3x - 4|$ will not overlap.

Hence, $|ax^2 + bx + c| = |x^2 + 3x - 4|$ have only two roots, $x = -4$ and 1.

Ex8. If $\dfrac{x^2 - 2|x| - 3}{x^2 - 4|x| - 5} > -1$, then x is equal to

(A) $(-\infty, -5) \cup (5, \infty)$

(B) $(-4, 4)$

(C) $(-\infty, -5) \cup (-4, 4) \cup (5, \infty)$

(D) R

Solution:

Let $|x| = t$

$$\dfrac{x^2 - 2|x| - 3}{x^2 - 4|x| - 5} > -1 \rightarrow \dfrac{t^2 - 2t - 3}{t^2 - 4t - 5} > -1 \rightarrow \dfrac{2t^2 - 6t - 8}{t^2 - 4t - 5} > 0$$

$$\rightarrow \dfrac{t^2 - 3t - 4}{t^2 - 4t - 5} > 0 \rightarrow \dfrac{(t-4)(t+1)}{(t-5)(t+1)} > 0 \rightarrow \dfrac{(|x|-4)(|x|+1)}{(|x|-5)(|x|+1)} > 0$$

As $|x| + 1 > 0 \ \forall \ x \in R$, so

$$\dfrac{(|x|-4)(|x|+1)}{(|x|-5)(|x|+1)} > 0 \rightarrow \dfrac{(|x|-4)}{(|x|-5)} > 0 \rightarrow \dfrac{(x-4)(x+4)}{(x-5)(x+5)} > 0$$

by applying inequality equivalence.

$\rightarrow x \in (-\infty, -5) \cup (-4, 4) \cup (5, \infty)$. Hence, option (C) is correct.

Ex9. If the solution set of $|x - k| < 2$ is a subset of the solution set of the inequality $\dfrac{2x-1}{x+2} < 1$, then the possible integral value(s) of 'k' is/are

(A) 0   (B) 1   (C) 2   (D) 3

Solution:

$|x - k| < 2 \rightarrow -2 < x - k < 2 \rightarrow x \in (k - 2, k + 2)$ ....(1)

$\dfrac{2x-1}{x+2} < 1 \rightarrow \dfrac{2x-1}{x+2} - 1 < 0 \rightarrow \dfrac{x-3}{x+2} < 0 \rightarrow x \in (-2, 3)$ ....(2)

From (1) and (2) and as per question $k - 2 \geq -2 \rightarrow k \geq 0$

And $k + 2 \leq 3 \rightarrow k \leq 1$, So $k \in [0, 1]$. Hence, (A) and (B) are correct options.

Ex10. Solution set for the inequation $\dfrac{x^2 - 1}{x} \leq 2 - x$ is

(A) $x \in \left(-\infty, \dfrac{1-\sqrt{3}}{2}\right] \cup \left(0, \dfrac{1+\sqrt{3}}{2}\right]$

(B) $x \in \left[\dfrac{1-\sqrt{3}}{2}, 0\right) \cup \left[\dfrac{1+\sqrt{3}}{2}, \infty\right)$

(C) $x \in \left[\dfrac{1-\sqrt{3}}{2}, \dfrac{1+\sqrt{3}}{2}\right]$

(D) None of these

Solution:

$$\frac{x^2-1}{x} \leq 2-x \rightarrow \frac{x^2-1}{x}-2+x \leq 0 \rightarrow \frac{2x^2-2x-1}{x} \leq 0 \rightarrow \frac{2\left(x-\frac{1-\sqrt{3}}{2}\right)\left(x-\frac{1+\sqrt{3}}{2}\right)}{x} \leq 0$$

$$x=\frac{1-\sqrt{3}}{2} \qquad x=0 \qquad x=\frac{1+\sqrt{3}}{2}$$

$$\rightarrow x \in \left(-\infty, \frac{1-\sqrt{3}}{2}\right] \cup \left(0, \frac{1+\sqrt{3}}{2}\right]$$ Hence, (A) is correct option.

MATHSARC EDUCATION

A learning place to fulfill your dream of success!

MATHEMATICS   **PRACTICE PROBLEMS**

SUBJECTIVE PROBLEMS

1. Minimum value of $f(x) = |x-1| + |2x-1| + |3x-1| + |4x-1|$ is p/q where p/q is in lowest form and $p, q \in I^+$ then p + q is _____

2. The number of integral solution of $\dfrac{x+1}{x^2+2} > \dfrac{1}{4}$ is/are _____

3. The greatest positive integers satisfies the in-equation $x < \dfrac{10}{x}$, is_____

4. Solve the inequality: $\dfrac{x+1}{2-x} \geq x + \dfrac{1}{x} + 2$

5. Solve the following Inequalities

    (i) $\dfrac{(x^2+1)(x^2+x+1)(x^2-2)}{(x^2-4)} \geq 0$ (ii) $\dfrac{(2x^2+3x+5)^5 x^3}{(x-1)^4(x-2)(x+1)} < 0$

    Hints: (i) $ax^2 + bx + c > 0 \; \forall \; x \in R$ if $a > 0$ and $D = b^2 - 4ac < 0$

             (ii) $ax^2 + bx + c < 0 \; \forall \; x \in R$ if $a < 0$ and $D = b^2 - 4ac < 0$

6. Solve the following In-equations

    (i) $|x^2 + x| - 5 < 0$ (ii) $3x^2 - 7x + 6 < 0$

    (iii) $x^2 - 5|x| + 6 \geq 0$ (iv) $|x^2 - 2x - 8| > 2x$

    (v) $\left|\dfrac{x-1}{2x-3}\right| \leq 5x - 6$ (vi) $(|x-1| - 3)(|x+2| - 5) < 0$

    (vii) $(x-1)(2x-1)(2x+3) \leq 90$ (viii) $\dfrac{-2x^2+3x-1}{x-3} \leq 0$

    (ix) $2x - 3 + \left|\dfrac{x-1}{x}\right| \leq x$ (x) $\left|x^2 - 4x + 3\right| \leq \dfrac{1}{2}$

    (xi) $\dfrac{|x|(x^2+6x-7)}{|x-4|} \geq 0$ (xii) $\dfrac{x^2 - |x| - 12}{x-3} \geq 2x$

7. Solve the following equations

    (i) $x^2 - |2x^2 - x| = 5$ (ii) $\left|\dfrac{3x-2}{x-3}\right| = \left|x + \dfrac{1}{2}\right|$

    (iii) $|3x^3 - 2x + 1| = 3x^3 - 2x + 1$ (iv) $\left|\dfrac{3x^2 - 3x - 1}{x}\right| = \dfrac{1}{|x|} + 3|x-1|$

    (v) $|2x^2 - 3x| + x|1 - x| = 5$ (vi) $x^3 - |x^2 - 6x + 8| = 8$

(vii) $|x^2 - 9| + |x^2 - 4| = 5$

(viii) $\left|\dfrac{x^2 - 5x + 4}{x^2 - 4}\right| = 1$

8. Number of integers x for which $5x - 1 < (x + 1)^2 < 7x - 3$, is _____

9. Solve the following inequalities

(i) $\dfrac{(x+4)x(x-1)^{2/3}(x+1)^{3/2}(e^x-1)}{|x-3|(x-4)(2x-3)^{1/3}} \le 0$

(ii) $\sqrt{2x-1} \le 3x-1$

(iii) $\dfrac{1}{\sqrt{x-2x^2}} \le 3$

(iv) $\sqrt{x^2-5x+8} - \sqrt{x^2-5x+5} \le 3$

(v) $\left|x+\dfrac{1}{x}\right| \le 2$

(vi) $\dfrac{x}{1+|x|} < \dfrac{2}{x}$

10. If $z \in C$ then number of roots of system of equations $|z-1| + |z+1| = 2\sqrt{2}$ and $|z| = 1$, is _____

OBJECTIVE PROBLEMS

1. The number of distinct real roots of equation $\left(|x|-1\right)^{|x-1|-3} = 1$

(A) 3　　　　　(B) 4　　　　　(C) 5　　　　　(D) None of these

2. Solution set of $\dfrac{|x-1|}{x(x-2)|x-3|} \ge 0$ is

(A) $x \in (-\infty, 0) \cup (2, \infty)$

(B) $x \in (-\infty, 0) \cup (2, \infty) \cup \{1\} - \{3\}$

(C) $x \in (0, 2)$

(D) none of these

3. Select the correct statement(s) for solution set of x

(A) $|2x-1| > -1 \to x \in R$

(B) $\dfrac{1}{x-1} < x \to x(x-1) > 1 \ \forall \ x > 1$

(C) $\dfrac{|x|-1}{x(x-2)} < 0 \equiv \dfrac{(x+1)(x-1)}{x(x-2)} < 0$

(D) $|x-1|(x-2)^2 \le 0 \to x \in \phi$

4. The set of real values of x satisfying $|x-1| \le 3$ & $|x-1| \ge 1$, is

(A) [2, 4]　　　(B) $(-\infty, 0] \cup [2, \infty)$　　　(C) $[-2, 0] \cup [2, 4]$　　　(D) [-2, 4]

5. The solution of the in-equation $x^2 - 3x + 2 \le 0$, is

(A) $1 < x < 2$　　　(B) $1 \le x \le 2$　　　(C) $x < 1$ or $x > 2$　　　(D) $x \le 1$ or $x \ge 2$

6. If x is real then greatest and least values of $\dfrac{x^2-x+1}{x^2+x+1}$ are respectively

(A) 3, -1/2　　　(B) 3, 1/3　　　(C) -1/3, -3　　　(D) 2, ½

7. If $\left|\dfrac{1}{x}\right| > -2$ then select the more appropriate statement

    (A) $x \in R$ (B) $x \in R - \{0\}$ (C) $-2 < x < 2$ (D) $x < -2$ or $x > 2$

8. If $(x - 1)^2 + (x - 2)^2 + (x - 3)^2 + (x - 4)^2 = 0$ then

    (A) $x = \dfrac{1+2+3+4}{4}$ (B) $x = 4!$ (C) $x = \dfrac{1}{4!}$ (D) Discriminant < 0

9. If $a < 0$ then the inequality $ax^2 - 2x + 4 > 0$ has the solutions represented by

    (A) $\dfrac{1+\sqrt{1-4a}}{a} > x > \dfrac{1-\sqrt{1-4a}}{a}$ (B) $x < \dfrac{1-\sqrt{1-4a}}{a}$

    (C) $\dfrac{1+\sqrt{1-4a}}{a} < x < \dfrac{1-\sqrt{1-4a}}{a}$ (D) $2 > x > \dfrac{1+\sqrt{1-4a}}{a}$

10. Consider the following statements for the $f(x) = \dfrac{\sqrt{(x-1)^2}}{x-1}$.

    (1) $f(x) = 1$ if $x > 1$ (2) $f(x) = -1$ if $x < 1$

    (3) $f(x) = 1 \ \forall \ x \in R$

    (A) Only 1 is correct   (B) Only 1 and 2 are correct

    (C) Only 1 and 3 are correct   (D) All are correct

11. Solution set of inequality $\left| 1 - \dfrac{|x|}{1+|x|} \right| \geq \dfrac{1}{2}$, is

    (A) $[0, 1]$ (B) $[-1, 0]$ (C) $[-1, 1]$ (D) R

12. Select the correct statement(s)

    (A) $x + \dfrac{1}{\sqrt{x-1}} = 1$ Has no real root   (B) $y = \sqrt{x-1}$ & $y = \sqrt{2-x}$ intersect at 2 points

    (C) $\dfrac{x}{x} = 1 \forall x \in R$ (D) $|a| = |b| \to a = b \forall a, b \in C$

EXERCISE - HINTS AND SOLUTIONS

Mind Refreshment Problem # 1: 90   Mind Refreshment Problem # 2: $^{18}C_3 = 816$

ANSWER KEY – EXERCISE 1

Q. NO.	ANSWER	Q. NO.	ANSWER	Q. NO.	ANSWER
1	C	3	B, C, D	5	A
2	B	4	C		

ANSWER KEY – EXERCISE 2

Q. NO.	ANSWER	Q. NO.	ANSWER	Q. NO.	ANSWER
1	C	5	D	9	D
2	D	6	B	10	C
3	D	7	C	11	A
4	C	8	C		

ANSWER KEY – EXERCISE · 3

1. (i) $x \in [1/2, 2]$

(ii) $x \in (-\infty, 2] \cup [3, \infty)$

(iii) $x \in (-1, 0] \cup [1, 2]$

(iv) $x \in \left(-\infty, -\dfrac{1+\sqrt{13}}{2}\right] \cup \left(-1, \dfrac{-1+\sqrt{13}}{2}\right]$

(v) $x \in \phi$

(vi) $x \in R$

(vii) $x \in [-3, 1) \cup [7, \infty)$

(viii) $x \in (-\infty, -3] \cup (1, 7] - \{3\}$

(ix) $x \in (0, 1) \cup [3, \infty) - \{5\}$

(x) $x \in (-\infty, 1) - \{-1, -3\}$

(xi) $x \in (-\infty, -4) \cup (-3, 3/2) \cup (5/2, \infty)$

(xii) $x \in (-2, -1) \cup (-2/3, -1/2)$

(xiii) $x \in (-\infty, -1) \cup (-1/2, 0) \cup (1/2, \infty)$

(xiv) $x \in (-\infty, 1) \cup (3/2, 2) \cup (3, \infty)$

2. (A)   3. (B)

4. (i) $(0, 1]$ (ii) $\{1, 2, 3\}$ (iii) $\{0, 1, 2\}$ (iv) $(-\infty, -1/2]$

5. (i) $(5, \infty)$ (ii) $(3, \infty)$ 6. $k \in (-3, 5)$

ANSWER KEY – EXERCISE 4

Q. NO.	ANSWER	Q. NO.	ANSWER	Q. NO.	ANSWER
1	B	4	B	7	C
2	A	5	A	8	B
3	C	6	A	9	C

SOLUTIONS – EXERCISE · 1

1.   **C**

As $|x| \geq 0 \ \forall \ x \in R$ but $|x + 1| + |x - 1| + |x + 2| + |x - 2| = 0 \to$ each $|x + 1|$, $|x - 1|$, $|x + 2|$ and $|x - 2|$ has to be zero simultaneously, but there is no any x for which situation to be possible. Hence, no solution

2.   **B**

If $x < 0$ then $\dfrac{1}{|x|} + \dfrac{1}{x} = 1 \rightarrow -\dfrac{1}{x} + \dfrac{1}{x} = 1 \rightarrow 0 = 1$ (Not possible) also $x \neq 0$ as $1/0 = $ Not Define.

So, let $x > 0$ then $\dfrac{1}{|x|} + \dfrac{1}{x} = 1 \rightarrow \dfrac{1}{x} + \dfrac{1}{x} = 1 \rightarrow \dfrac{2}{x} = 1 \rightarrow x = 2$

3.   B, C, D

The possible pairs are $(1, 2)$, $(1, 0)$, $\left(-\dfrac{1}{3}, 2\right)$ & $\left(-\dfrac{1}{3}, 0\right)$.

Maximum area will be formed in the case of rectangle only.

So, required area = $\left| 1 - \left(-\dfrac{1}{3}\right) \right| \times \left| 2 - 0 \right| = \dfrac{8}{3}$ sq. unit.

There are 2 choices for each x and y hence total number of such matrices = $2 \times 2 = 4$

4.   C

The minimum value will be meet at $x = 3$.

5.   A

$|x| > x \rightarrow x < 0$

Similarly, $|(x - 1)(x - 7)| > x^2 - 8x + 7 \rightarrow (x - 1)(x - 7) < 0$

The product of two numbers is less than zero $\rightarrow$ both are of opposite sign

Case: (I) $x - 1 < 0$ and $x - 7 > 0 \rightarrow x < 1$ and $x > 7$ which is not possible.

Case: (II) $x - 1 > 0$ and $x - 7 < 0 \rightarrow x > 1$ and $x < 7 \rightarrow 1 < x < 7$ and $x \in I$, so $x = \{2, 3, 4, 5, 6\}$.

Hence, the total number of values of $x = 5$.

ANSWER KEY – PRACTICE PROBLEMS

SUBJECTIVE PROBLEMS

1.   7   2.   5   3.   3

4. $x \in (-\infty, -\sqrt{2}] \cup [-1, 0) \cup [\sqrt{2}, 2)$

5   (i) $x \in (-\infty, -2) \cup [-\sqrt{2}, \sqrt{2}] \cup [2, \infty)$ (ii) $x \in (-\infty, -1) \cup (0, 2) - \{1\}$

6.   (i) $x \in \left[-\dfrac{1 + \sqrt{21}}{2}, \dfrac{-1 + \sqrt{21}}{2} \right]$ (ii) $x \in \phi$

       (iii) $x \in (-\infty, -3] \cup [-2, 2] \cup [3, \infty)$ (iv) $x \in (-\infty, 2\sqrt{2}) \cup (2 + 2\sqrt{3}, \infty)$

       (v) $x \in \left[\dfrac{14 + \sqrt{6}}{10}, \infty \right)$ (vi) $x \in (-7, -2) \cup (3, 4)$

       (vii) $x \in (-\infty, 3]$ (viii) $x \in \left[\dfrac{1}{2}, 1 \right] \cup (3, \infty)$

       (ix) $x \in (-\infty, 1 - \sqrt{2}] \cup [2 - \sqrt{3}, 1 + \sqrt{2}]$ (x) $\left[2 - \sqrt{\dfrac{3}{2}}, 1 \right) \cup \left[2 - \dfrac{1}{\sqrt{2}}, 2 + \dfrac{1}{\sqrt{2}} \right] \cup \left(3, 2 + \sqrt{\dfrac{3}{2}} \right]$

       (xi) $x \in (-\infty, -7] \cup [1, \infty) \cup \{0\} - \{4\}$ (xii) $x \in (-\infty, 3)$

7. (i) $x \in \phi$ (ii) $x = \dfrac{-1 \pm \sqrt{57}}{4}, \dfrac{11 \pm \sqrt{113}}{4}$ (iii) $x \in [-1, \infty)$

(iv) $x \in (0, 1]$ (v) $x = 1 - \sqrt{6}$ (vi) $x = 2$

(vii) $x \in [-3, -2] \cup [2, 3]$ (viii) $x = \dfrac{8}{5}, 0, \dfrac{5}{2}$

8. 1

9. (i) $x \in \left(\dfrac{3}{2}, 4\right) \cup \{-1, 0, 1\} - \{3\}$ (ii) $x \in \left[\dfrac{1}{2}, \infty\right)$

(iii) $x \in \left[\dfrac{1}{6}, \dfrac{1}{3}\right]$ (iv) $x \in \left(-\infty, \dfrac{10 - 2\sqrt{13}}{3}\right] \cup \left[\dfrac{10 + 2\sqrt{13}}{3}, \infty\right)$

(v) $x = \pm 1$ (vi) $x \in (-\infty, -(1 + \sqrt{3})) \cup (0, 1 + \sqrt{3})$

10. 2

OBJECTIVE PROBLEMS

1. B   2. B   3. A, B, C   4. C

5. B   6. B   7. B   8. D

9. C   10. B   11. C   12. A, B

	NOTES MAKING
Que / Page No.	**NOTES**

POINTS TO REMEMBER:

❖ .

❖ .

❖ .

❖ .

❖ .

OTHERS: --

QUADRATIC EQUATIONS AND EXPRESSIONS

"One best book is equal to hundred good friends but one good friend is equal to a library"

Dr. A. P. J. Abdul Kalam

ABOUT QUADRATIC EQUATION

It is often claimed that the Babylonians (about 400 BC) were the first to solve quadratic equations. This is an over simplification, for the Babylonians had no notion of 'equation'. What they did develop was an algorithmic approach to solving problems which, in our terminology, would give rise to a quadratic equation. The method is essentially one of completing the square. However all Babylonian problems had answers which were positive (more accurately unsigned) quantities since the usual answer was a length.

In about 300 BC Euclid developed a geometrical approach which, although later mathematicians used it to solve quadratic equations, amounted to finding a length which in our notation was the root of a quadratic equation. Euclid had no notion of equation, coefficients etc. but worked with purely geometrical quantities.

Hindu mathematicians took the Babylonian methods further so that Brahmagupta (598-665 AD) gives an, almost modern, method which admits negative quantities. He also used abbreviations for the unknown, usually the initial letter of a colour was used, and sometimes several different unknowns occur in a single problem.

$$ax^2 + bx + c = 0 \rightarrow x = \frac{-b \pm \sqrt{b^2 - 4ac}}{2a}, a \neq 0 \text{ (Sanskrit Pandit, Sridhar Acharya Method)}$$

References: University of Virginia USA

Article by: J J O'Connor and E F Robertson.

"Suffering is the essence of success!"

REAL POLYNOMIAL:

The expression of type $a_nx^n + a_{n-1}x^{n-1} + a_{n-2}x^{n-2} + ... + a_1x + a_0$ where $a_n, a_{n-1}, a_{n-2}, ..., a_1, a_0 \in R$, $a_n \neq 0$ and $n \in W$, known as real polynomial in one variable of degree 'n' & $a_n, a_{n-1}, a_{n-2}, ..., a_1, a_0$ are known as coefficients of $x^n, x^{n-1}, ..., x^1, x^0$ respectively.

Note: $a_n \neq 0$ for retaining degree of polynomial. It's also known as leading coefficient.

Ex1. (i) $3x^2 + 5x - 6$: real polynomial of degree two (Quadratic Expression)

(ii) $3x^{3/2} - 5x + 1$: Not a real polynomial, as $3/2$ not a whole number

(iii) $u^2 - 3uv + 4$: is a real polynomial of degree two in two variable u and v

(iv) $u^2 + 3u - 3v^3 + uvw$: is a real polynomial of degree 3 in three variable u, v and w

(v) $3x^3 - (2 + 3i)x^2 + ix - 3$: is a complex polynomial of degree 3

(vi) $2x^{3/5} + x^{2/5} - 3x^{1/5} + 2$: is not a real polynomial, since powers of variables x are not whole nos.

Degree: The highest power of variable in a polynomial known as degree of that polynomial.

Ex2. Select the correct statements

(A) $p(x) = 2$, is a polynomial of degree zero

(B) $p(x) = x^3 - x + 1$, is a polynomial of degree 3

(C) $p(x) = 0$, is a polynomial with degree Not Define

(D) $p(x) = x^{2/3} + x^{1/3} + 1$ is a polynomial of degree 2/3

Ans. A, B, C

Ex3. Given that P(x) and Q(x) are polynomials of degree 3 with real coefficients, which one of the following is not true?

(A) $\deg[P(x) \times Q(x)] = 6$ (B) $\deg[P(x) + Q(x)] = 3$

(C) $\deg[P(x) - Q(x)] \leq 3$ (D) $\deg[[P(x)]^2 \times [Q(x)]^3] = 15$

Ans. B.

Look at the degree of $[(x^3 + x - 1) + (-x^3 - 2x^2 - x + 1)] = 2$

ZERO OF A POLYNOMIAL

Consider a polynomial $p(x) = 2x^3 - x - 1$, here $p(1) = 2(1)^3 - 1 - 1 = 0$. i.e. $p(x)$ is becoming zero at $x = 1$ hence '1' is called 'zero' of polynomial $p(x)$.

In general, if $p(x_1) = 0$ then $x = x_1$ is a zero of polynomial $p(x)$.

Ex1. Find the zeroes of following polynomials

(i) $p(x) = x^2 - 5x + 6$ (ii) $p(x) = x^3 - 1$

(iii) $p(x) = 2x - 3$ (iv) $p(x) = (x - 1)(x - 2)(2x - 3)$

Solutions:

(i) $p(x) = x^2 - 5x + 6 \rightarrow p(x) = (x - 2)(x - 3)$.

Observe $p(2) = 0$, $p(3) = 0$ hence $x = 2$ and $x = 3$ are zero's of $p(x) = x^2 - 5x + 6$

The curve of $y = x^2 - 5x + 6$ is as shown in figure.

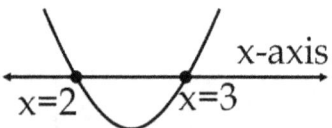

(ii) $p(x) = x^3 - 1 = (x-1)\left(x-\left(\dfrac{-1+i\sqrt{3}}{2}\right)\right)\left(x-\left(\dfrac{-1-i\sqrt{3}}{2}\right)\right)$

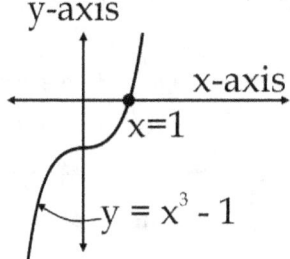

$p(x)$ has three zeroes $x = 1$, $\dfrac{-1+i\sqrt{3}}{2}$ and $\dfrac{-1-i\sqrt{3}}{2}$

If we plot the curve of $y = x^3 - 1$ then we will get real zero as $x = 1$ on number line as shown in figure.

(iii) $x = 3/2$ is zero.

(iv) $x = 1, 2$ and $3/2$ are zeroes.

Look at the real zeroes x_1, x_2 and x_3 of polynomial $p(x)$ having curve as shown in figure

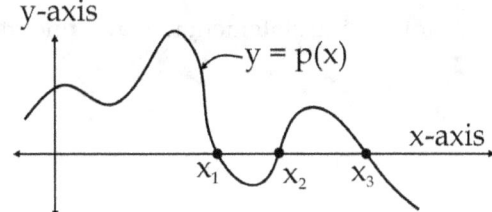

ROOT OF AN EQUATION

Consider an equation $(x - 1)(x - 2) = 0$. Here you can observe that $x = 1$ and 2 are satisfying the equation so 1 and 2 are roots of this equation.

Note: we meet roots correspond to an equation and zeroes correspond to expressions.

Look at the figure to know the difference between root and zero

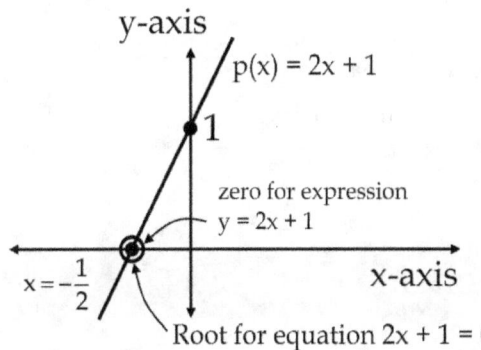

FUNDAMENTAL THEOREM OF ALGEBRA

Every polynomial $p(x) \in C(x)$ with complex coefficients can be factored into linear factors over the complex number.

Ex. (i) $x^2 - 7x + 12 = (x - 4)(x - 3)$

(ii) $x^2 - (1 + i)x + i = (x - 1)(x - i)$

(iii) $x^3 + 1 = (x+1)\left(x+\left(\dfrac{-1+i\sqrt{3}}{2}\right)\right)\left(x+\left(\dfrac{-1-i\sqrt{3}}{2}\right)\right)$

(iv) $2i x^3 - (1 + 3i) x^2 + (1 + i) x = x(2i - 1 - i)(x - 1)$

GAUSS ALTERNATIVE FORMULATION

Every polynomial $p(x) \in R(x)$ with real coefficients can be factored into linear and quadratic factors where degree of polynomial $p(x)$ must be greater than or equal to 1.

Ex. (i) $x^3 - x^2 + x - 1 = (x - 1)(x^2 + 1)$ (ii) $x^3 - 1 = (x - 1)(x^2 + x + 1)$

(iii) $x^4 - x^3 + 4 x^2 - x + 3 = (x^2 + 1)(x^2 - x + 3)$ (iv) $x^4 + x^2 + 1 = (x^2 + x + 1)(x^2 - x + 1)$

Note: (i) A real linear factor can be written as $x \pm r$.

(ii) An irreducible quadratic factor over the real can be written as $x^2 - 2x\, r \cos(\phi) + r^2, r > 0$.

(iii) The roots of equation $x^2 - 2x\, r \cos(\phi) + r^2 = 0$ are $x = r(\cos \phi \pm i \sin \phi)$, where $i = \sqrt{(-1)}$.

POLYNOMIAL REMAINDER THEOREM

Let's have some preliminary work before actual statement to better understanding

Ex1. Find the remainder when

(i) 14 divided by 3 (ii) - 2 divided by 5

(iii) $x^3 - 2x + 3$ divided by $x - 1$ (iv) $x^4 - 3x + 5$ divided by $x - 2$

Solution:

(i)
$$3\overline{)14} \quad \rightarrow 14 = 3 \times 4 + 2$$

quotient 4, 12, remainder 2

$$\begin{array}{r} \text{Quotient} \\ \text{Divisor}\,\overline{)\,\text{Dividend}} \\ \text{------} \\ \text{------} \\ \overline{\text{Remainder}} \end{array}$$

In general we can have:

$$\text{Dividend} = \text{Divisor} \times \text{Quotient} + \text{Remainder}$$

For the case, Dividend is 14, divisor is 3, quotient is 4 and remainder is 3

(ii) As per Euclid division lemma for integers p, d, q, r.

$$d\overline{)\,p} \quad \rightarrow p = d \times q + r \text{ where } 0 \leq r < d,\ d > 0.$$

$- 2 = 5 \times (- 1) + 3$, so remainder is 3.

(iii)
$$x - 1\overline{)\,x^3 - 2x + 3} \quad (\text{quotient } x^2 + x - 1)$$

As per synthetic division Method:

$$\begin{array}{r} x^3 - x^2 \\ \hline x^2 - 2x \\ x^2 - x \\ \hline -x + 3 \\ -x + 1 \\ \hline R = 2 \end{array}$$

$\rightarrow x^3 - 2x + 3 = (x - 1)(x^2 + x - 1) + 2$

So, remainder is 2.

SYNTHETIC DIVISION
$\begin{array}{r} 1\,\vert\ 1\ \ 0\ \ -2\ \ \ 3 \\ \quad\ \ 1\ \ \ 1\ \ -1 \\ \hline \ \ 1\ \ \ 1\ \ -1\ \ \boxed{2}\ \text{Remainder} \end{array}$
$x^3 - 2x + 3 = (x - 1)(x^2 + x - 1) + 2$

(iv)   Look at the synthetic division method as shown.

$$x - 2 \overline{) \begin{array}{c} x^3 + 2x^2 + 4x + 5 \\ x^4 - 3x + 5 \end{array}}$$

$$\underline{x^4 - 2x^3}$$
$$2x^3 - 3x$$
$$\underline{2x^3 - 4x^2}$$
$$4x^2 - 3x$$
$$\underline{4x^2 - 8x}$$
$$5x + 5$$
$$\underline{5x - 10}$$
$$R = 15$$

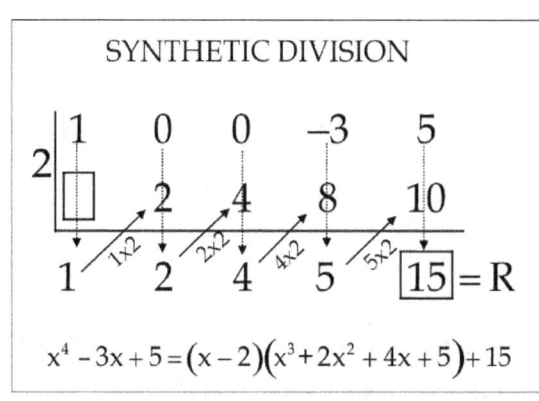

$x^4 - 3x + 5 = (x - 2)$ $(x^3 + 2x^2 + 4x + 5) + 15$. Hence, remainder is 2.

I hope you have analyzed and observe both the methods for polynomial division.

REMAINDER THEOREM STATEMENT:

If $p(x)$ & $g(x)$ are two polynomials such that degree of $p(x) \geq$ degree of $g(x)$ where $g(x) \neq 0$, then we can find polynomial $q(x)$ and $r(x)$ such that $p(x) = g(x) \times q(x) + r(x)$, where degree of $r(x)$ is less than the degree of $g(x)$.

Ex1.   Find the quotient and remainder when polynomial $3x^2 + x - 1$ is divided by $x + 1$.

Sol.   As per long division method

$$x + 1 \overline{) \begin{array}{c} 3x - 2 \\ 3x^2 + x - 1 \end{array}}$$
$$\underline{3x^2 + 3x}$$
$$-2x - 1$$
$$\underline{-2x - 2}$$
$$1 = R$$

Quotient = $3x - 2$, Remainder = 1

$3x^2 + x - 1 = (x + 1)(3x - 2) + 1$.

OBSERVATION: $3x^2 + x - 1 = (x + 1)(3x - 2) + 1 \; \forall \; x \in R$

   Let $x = 0 \rightarrow 3(0)^2 + (0) - 1 = (0 + 1) (3 \times 0 - 2) + 1 \rightarrow -1 = -2 + 1 = -1$.

   Let $x = 1 \rightarrow 3(1)^2 + (1) - 1 = (1 + 1) (3 \times 1 - 2) + 1 \rightarrow 3 = 3$.

   So, if $p(x) = 3x^2 + x - 1 = (x + 1)(3x - 2) + 1 \rightarrow p(-1) = 0 + 1 = R$

STATEMENT – II

Let $p(x)$ be any polynomial of degree greater than or equal to one. If $p(x)$ is divided by linear polynomial $x - a$ then the remainder is $p(a)$.

Proof: $p(x) = (x - a)q(x) + R$, where remainder R is any constant

   $p(a) = 0 + R \rightarrow R = p(a)$.

Ex2.   Find the remainder when $p(x) = x^3 + x - 1$ is divided by $(x - 3)$

Sol.   Remainder = $p(3) = 3^3 + 3 - 1 = 29$

   By long division method

$$x - 3 \overline{)\begin{array}{l} x^2 + 3x + 10 \\ x^3 + x - 1 \end{array}}$$

$$\begin{array}{l} \underline{x^3 - 3x^2} \\ \quad\ 3x^2 + x \\ \quad\ \underline{3x^2 - 9x} \\ \qquad\ 10x - 1 \\ \qquad\ \underline{10x - 30} \\ \qquad\qquad 29 = R \end{array}$$

SYNTHETIC DIVISION

$$3\begin{array}{|rrrr} 1 & 0 & 1 & -1 \\ \square & 3 & 9 & 30 \\ \hline 1 & 3 & 10 & \boxed{29} = R \end{array}$$

Ex3. Find the remainder when $f(x) = x^3 + 7x + 6$ is divided by $2x + 1$

Sol. $f(x) = x^3 + 7x + 6 = (2x + 1)q(x) + R$

So, remainder (R) is $f\left(-\dfrac{1}{2}\right) = \left(-\dfrac{1}{2}\right)^3 + 7\left(-\dfrac{1}{2}\right) + 6 = \dfrac{19}{8}$, so remainder is 19/8.

Ex4. If $f(x) = x^4 - 2x^3 + 3x^2 - ax + b$ leaves remainder 5 and 19 on division by $(x - 1)$ and $(x + 1)$ respectively. Find the remainder when $f(x)$ is divided by $x - 2$.

Sol. As per question, $f(1) = 5$ and $f(-1) = 19$.

$f(1) = 5 = 1^4 - 2(1)^3 + 3(1)^2 - a(1) + b \rightarrow b - a = 3$ ....(1)

$f(-1) = 19 = (-1)^4 - 2(-1)^3 + 3(-1)^2 - a(-1) + b \rightarrow a + b = 13$ ...... (2)

from (1) & (2) $a = 5$, $b = 8$.

So, polynomial is $f(x) = x^4 - 2x^3 + 3x^2 - 5x + 8$

Hence, remainder when divided $f(x)$ by $x - 2$ is $f(2) = 2^4 - 2(2)^3 + 3(2)^2 - 5(2) + 8 = 10$.

Ex5. An unknown polynomial yields a remainder 2 upon division by $x - 1$ and remainder 1, upon division by $x - 2$. What remainder is obtained if this polynomial is divided by $(x - 1)(x - 2)$?

Sol. The unknown polynomial $p(x)$ must be $= q(x) (x - 1)(x - 2) + R(x)$

Let $p(x) = q(x) (x - 1)(x - 2) + ax + b$.

As per question, $p(1) = 2 \rightarrow a + b = 2$ ....(1)

$p(2) = 1 \rightarrow 2a + b = 1$ ....(2)

from (1) & (2) $a = -1$, $b = 3$. So, remainder is $-x + 3$.

Ex6. A cubic polynomial $p(x)$ yields remainders 2, 3 and 4 upon division by $x - 1$, $x - 2$ and $x - 3$ respectively. If $p(5) = 30$ then the value of $p(6)$ is _____

Sol. As per question $p(1) = 2$, $p(2) = 3$ and $p(3) = 4$.

Let cubic polynomial $p(x) = k(x - 1)(x - 2)(x - 3) + x + 1$

Now $p(5) = 30 = k (4)(3)(2) + 6 \rightarrow k = 1$.

So, polynomial $p(x) = (x - 1)(x - 2)(x - 3) + x + 1$

Hence, $p(6) = 5 \times 4 \times 3 + 7 = 67$.

Ex7. a cubic polynomial P(x) when divided by (x - 1), (x - 2) and (x - 3) leaves remainder 3, 8 and 15 respectively. If P(4) = 30 then the remainder, when P(x) is divided by (x + 1) is

(A) - 25   (B) - 20   (C) - 16   (D) none of these

Sol. As per question, p(1) = 3, p(2) = 8 and p(3) = 15.

Now challenge is to find a polynomial expression of degree less than 3 which follow p(1) = 3, p(2) = 8 and p(3) = 15. Think!
I find a required expression as $x^2 + 2x$.
So, let P(x) = k(x - 1)(x - 2)(x - 3) + x^2 + 2x.
Now, P(4) = 30 = k(4 - 1)(4 - 2)(4 - 3) + 4^2 + 2×4 → k = 1.
Hence, P(x) = (x - 1)(x - 2)(x - 3) + x^2 + 2x.
Required remainder = P(- 1) = 25. Therefore option (A) is correct.

Ex8. Find the remainder when $x^7 + 2x^6 + 3x^5 + 2x^4 + x^3$ is divided by $x^2 + x + 1$.

Sol. As per long division method

$$x^2 + x + 1 \overline{)x^7 + 2x^6 + 3x^5 + 2x^4 + x^3} \quad \dfrac{x^5 + x^4 + x}{}$$

$$\underline{x^7 + x^6 + x^5}$$
$$x^6 + 2x^5 + 2x^4$$
$$\underline{x^6 + x^5 + x^4}$$
$$x^5 + x^4 + x^3$$
$$\underline{x^5 + x^4 + x^3}$$
$$0 = \text{Remainder}$$

Method – II
If we observe: $x^7 + 2x^6 + 3x^5 + 2x^4 + x^3 = (x^7 + x^6 + x^5) + (x^6 + x^5 + x^4) + (x^5 + x^4 + x^3)$
$= x^5 (x^2 + x + 1) + x^4 (x^2 + x + 1) + x^3 (x^2 + x + 1) = (x^5 + x^4 + x^3)(x^2 + x + 1) + 0$
So, remainder is zero.

Ex9. If the remainders of $ax^3 + 3x^2 - 13$ & $2x^3 - 5x + a$ are same when divided by x + 2. Then find the value of a.

Sol. Let p(x) = $ax^3 + 3x^2 - 13$ and q(x) = $2x^3 - 5x + a$.
As per question, remainders are same → p (- 2) = q (- 2)
→ - 8a – 1 = - 16 + 10 + a → 9a = 5 → a = 5/9.

Ex10. The remainder when $(x + 3)^5 + (x + 2)^8 + (5x + 9)^{1997}$ is divided by x + 2 is ____

(A) 1   (B) - 1   (C) 0   (D) None of these

Sol. Let p(x) = $(x + 3)^5 + (x + 2)^8 + (5x + 9)^{1997}$

Remainder R = p(- 2) = 1 – 1 = 0. So, (C) is correct option.

Let's have a mind refreshing Geometry problem.

GEO PROBLEM # 1

Consider the rectangle ABCD as shown in figure. If area of square is 3 cm^2, then area of rectangle in shaded region is _____

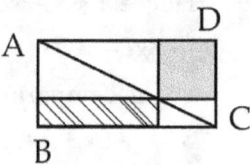

FACTOR THEOREM

It's a special case of remainder theorem. The theorem connects the factors or zeroes of polynomial. It's applied commonly at two places.

    (1) Factorization   (2) finding zeroes of polynomial

STATEMENT:

If $P(x)$ is a polynomial of degree n (≥ 1) then $x - a$ is a factor of $P(x)$ if and only if $P(a) = 0$.

Proof: as per remainder theorem.

If we divide $P(x)$ by linear factor $x - a$ then we will have a remainder of zero degree.

$$P(x) = (x - a)Q(x) + R$$

$\dfrac{Q(x)}{x - a \overline{)\, P(x)}}$ (i)   If $P(a) = 0$, then

.... $P(a) = (a - a)Q(a) + R \rightarrow 0 = R$

                So, $P(x) = (x - a)Q(x) + 0 \rightarrow (x - a)$ is a factor of $P(x)$.

     R   (ii)   If $(x - a)$ is a factor of $P(x)$

Then, $P(x) = (x - a)Q(x)$

$\rightarrow P(a) = 0$.

Ex1. Factorize $3x^2 - 5x + 2$ using factor theorem.

Sol. Let $P(x) = 3x^2 - 5x + 2$

$P(2) = 3(2)^2 - 5(2) + 2 = 12 - 10 + 2 = 4 \neq 0$ so $(x - 2)$ is not a factor.

$P(1) = 3(1)^2 - 5(1) + 2 = 0 \rightarrow (x - 1)$ is a factor of $P(x)$.

$P(2/3) = 3(2/3)^2 - 5(2/3) + 2 = 0 \rightarrow (x - (2/3))$ is another factor of $P(x)$.

Since, $P(x)$ is quadratic so $P(x) = 3x^2 - 5x + 2 = 3(x - 1)(x - (2/3))$.

Why, we multiply factors by 3? (to balance x^2 coefficients).

Here, guessing the zero a (like a =2/3) is very hard but for learning in text books we use easily thinkable zero.

METHOD – II

I know, you are thinking of splitting method to factorize.

(Mainly used in quadratic and not in cubic and higher degree polynomial)

$3x^2 - 5x + 2 \rightarrow 3x^2 - 3x - 2x + 2 \rightarrow 3x(x - 1) - 2(x - 1) \rightarrow (3x - 2)(x - 1)$.

$3x^2 - 5x + 2 = (3x - 2)(x - 1)$

METHOD – III

Let $P(x) = 3x^2 - 5x + 2$

$P(1) = 3(1)^2 - 5(1) + 2 = 0 \rightarrow (x - 1)$ is a factor of $P(x)$.

Since, $P(x)$ is a quadratic polynomial so it will have one more linear factor.

Hence, $P(x) = 3x^2 - 5x + 2 = k(x - 1)(x - \alpha)$

$\rightarrow 3x^2 - 5x + 2 = k(x^2 - (1 + \alpha)x + \alpha) \rightarrow 3x^2 - 5x + 2 = kx^2 - k(1 + \alpha)x + k\alpha$

So, on comparing, $k = 3$, $2 = 3\alpha \rightarrow \alpha = 2/3$

$\therefore 3x^2 - 5x + 2 = 3(x - 1)(x - (2/3))$.

METHOD – IV

$P(x) = 3x^2 - 5x + 2$ and $P(1) = 0 \rightarrow (x - 1)$ is a factor

Now, $P(x)$ is a quadratic so write $(x - 1)$ twice as shown

$P(x) = 3x^2 - 5x + 2 = a (x - 1) + b (x - 1)$

What should be 'a' to get $3x^2$? Obviously $a = 3x$

Now, $3x(x - 1) = 3x^2 - 3x$. we need $- 2x$ more to get $P(x)$. so, what should be 'b'?

Again answer is, $b = - 2$.

$\rightarrow 3x^2 - 5x + 2 = 3x (x - 1) - 2 (x - 1) \rightarrow (3x - 2) (x - 1)$

METHOD · V

$P(x) = 3x^2 - 5x + 2$ and $P(1) = 0 \rightarrow (x - 1)$ is a factor

Now, using synthetic division, we can get.

$3x^2 - 5x + 2 = (x - 1)(3x - 2)$

$$
\begin{array}{r|rrr}
 & 3 & -5 & 2 \\
1 & \square & 3 & -2 \\
\hline
 & 3 & -2 & 0 \\
\end{array}
$$

$$
\begin{array}{r}
3x - 2 \\
x - 1 \overline{\smash{)}3x^2 - 5x + 2} \\
\underline{3x^2 - 3x} \\
-2x + 2 \\
\underline{-2x + 2} \\
0
\end{array}
$$

METHOD – VI

$P(x) = 3x^2 - 5x + 2$ and $P(1) = 0 \rightarrow (x - 1)$ is a factor.

Now use long division method to find quotient

Here, remainder is zero.

So, $3x^2 - 5x + 2 = (x - 1)(3x - 2)$

Ex2. Find k if $(x - 3)$ is a factor of $f(x) = 2x^3 - 4kx + k$

Sol. $(x - 3)$ is a factor of $f(x) \rightarrow f(k) = 0$

$\rightarrow 2k^3 - 4 k^2 + k = 0 \rightarrow k (2k^2 - 4k + 1) = 0$

$\rightarrow k = 0, \dfrac{2 \pm \sqrt{2}}{2}$.

Ex3. Show that $(x - 2)$ is a factor of $x^3 - 3x^2 - 10x + 24$.

Sol. Let $P(x) = x^3 - 3x^2 - 10x + 24$

$P(2) = (2)^3 - 3(2)^2 - 10(2) + 24 = 8 - 12 - 20 + 24 = 0$

$\therefore (x - 2)$ is a factor.

Ex4. The values of a & b if $x^3 - ax^2 - 13x + b$ have $(x - 1)$ and $(x + 3)$ as factors

(A) 3, 12   (B) 3, 15   (C) 4, 12   (D) 6, 15

Sol. Let $p(x) = x^3 - ax^2 - 13x + b$

Since, $(x - 1)$ and $(x + 3)$ are factors, so $p (1) = 0$ and $p (- 3) = 0$

$p (1) = 0 \rightarrow 1 - a - 13 + b = 0 \rightarrow b - a = 12 \ldots\ldots\ldots (1)$

$p (- 3) = 0 \rightarrow - 27 - 9a + 39 + b = 0 \rightarrow b - 9a = - 12 \ldots\ldots(2)$

From (1) and (2) $a = 3$ and $b = 15$. Option (B) is correct.

Ex5. Factorize: $P(x) = x^3 - 2x^2 + 2x - 1$.

Sol. Clearly $P(1) = 0 \to x - 1$ is a factor of $P(x)$.

$$x - 1 \overline{\smash{)}x^3 - 2x^2 + 2x - 1} \quad \frac{x^2 - x + 1}{}$$

$$\underline{x^3 - x^2}$$
$$-x^2 + 2x$$
$$\underline{-x^2 + x}$$
$$x - 1$$
$$\underline{x - 1}$$
$$0$$

Factorization of $x^3 - 2x^2 + 2x - 1 = (x - 1)(x^2 - x + 1)$.

METHOD – II

Clearly $P(1) = 0$ *(Observation)*

$\to x - 1$ is a factor of $P(x)$.

Now, $P(x)$ is cubic polynomial so we will write $(x - 1)$ three time as shown below.

$\to P(x) = x^3 - 2x^2 + 2x - 1 = a(x - 1) + b(x - 1) + c\,(x - 1)$

What should be 'a' to get x^3? It should be x^2

Now $x^2(x - 1) = x^3 - x^2$ but we need $- 2x^2$. So what should be b? It should be $- x$.

Now $x^2(x - 1) - x(x - 1) = x^3 - 2x^2 + x$. now we remains to get $x - 1$. So what should be c?

We must have $c = 1$.

Hence, $x^3 - 2x^2 + 2x - 1 = x^2(x - 1) - x(x - 1) + 1\,(x - 1)$

$\to x^3 - 2x^2 + 2x - 1 = (x - 1)(x^2 - x + 1)$. *(This is a one line method)*

Ex6. Find the value of k if the polynomial $P(x) = 2x^3 + kx^2 - 7x - 12$ is divisible by $x + 4$. Hence, find all the factors of $P(x)$.

Sol. $P(x)$ is divisible by $x + 4 \to P(- 4) = 0$

$\to 2(- 4)^3 + k\,(- 4)^2 - 7\,(- 4) - 12 = 0 \to - 128 + 16\,k + 28 - 12 = 0$

$\to 16\,k = 112 \to k = 7$.

$P(x) = 2x^3 + 7x^2 - 7x - 12$ having $(x + 4)$ as factor.

$\to 2x^3 + 7x^2 - 7x - 12 = 2x^2\,(x + 4) - x\,(x + 4) - 3\,(x + 4)$

$\to \quad 2x^3 + 7x^2 - 7x - 12 = (x + 4)(2x^2 - x - 3)$

$\to \quad 2x^3 + 7x^2 - 7x - 12 = (x + 4)(2x - 3)(x + 1)$

$(x + 4)$, $(2x - 3)$ and $(x + 1)$ are the factors of $2x^3 + kx^2 - 7x - 12$.

Ex7. Factorize: $x^4 + x^3 - 7x^2 - x + 6$

Sol. Let $P(x) = x^4 + x^3 - 7x^2 - x + 6$

$P(1) = 0$ *(observation)*

$(x - 1)$ is a factor of $P(x) = (x - 1)(x^3 + 2x^2 - 5x - 6)$ (One line factorization) Think!

Now, let $Q(x) = x^3 + 2x^2 - 5x - 6 \to Q(- 1) = 0$. So $(x + 1)$ is a factor of $Q(x)$.

$Q(x) = x^3 + 2x^2 - 5x - 6 = (x + 1)(x^2 + x - 6)$ and $x^2 + x - 6 = (x + 3)(x - 2)$

So, $P(x) = x^4 + x^3 - 7x^2 - x + 6 = (x - 1)(x + 1)(x + 3)(x - 2)$.

Ex8. Find the number of real roots of equation $3x^4 + 8x^3 - 6x^2 - 24x - 10 = 0$.

Sol. Let $f(x) = 3x^4 + 8x^3 - 6x^2 - 24x - 10$

In this case finding zero is a difficult task, that's why we have question like "numbers of zeroes" instead of finding zeroes. Some time we need to take help of calculus to observe the behavior of polynomial. I will not go in to details but have sense of such kind of questions.

$f'(x) = 12x^3 + 24x^2 - 12x - 24 \rightarrow f'(x) = 12(x^3 + 2x^2 - x - 2)$

$\rightarrow f'(x) = 12(x^2(x + 2) - (x + 2)) \rightarrow f'(x) = 12(x + 2)(x^2 - 1) \rightarrow f'(x) = 12(x - 1)(x + 1)(x + 2)$

Let's plot the sign scheme of $f'(x)$ and a rough working sketch of $f(x)$.

Note: $f'(x) > 0$, $f'(x) < 0 \rightarrow f(x)$ is strictly increasing, strictly decreasing respectively.

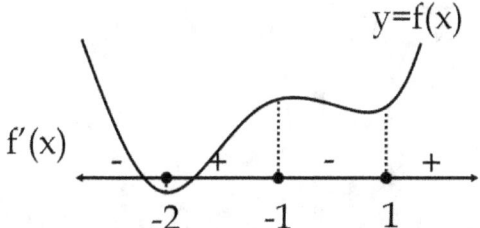

As $f'(x) > 0$ in $x \in (-2, -1) \cup (1, \infty)$ so it's increasing in the interval $x \in (-2, -1) \cup (1, \infty)$.

And $f'(x) < 0$ in $x \in (-\infty, -2) \cup (-1, 1)$ so it's decreasing in the interval $x \in (-\infty, -2) \cup (-1, 1)$.

Above figure is not an actual figure. It's our working diagram.

For actual diagram, we need to find $f(-2)$, $f(-1)$ and $f(1)$ for exact value.

$f(-2) = 3(-2)^4 + 8(-2)^3 - 6(-2)^2 - 24(-2) - 10 = 48 - 64 - 24 + 48 - 10 = -2$. So $f(-2) = -2$.

$f(-1) = 3(-1)^4 + 8(-1)^3 - 6(-1)^2 - 24(-1) - 10 = 3$. So $f(-1) = 3$

$f(1) = 3(1)^4 + 8(1)^3 - 6(1)^2 - 24(1) - 10 = -29$. So, $f(1) = -29$.

Here is the actual curve.

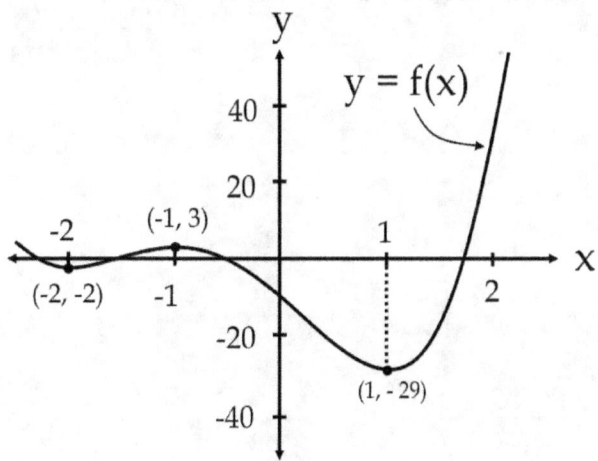

Hence, there are four real roots of given equation.

Ex9. The value of $\sqrt[3]{20+14\sqrt{2}}+\sqrt[3]{20-14\sqrt{2}}$ is _____

Sol. Let $x=\sqrt[3]{20+14\sqrt{2}}+\sqrt[3]{20-14\sqrt{2}}$ is a real number

Now take cube both side we get

$$x^3 = 20 + 14\sqrt{2} + 20 - 14\sqrt{2} + 3\sqrt[3]{\left(20+14\sqrt{2}\right)\left(20-14\sqrt{2}\right)}\left(\sqrt[3]{20+14\sqrt{2}}+\sqrt[3]{20-14\sqrt{2}}\right)$$

$\rightarrow x^3 = 40 + 3x\sqrt[3]{400-392} = 40 + 6x$

Now we see that x satisfies the equation $x^3 - 6x - 40 = 0$

Let $P(x) = x^3 - 6x - 40 \rightarrow P(4) = 0$

$\rightarrow$ (x - 4) is a factor of $P(x)$

$\rightarrow x^3 - 6x - 40 = (x - 4)(x^2 + 4x + 10) = 0$

Since the equation $x^2 + 4x + 10 = 0$ has imaginary roots so x must be 4

Hence, $\sqrt[3]{20+14\sqrt{2}}+\sqrt[3]{20-14\sqrt{2}} = 4$

☺ ☺ *Let's have a mind refreshment problem to relax your mind.* ☺ ☺

MIND REFRESHMENT # 1

A Coin of diameter 2 unit rolls over a triangle of perimeter 3, as shown in figure. Then the perimeter of outer boundary trace is?

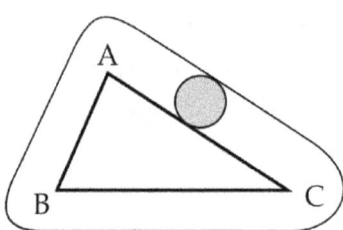

EXERCISE - 1

1. Find the remainder when x^{100} is divided by $x^2 - x$.

2. If $2x^3 - 5x^2 + x + 2 = (x - 2)(ax^2 - bx - 1)$ then a = _____ and b = _____

3. Select the quadratic equations

   (A) $2(x - 1)^2 = 4x^2 - 2x + 1$ (B) $(x^2 + 1)^2 = x^2 + 3x + 9$

   (C) $(x^2 + 2x)^2 = x^4 + 3 + 4x^3$ (D) $x^2 + 9 = 3x^2 - 5x$

4. Factorize the following:

   (i) $x^4 + x^2 + 1$ (ii) $2x^3 - 7x^2 + 2x + 3$ (iii) $6x^3 - x^2 - 5x + 2$ (iv) $x^8 - y^8$

   (v) $x^{10} + x^5 + 1$ (vi) $x^8 + x^4 + 1$

5. If $f(x) = x^3 - 4x + p$, and $f(0)$ and $f(1)$ are of opposite sign, then which of the following is necessarily true?

   (A) $-1 < p < 2$ (B) $0 < p < 3$

   (C) $-2 < p < 1$ (D) $-3 < p < 0$

6. Let $f(x) = x^4 + x^3 + x^2 + x + 1$ then remainder of $f(x)$ on dividing $f(x^5)$ by $f(x)$ is

   (A) 1   (B) 3   (C) 4   (D) 5

7. $P(x)$ is a polynomial with integral coefficients such that the absolute value of the constant term of $P(x)$ is smaller than 1000. Given $P(19) = P(94) = 1994$, then the constant term of $P(x)$ is

   (A) 207   (B) 208   (C) 209   (D) 220

8. If the two numbers are selected from the set $\{1, 2, 3, \ldots, 100\}$ without replacement and $x < y$ then the number of ways such that $x^3 + y^3$ when divided by '9' leaves odd remainder, is

   (A) 1650   (B) 1750   (C) 2225   (D) 2805

9. When the polynomial $5x^3 + Mx + N$ is divided by $x^2 + x + 1$, the remainder is 0. Then the value of $|M + N|$ is _____

10. Let $f(x)$ be a polynomial of degree 4 such that $f([i]) = \dfrac{1}{1+[i]}$ for $0 \le i \le 4$,

    Where $[x]$ = Greatest integer less than or equal to x, then
    (A) $f(-2) < 0$ (B) $f(6) = 1$

    (C) $f(5) \in (0, 1)$ (D) $f(-2) + f(5) + f(6) \in (6, 7)$

11. Find the set of values of λ for which $x^3 - 6x + \lambda = 0$ have three distinct real roots.

QUADRATIC EXPRESSION

A polynomial of degree 2 is known as a quadratic expression.

Ex1. $ax^2 + bx + c$, where $a \neq 0$ and $a, b, c \in C$ is a quadratic expression. Where a, b & c are the coefficients of x^2, x and x^0 respectively.

Ex2. $f(x) = a x^2 + bx + c$ or $y = a x^2 + bx + c$, where $a \neq 0$ and $a, b, c \in R$, is a real quadratic expression.

Ex3. $ax^2 + 2hxy + by^2 + 2gx + 2fy + c = 0$ is a quadratic equation in two variables x and y.

Here is example of some more quadratic equations.

(i) $x^2 + x + 1 = 0$

(ii) $2x^2 - 1 = 0$

(iii) $x^2 - x = 0$

(iv) $x^2 + 5x - 6 = 0$

(v) $2x^2 - (5 + 2i)x + 3 + 2i = 0$

(vi) $h = ut - \dfrac{gt^2}{2}$

QUADRATIC IN DAILY LIFE:

I know, the picture is a big question at the moment. Soon you will understand the meaning of it. ☺

Consider a general quadratic expression $y = Q(x) = ax^2 + bx + c$, where $a, b, c \in C$ and $a \neq 0$, then

$$Q(x) = a\left(x^2 + \frac{b}{a}x \right) + c = a\left(x^2 + 2\left(\frac{b}{2a} \right)x + \left(\frac{b}{2a} \right)^2 - \left(\frac{b}{2a} \right)^2 \right) + c$$

$$= a\left(\left(x + \frac{b}{2a} \right)^2 - \left(\frac{b}{2a} \right)^2 \right) + c = a\left(x + \frac{b}{2a} \right)^2 - \frac{b^2}{4a} + c$$

$$= a\left(x + \frac{b}{2a} \right)^2 - \left(\frac{b^2 - 4ac}{4a} \right)$$

Now if we have quadratic equation $ax^2 + bx + c = 0$

$$\rightarrow a\left(x+\frac{b}{2a}\right)^2 - \left(\frac{b^2-4ac}{4a}\right) = 0 \rightarrow \left(x+\frac{b}{2a}\right)^2 = \frac{b^2-4ac}{4a^2}$$

$$\rightarrow x+\frac{b}{2a} = \pm\sqrt{\frac{b^2-4ac}{4a^2}} \rightarrow x = -\frac{b}{2a} \pm \frac{\sqrt{b^2-4ac}}{2a} \rightarrow x = \frac{-b \pm \sqrt{b^2-4ac}}{2a}$$

So, roots of equation $ax^2 + bx + c = 0$ are $x_1 = \dfrac{-b+\sqrt{b^2-4ac}}{2a}$ & $x_2 = \dfrac{-b-\sqrt{b^2-4ac}}{2a}$

We called $D = b^2 - 4ac$, as Discriminant of quadratic equation $ax^2 + bx + c = 0$.

Hence, roots can also be written as $x_1 = \dfrac{-b+\sqrt{D}}{2a}$ & $x_2 = \dfrac{-b-\sqrt{D}}{2a}$

ALITER: $ax^2 + bx + c = 0$ as $a \neq 0 \rightarrow 4a^2x^2 + 4abx + 4ac = 0$ (Multiply by 4a)

$\rightarrow 4a^2x^2 + 4abx + b^2 - b^2 + 4ac = 0 \rightarrow (2ax + b)^2 = b^2 - 4ac$

$\rightarrow 2ax + b = \pm\sqrt{b^2-4ac} \rightarrow x = \dfrac{-b \pm \sqrt{b^2-4ac}}{2a}$.

Ex4. Find the Discriminant of following quadratic equations and hence find its roots.

(i) $x^2 + x + 1 = 0$

(ii) $x^2 - 2x + 1 = 0$

(iii) $x^2 - 5x + 6 = 0$

(iv) $ix^2 - 3x - 2i = 0$

(v) $2(1 + i) x^2 - 4 (2 - i) x - 5 - 3i = 0$

(vi) $3x^2 - 2x = 0$

(vii) $x^2 - 2 = 0$

(viii) $x^2 - 4x + 1 = 0$

Where $i = \sqrt{-1}$, $i^2 = -1$, $i^3 = -i$ & $i^4 = 1$

Solution

(i) for $x^2 + x + 1 = 0$, $a = b = c = 1$

So, $D = b^2 - 4ac \rightarrow D = (1)^2 - 4(1)(1) = -3$

Discriminant $D = -3$ which is less than zero.

Roots are x_1 and $x_2 = \dfrac{-1 \pm \sqrt{-3}}{2} = \dfrac{-1 \pm \sqrt{-1} \times \sqrt{3}}{2} = \dfrac{-1 \pm i\sqrt{3}}{2}$

Here, you can observe that if $D < 0$ then you will have negative quantity under, root which leads to makes roots imaginary.

Here we are getting imaginary roots $\dfrac{-1+i\sqrt{3}}{2}$ and $\dfrac{-1-i\sqrt{3}}{2}$

(ii) For quadratic equation $x^2 - 2x + 1 = 0$, we have $a = 1$, $b = -2$ and $c = 1$.

So, Discriminant $D = (-2)^2 - 4 \times 1 \times 1 = 0$. This is zero

Roots are $x_1, x_2 = \dfrac{-(-2) \pm \sqrt{0}}{2 \times 1} = \dfrac{2 \pm 0}{2} = \dfrac{2+0}{2} = 1, \dfrac{2-0}{2} = 1$

So, roots are $x_1, x_2 = 1, 1$ (real and equal, we also say that repeated roots)

Here, you can observe that if $D = 0$ then we are getting real and equal roots $(a, b, c \in R)$.

And $x^2 - 2x + 1 = (x - 1)^2$ (quadratic expression is becoming a perfect square)

(iii) For $x^2 - 5x + 6 = 0$, $a = 1$, $b = -5$ and $c = 6$

So, Discriminant $D = (-5)^2 - 4(1)(6) = 1 > 0$

Roots are $x_1, x_2 = \dfrac{-(-5) \pm \sqrt{1}}{2} = \dfrac{5 \pm 1}{2} \to x_1, x_2 = 3, 2$

Here you can observe that a, b, c $\in$ R. if D > 0 then $\sqrt{D} \in$ R, so we are getting real and distinct roots in this case. We also needs to observe that a = 1 and b, c $\in$ I and D = 1, is a perfect square so, we are getting integral roots too.

(iv) For $ix^2 - 3x - 2i = 0$, $a = i$, $b = -3$ and $c = -2i$

So, Discriminant $D = (-3)^2 - 4(i)(-2i) = 9 + 8i^2 = 9 - 8 = 1$

So, roots are $x_1, x_2 = \dfrac{-(-3) \pm \sqrt{1}}{2i} = \dfrac{3 \pm 1}{2i} \to x_1, x_2 = \dfrac{2}{i}, \dfrac{1}{i} = -2i, -i$

Here you needs to observe that D = 1 > 0 but still we are not getting real roots. Why? Think!

As you can see, here a, c $\in$ Imaginary numbers

So, if a, b, c $\in$ R and D > 0, only then we can have real and distinct roots.

(v) For $2(1 + i) x^2 - 4(2 - i) x - 5 - 3i = 0$

$D = (-4(2 - i))^2 - 4 \times 2(1 + i) \times (-5 - 3i) = 16(2 - i)^2 + 8(1 + i)(5 + 3i)$

$\to D = 16(4 - 1 - 4i) + 8(5 + 5i + 3i - 3) \to D = 16(3 - 4i) + 8(2 + 8i) \to D = 64$

Note: *if you didn't get the calculation part, then do it your own to reach the final point.*

So, roots are $x_1, x_2 = \dfrac{4(2 - i) \pm 8}{4(1 + i)} = \dfrac{(2 - i) \pm 2}{1 + i} \to x_1, x_2 = \dfrac{4 - i}{1 + i}, -\dfrac{i}{1 + i}$

(vi) For $3x^2 - 2x = 0$, $a = 3$, $b = -2$ and $c = 0$

$D = 4$, hence roots are $x_1, x_2 = 0$ and $2/3$

(vii) For $x^2 - 2 = 0$, $a = 1$, $b = 0$ and $c = -2$, so $D = 8$

Hence roots are $-\sqrt{2}$ and $\sqrt{2}$

(viii) For $x^2 - 4x + 1 = 0$, $D = 12 \to$ Roots are, $2 + \sqrt{3}$ and $2 - \sqrt{3}$

NATURE OF ROOTS

Let α and β are roots of the real quadratic equation $ax^2 + bx + c = 0$ where $a \neq 0$, a, b, c $\in$ R.

Then $\alpha, \beta = \dfrac{-b \pm \sqrt{D}}{2a}$

(i) If D > 0 then roots are real and distinct. (i.e. $\alpha, \beta = \dfrac{-b \pm \sqrt{D}}{2a}$)

In this case we can express $ax^2 + bx + c = a(x - \alpha)(x - \beta)$

Ex. $2x^2 - 5x + 2 = 0$ have roots x = 2 and 1/2 so, $2x^2 - 5x + 2 = 2(x - 2)(x - 1/2)$.

(ii) If D = 0 then roots are real and equal and quadratic expression is a perfect square. (i.e. $\alpha = \beta = -\dfrac{b}{2a}$)

Roots of $ax^2 + bx + c = 0$ are $\alpha = \beta = -\dfrac{b}{2a}$ so, $ax^2 + bx + c = a(x - \alpha)^2 = a\left(x + \dfrac{b}{2a}\right)^2$

Ex. $x^2 - 4x + 4 = 0$. D = 0 and roots are α, β = 2, 2 and $x^2 - 4x + 4 = (x - 2)^2$

(iii)   If D < 0 then roots are imaginary and occurs in conjugate pairs. (i.e p + iq, p – iq where p, q$\in$ R)

Ex. $x^2 - 2x + 2 = 0$, D = - 4 < 0 so have imaginary roots 1 + i and 1 – i (conjugate pairs).

Note: a number a + i b is said to be imaginary if b $\neq$ 0 and a, b $\in$ R, i = $\sqrt{-1}$

Que.   Is there any quadratic equation having one root real and other imaginary?

Sol.   Yes

You can think of equation (x - 1)(x - i) = 0

$\rightarrow x^2 - (1 + i)x + i = 0$. Here you can observe a, b, c $\in$ C.

If a, b, c $\in$ R then there is no as such quadratic equations.

SOME IMPORTANT STATEMENTS

Think and prove the following statement your own.

Statement - 1

If a, b, c $\in$ Q (rational numbers) & D is a perfect square of a rational number then the roots are rational and in case if it's not a perfect square of rational then roots are irrational.

Ex1. $x^2 - 2x - 2 = 0$. D = 9 = (3)^2 and roots are 2 and – 1.

Ex2. $x^2 - x - 1 = 0$. D = 5 (not a perfect square of rational)

Roots are α, $\beta = \dfrac{1 \pm \sqrt{5}}{2}$ (irrational in conjugate pairs)

Statement - 2

If p + $\sqrt{q}$ (where p, q $\in$ Q) is an irrational root of real quadratic equation $ax^2 + bx + c = 0$ then p - $\sqrt{q}$ must be the other root of it provided coefficients are rational.

Ex.   3 - $\sqrt{2}$ and 3 + $\sqrt{2}$ are roots of $x^2 - 6x + 7 = 0$

Statement – 3

If a = 1 & b, c $\in$ I and D is a perfect square of rational number then roots must be integral.

Ex. $x^2 - 5x + 6 = 0$, D = 1 = (1)^2

Roots are x = 2, 3 $\in$ I

Let's solve some problems to apply our learning.

Ex1.   Find the value of m for which the equation $(1 + m)x^2 - 2(1 + 3m)x + (1 + 8m) = 0$ has equal roots.

Sol.   As per condition of equal roots, D = 0

$\rightarrow 4(1 + 3m)^2 - 4 \times (1 + m)(1 + 8m) = 0$

$\rightarrow (1 + 3m)^2 - (1 + m)(1 + 8m) = 0$

$\rightarrow 1 + 9m^2 + 6m - (8m^2 + 9m + 1) = 0$

$\rightarrow m^2 - 3m = 0 \rightarrow m(m - 3) = 0 \rightarrow m = 0, 3$.

Can we proceed like $m^2 - 3m = 0 \rightarrow m^2 = 3m \rightarrow \cancel{m} \times m = 3\cancel{m} \rightarrow m = 3$? Answer is No.

As the process causes loss of genuine root m = 0.

Think about the situation. $0 = 0 \rightarrow 0 \times 2 = 0 \times 3 \rightarrow \cancel{0} \times 2 = \cancel{0} \times 3 \rightarrow 2 = 3$ (which is wrong)

So cancelation not allowed for expression becoming zero.

$6 = 6 \rightarrow \cancel{2} \times 3 = \cancel{2} \times 3 \rightarrow 3 = 3$ *(True). Cancellation can apply on non zero quantities.*

Ex2. Prove that roots of $ax^2 - 3b\, x - 4a = 0$ are real and distinct $\forall\ a, b \in R, a \neq 0$

Sol. $D = 9b^2 + 16\, a^2 > 0$ so roots are real and distinct.

Ex3. Find all the integral values of a for which the quadratic equation $(x - a)(x - 10) + 1 = 0$ have integral roots.

Sol. $(x - a)(x - 10) + 1 = 0 \rightarrow x^2 - (a + 10)x + 10a + 1 = 0$.

Since, coefficients are integers and x^2 coefficient is 1, then as per condition of integral roots D must be a perfect square of a rational number.

$D = (a + 10)^2 - 4(10a + 1) \rightarrow D = a^2 - 20\, a + 96 = k^2$ where $k \in I$ as $a \in I$

$\rightarrow a^2 - 20\, a + 100 = k^2 + 4 \rightarrow (a - 10)^2 - k^2 = 4$

$\rightarrow (a - 10 - k)(a - 10 + k) = 4 = 1 \times 4 = 4 \times 1 = 2 \times 2 = (-1) \times (-4) = (-4) \times (-1) = (-2) \times (-2)$

So, consider first situation

$(a - 10 - k)(a - 10 + k) = 1 \times 4$

$\rightarrow a - 10 - k = 1$ and $a - 10 + k = 4 \rightarrow a = 12/5 \notin I$

Similarly, $(a - 10 - k)(a - 10 + k) = 4 \times 1 \rightarrow a - 10 - k = 4$ and $a - 10 + k = 1 \rightarrow a = 12/5 \notin I$

Consider $(a - 10 - k)(a - 10 + k) = 2 \times 2$

$\rightarrow a - 10 - k = 2$ and $a - 10 + k = 2 \rightarrow a = 12$ and $k = 0$, so $a = 12$ is an answer.

Now, consider $(a - 10 - k)(a - 10 + k) = (-1) \times (-4)$

$\rightarrow a - 10 - k = -1$ and $a - 10 + k = -4 \rightarrow a = 15/2 \notin I$, similarly there is no answer for $(-4) \times (-1)$.

Now look at the situation

$(a - 10 - k)(a - 10 + k) = (-2) \times (-2)$

$\rightarrow a - 10 - k = -2$ and $a - 10 + k = -2 \rightarrow a = 8$ and $k = 0$

So, $a = 8$ and 12 are the only integral values of a.

Ex4. If $a, b \in R$ then roots of the equation $(x - a)(x - b) = b^2$ are

(A) Real & Equal   (B) Real & Distinct   (C) Imaginary   (D) none of these

Sol. $(x - a)(x - b) = b^2 \rightarrow x^2 - (a + b)\, x + ab - b^2 = 0$

$D = (a + b)^2 - 4\, (ab - b^2) \rightarrow D = (a - b)^2 + 4\, b^2 > 0$

So roots are real and distinct. (B) is correct option.

Ex5. If "a" can take values from set $\{1, 2, 3, 4, 5, 6, 7\}$ then number of distinct values of "a" for which equation $x^2 - ax + 3 = 0$ has real and distinct roots?

Sol. For real and distinct roots: $D > 0$

$\rightarrow a^2 - 4 \times 3 > 0 \rightarrow a^2 - 12 > 0$ is true for $a = 4, 5, 6$ and 7.

So, number of values of a is 4.

VIETA'S THEOREM

If α and β are roots of quadratic equation $ax^2 + bx + c = 0$ then

(i) *Sum of roots* = $\alpha + \beta = -b/a$ (ii) *Product of roots* = $\alpha\beta = c/a$

Proof:

Let $Q(x) = ax^2 + bx + c$. As α and β are roots so $Q(\alpha) = 0$, $Q(\beta) = 0$

$\rightarrow Q(x) = ax^2 + bx + c = a(x - \alpha)(x - \beta)$

$\rightarrow ax^2 + bx + c = a(x^2 - (\alpha + \beta)x + \alpha\beta)$

On comparing the coefficients of x and x^0 we get

$b = -a(\alpha + \beta) \rightarrow \alpha + \beta = -b/a$ and $c = a\alpha\beta \rightarrow \alpha\beta = c/a$

Sum of roots = $\alpha + \beta = -b/a$ product of roots = $\alpha\beta = c/a$

Note: Difference of roots $\alpha - \beta = \pm\dfrac{\sqrt{D}}{a}$.

Ex1. Find the sum and product of roots of following quadratic equations

(i) $x^2 + 4x + 1 = 0$ (ii) $2x^2 - 3x - 1 = 0$ (iii) $x^2 + x = 0$

Sol. (i)   Sum of roots = $-4/1 = -4$, product of roots = $1/1 = 1$.

(ii) Sum of roots = $-(-3)/2 = 3/2$, product of roots = $-1/2$.

(iii) Sum of roots = $-1/1 = -1$, product of roots = $0/1 = 0$.

OBSERVATION

$ax^2 + bx + c = 0 \rightarrow x^2 + \dfrac{b}{a}x + \dfrac{c}{a} = 0$ as $a \neq 0$

$\rightarrow x^2 + \dfrac{b}{a}x + \dfrac{c}{a} = x^2 - \left(-\dfrac{b}{a}\right)x + \left(\dfrac{c}{a}\right) = x^2 - (\text{Sum of roots})x + (\text{product of roots}) = 0$

$\rightarrow$ a quadratic equation having roots α and β can be written as $(x - \alpha)(x - \beta) = 0$

$\rightarrow x^2 - (\alpha + \beta)x + \alpha\beta = 0 \rightarrow x^2 - (\text{sum of roots})x + (\text{product of roots}) = 0$.

Ex2. Find the quadratic equation having roots 2 & 3

Sol. $\alpha + \beta = 2 + 3 = 5$, $\alpha\beta = 2\times3 = 6$

So required quadratic equation is $x^2 - 5x + 6 = 0$

Ex3. If α and β are roots of $x^2 - 5x + a = 0$, then find the value of following expressions

(i) $\alpha^2 + \beta^2$ (ii) $\dfrac{1}{\alpha^2} + \dfrac{1}{\beta^2}$ (iii) $\alpha^3 + \beta^3 - 125$ (iv) $\dfrac{\alpha}{\beta} + \dfrac{\beta}{\alpha}$

Sol. (i) $\alpha + \beta = 5$ and $\alpha\beta = a$

$\alpha^2 + \beta^2 = (\alpha + \beta)^2 - 2\alpha\beta \rightarrow \alpha^2 + \beta^2 = 5^2 - 2a = 25 - 2a$

(ii) $\dfrac{1}{\alpha^2} + \dfrac{1}{\beta^2} = \dfrac{\alpha^2 + \beta^2}{(\alpha\beta)^2} = \dfrac{25 - 2a}{a^2}$

(iii) $\alpha^3 + \beta^3 = (\alpha + \beta)^3 - 3\alpha\beta(\alpha + \beta) \rightarrow \alpha^3 + \beta^3 = 5^3 - 3a \times 5 \rightarrow \alpha^3 + \beta^3 - 125 = -15a$

(iv) $\dfrac{\alpha}{\beta} + \dfrac{\beta}{\alpha} = \dfrac{\alpha^2 + \beta^2}{\alpha\beta} = \dfrac{25 - 2a}{a}$

Ex4. Select the wrong option for $ax^2 + bx + c = 0$, $a \neq 0$, $a, b, c \in R$

(A) If $b = 0$ then roots are equal in magnitude but opposite in sign

(B) If $c = 0$ then one root is zero and other is $x = -b/a$.

(C) The dimensions of a rectangle of area 6 are $x - 2$ and $x - 3$ then $x^2 - 5x = 0 \rightarrow x = 0, 5$

(D) If $a = c \neq 0$ then roots are reciprocal of each other

Sol. (A) If $b = 0$ then equation become $ax^2 + c = 0 \rightarrow x^2 = -\dfrac{c}{a} \rightarrow x = \pm\sqrt{-\dfrac{c}{a}}$

(B) If $c = 0$ then equation become $ax^2 + bx = 0 \rightarrow x(ax + b) = 0 \rightarrow x = 0$ or $-b/a$

(C) Area of rectangle is $(x - 2)(x - 3) = 6 \rightarrow x^2 - 5x + 6 = 6 \rightarrow x^2 - 5x = 0 \rightarrow x = 0, 5$

But for $x = 0$ dimensions of rectangle becomes negative which is not possible, hence $x \neq 0$.

For $x = 5$, dimensions are positive and satisfies the given conditions so $x = 5$ is only solution.

(D) If $a = c \neq 0 \rightarrow$ product of roots $\alpha\beta = 1 \rightarrow \alpha = \dfrac{1}{\beta}$ or $\beta = \dfrac{1}{\alpha}$.

From above discussion we can say that option (C) is wrong.

Ex5. Roots of the quadratic equation $(x - 1)(x - 1) + (x - 1)(x - 2) + (x - 1)(x - 3) + \ldots + (x - 1)(x - 100) = 0$ are 1 and _____?

Sol. $(x - 1)(x - 1) + (x - 1)(x - 2) + (x - 1)(x - 3) + \ldots + (x - 1)(x - 100) = 0$

$\rightarrow (x - 1)(100x - (1 + 2 + 3 + 4 + \ldots + 100)) = 0$

$\rightarrow (x - 1)(100x - 5050) = 0 \rightarrow x = 1$ and $101/2$.

Ex6. If the difference between the corresponding roots of $x^2 + ax + b = 0$ and $x^2 + bx + a = 0$ is same and $a \neq b$, then

(A) $a + b + 4 = 0$ (B) $a + b - 4 = 0$ (C) $a - b - 4 = 0$ (D) $a - b + 4 = 0$

Sol. Let α and β are roots of $x^2 + ax + b = 0$ and γ and δ are the roots of $x^2 + bx + a = 0$

$\rightarrow \alpha + \beta = -a$ & $\alpha\beta = b \rightarrow (\alpha - \beta)^2 = (\alpha + \beta)^2 - 4\alpha\beta = a^2 - 4b$

$\rightarrow (\alpha - \beta)^2 = a^2 - 4b$ ......(1)

Similarly $(\gamma - \delta)^2 = b^2 - 4a$ ......(2)

From (1) and (2) we get $a^2 - 4b = b^2 - 4a$

$\rightarrow a^2 - b^2 + 4a - 4b = 0 \rightarrow (a - b)(a + b + 4) = 0$ but $a \neq b$ so $a + b + 4 = 0$

Hence, (A) option is correct.

Ex7. If α, β are roots of $x^2 - p(x + 1) - c = 0$ then $(\alpha + 1)(\beta + 1)$ is equal to

(A) c (B) $c - 1$ (C) $1 - c$ (D) $1 + c$

Sol. If α, β are roots of $x^2 - p(x + 1) - c = 0$ then $x^2 - p(x + 1) - c = (x - \alpha)(x - \beta) \; \forall \; x \in R$ ......(1)

So, to get $(\alpha + 1)(\beta + 1)$, put $x = -1$ in equation (1)

$\to 1 - p(-1 + 1) - c = (\alpha + 1)(\beta + 1)$

$\to (\alpha + 1)(\beta + 1) = 1 - c$. Hence (C) is the correct option.

Ex8. If α and β are the roots of $2x^2 - 5x + 7 = 0$ then equation whose roots are $\alpha + 2\beta$, $2\alpha + \beta$ is

(A) $x^2 - 5x + 81 = 0$ (B) $-7x^2 + 15x + 82 = 0$ (C) $7x^2 - 15x + 64 = 0$ (D) $98x^2 - 210x + 443 = 0$

Sol. $\alpha + \beta = 5/7$ and $\alpha\beta = 7/2$

So, sum of roots $= (\alpha + 2\beta) + (2\alpha + \beta) = 3(\alpha + \beta) = 15/7$

Product of roots $= (\alpha + 2\beta) \times (2\alpha + \beta) = 2(\alpha^2 + \beta^2) + 5\alpha\beta = 2(\alpha + \beta)^2 + \alpha\beta = 2(5/7)^2 + 7/2$

So product of roots $= \dfrac{443}{98}$

So required quadratic equation is $x^2 - (15/7)x + \dfrac{443}{98} = 0$

$= 98x^2 - 210x + 443 = 0$. So option (D) is correct.

Ex9. Let α, β are roots of $x^2 + (3 - \lambda)x - \lambda = 0$. The value of λ for which $\alpha^2 + \beta^2$ is minimum, is

(A) 0   (B) 1   (C) 2   (D) 3

Sol. $\alpha + \beta = \lambda - 3$, $\alpha\beta = -\lambda$.

$\alpha^2 + \beta^2 = (\alpha + \beta)^2 - 2\alpha\beta = (\lambda - 3)^2 - 2(-\lambda) = \lambda^2 - 4\lambda + 9$

$\to \alpha^2 + \beta^2 = (\lambda - 2)^2 + 5$.

So, $\alpha^2 + \beta^2|_{min} = 5$ at $\lambda = 2$. Hence, (C) is correct option.

Ex10. Answer these two questions based on the following information.

Let $f(x) = ax^2 + bx + c$, where a, b and c are certain constants and $a \neq 0$. Its known that $f(5) = -3f(2)$ and that 3 is a root of $f(x) = 0$.

1. What is the other root of $f(x) = 0$?

(A) -7   (B) -4   (C) 6   (D) Can't be determined

2. What is the value of $a + b + c$?

(A) 9   (B) 14   (C) 37   (D) Can't be determined

Solution:

1. $f(5) = -3f(2) \to 25a + 5b + c = -3(4a + 2b + c) \to 37a + 11b + 4c = 0$ ....(1)

   Since 3 is a root $\to 9a + 3b + c = 0$....................(2)

   If we perform the operation (1) – 4(2) we get a = b and c = -12a.

   Now $f(x) = ax^2 + ax - 12a \to f(x) = a(x + 4)(x - 3)$. So $f(x) = 0 \to x = 3 \ \& -4$. So (B) is correct option.

2. $a + b + c = a + a - 12a = -10a$, so it depend on 'a'. So, (D) is correct option.

Ex11. If α, β are roots of $x^2 - 2x + 3 = 0$ then find the equations having roots

(i) α^2 and β^2 (ii) $\dfrac{1}{\alpha} \ \& \ \dfrac{1}{\beta}$ (iii) $\alpha^2\beta$ and $\beta^2\alpha$ (iv) $\dfrac{\alpha}{\beta} \ \& \ \dfrac{\beta}{\alpha}$

Sol. (i)   The equation having roots α^2 and β^2 is $x^2 - (\alpha^2 + \beta^2)x + \alpha^2\beta^2 = 0$

Here $\alpha + \beta = 2$ and $\alpha\beta = 3$. So $\alpha^2 + \beta^2 = (\alpha + \beta)^2 - 2\alpha\beta = 4 - 6 = -2$.
$\alpha^2\beta^2 = 3^2 = 9$. So required equation is $x^2 + 2x + 9 = 0$

(ii) The equation having roots $1/\alpha$ and $1/\beta$ is $x^2 - (1/\alpha + 1/\beta)x + 1/\alpha\beta = 0$

$\dfrac{1}{\alpha} + \dfrac{1}{\beta} = \dfrac{\alpha + \beta}{\alpha\beta} = \dfrac{2}{3}$ and $\dfrac{1}{\alpha\beta} = \dfrac{1}{3}$ so, required equation is $x^2 - \dfrac{2}{3}x + \dfrac{1}{3} = 0 \rightarrow 3x^2 - 2x + 1 = 0$

(iii) The equation having roots $\alpha^2\beta$ and $\beta^2\alpha$ is $x^2 - \alpha\beta(\alpha + \beta)x + (\alpha\beta)^3 = 0$
$\rightarrow$ Equation is $x^2 - 6x + 27 = 0$

(iv) The quadratic equation having roots $\dfrac{\alpha}{\beta}$ & $\dfrac{\beta}{\alpha}$ is $x^2 - \left(\dfrac{\alpha^2 + \beta^2}{\alpha\beta}\right)x + 1 = 0$.

$\rightarrow$ Equation is $x^2 + \dfrac{2}{3}x + 1 = 0 \rightarrow 3x^2 + 2x + 3 = 0$.

Ex12. α is a root of $4x^2 + 2x - 1 = 0$. Prove that $4\alpha^3 - 3\alpha$ is the other root.

Sol. Let β be the other root $\rightarrow \alpha + \beta = -1/2 \rightarrow \beta = -\dfrac{1}{2} - \alpha$

Since α is a root of $4x^2 + 2x - 1 = 0 \rightarrow 4\alpha^2 + 2\alpha - 1 = 0$ ......(1)

Now multiply by $\alpha \ (\neq 0)$ both side $\rightarrow 4\alpha^3 + 2\alpha^2 - \alpha = 0$

as $4\alpha^2 + 2\alpha - 1 = 0 \rightarrow \alpha^2 = \left(\dfrac{1 - 2\alpha}{4}\right)$ so $4\alpha^3 + 2\alpha^2 - \alpha = 0 \rightarrow 4\alpha^3 + 2\left(\dfrac{1 - 2\alpha}{4}\right) - \alpha = 0$

$\rightarrow 4\alpha^3 + \dfrac{1}{2} - 2\alpha = 0 \rightarrow 4\alpha^3 - 3\alpha = -\dfrac{1}{2} - \alpha = \beta$

So, other root is $4\alpha^3 - 3\alpha$.

Ex13. If $\alpha \neq \beta$ & $\alpha^2 = 5\alpha - 3$ & $\beta^2 = 5\beta - 3$. Find the equation having roots $\dfrac{\alpha}{\beta}$ & $\dfrac{\beta}{\alpha}$.

Sol. α and β both are satisfying the equation $x^2 = 5x - 3$

$\rightarrow \alpha$ and β are roots of $x^2 = 5x - 3 \rightarrow x^2 - 5x + 3 = 0$

So, $\alpha + \beta = 5$ and $\alpha\beta = 3$.

Hence desire quadratic equation is $x^2 - \left(\dfrac{\alpha^2 + \beta^2}{\alpha\beta}\right)x + 1 = 0$

Where $\alpha^2 + \beta^2 = 25 - 6 = 19$. So equation is $3x^2 - 19x + 3 = 0$.

Ex14. If the roots of the equation $(x - a)(x - b) - k = 0$ are c and d then prove that roots of the equation $(x - c)(x - d) + k = 0$ are a and b.

Sol. If c and d are roots of $(x - a)(x - b) - k = 0 \rightarrow (x - a)(x - b) - k = (x - c)(x - d)$

$\rightarrow (x - a)(x - b) = (x - c)(x - d) + k$. Hence $(x - c)(x - d) + k = 0 \rightarrow (x - a)(x - b) = 0$

So, roots are a and b.

Ex15. Consider $4x^2 - 4(\alpha - 2)x + \alpha - 2 = 0$, $\alpha \in$ R. Find the set of values of α for which

    (i)   Both roots are real and distinct   (ii) both roots are equal

    (iii) Both roots are imaginary   (iv) both roots are opposite in sign

    (v)   Both roots are non zero, equal in magnitude but opposite in sign

Sol.

(i)   Condition for both roots real and distinct is D > 0

      $\rightarrow 16(\alpha - 2)^2 - 4 \times 4 \times (\alpha - 2) > 0 \rightarrow 16(\alpha - 2)(\alpha - 3) > 0 \rightarrow \alpha \in (-\infty, 2) \cup (3, \infty)$.

(ii)   Condition for both roots equal is D = 0

      $\rightarrow D = 16(\alpha - 2)^2 - 4 \times 4 \times (\alpha - 2) = 16(\alpha - 2)(\alpha - 3) = 0 \rightarrow \alpha = 2, 3$

(iii)   Condition for both roots imaginary is D < 0

      $\rightarrow D = 16(\alpha - 2)^2 - 4 \times 4 \times (\alpha - 2) < 0 \rightarrow 16(\alpha - 2)(\alpha - 3) < 0 \rightarrow \alpha \in (2, 3)$

(iv)   Both roots are opposite in sign if product of roots < 0

      (D = $b^2 - 4ac > 0$ is an inbuilt situation, as a and c are of opposite sign)

      Product of roots = $(\alpha - 2)/4 < 0 \rightarrow \alpha < 2 \rightarrow \alpha \in (-\infty, 2)$.

(v)   For both roots equal in magnitude but opposite in sign

      Sum of roots must be zero $\rightarrow \alpha - 2 = 0 \rightarrow \alpha = 2$

      But for $\alpha = 2$, the equation $4x^2 - 4(\alpha - 2)x + \alpha - 2 = 0$ become $4x^2 = 0 \rightarrow x = 0, 0$

      Hence, $\alpha \in \phi$

Ex16. The graph of $y = ax^2 + bx + c$ is as shown in the figure. α and β are roots of $ax^2 + bx + c = 0$. Select the correct statements

    (i)   a < 0

    (ii)   D > 0

    (iii) $S = \alpha + \beta > 0$ (S : Sum of roots)

    (iv) $P = \alpha\beta < 0$ (P: Product of roots)

    (v) $-\dfrac{b}{a} > 0$ where (b > 0)

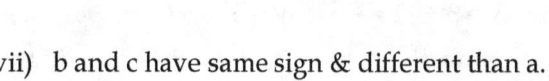

    (vi) $\dfrac{c}{a} < 0$ where (c > 0)   (vii) b and c have same sign & different than a.

Note: do not assume a < 0

Sol.   As graph of $y = ax^2 + bx + c$ is intersecting x – axis at two distinct points so roots are real and distinct $\rightarrow$ D > 0 so option (ii) is correct.

      Let $\alpha < \beta \rightarrow \alpha < 0 < \beta$ so product of roots P < 0 so option (iv) is correct.

      As per question figure $\alpha < 0 < |\alpha| < \beta$ (assume $\alpha < \beta$) so sum of roots S > 0 so option (iii) is correct.

      The graph intersect y – axis at (0, c), above x - axis $\rightarrow$ c > 0

      So, product of roots $P = \dfrac{c}{a} < 0$ where (c > 0) is correct and imply a < 0 so option (i) & (vi) are correct.

      Now sum of roots $S = -\dfrac{b}{a} > 0 \rightarrow b > 0$ as a < 0, So option (v) is correct.

Now, we have a < 0, b > 0 and c > 0 → option (vii) is correct.

Ex17. If $(3a - b)x^2 + (a - 2)x + 4 = 0$ has both roots at infinity then find a and b.

Sol. $(3a - b)x^2 + (a - 2)x + 4 = 0$ has both roots at infinity

$$\to (3a-b)\frac{1}{x^2}+(a-2)\frac{1}{x}+4=0 \to 4x^2+(a-2)x+(3a-b)=0 \text{ has both roots at origin.}$$

$\to a - 2 = 0 \to a = 2$ and $3a - b = 0 \to b = 3a = 6$ so a = 2, b = 6.

Ex18. Find the conditions on a, b and c for which $ax^2 + bx + c = 0$ has

   (i)   At least one root is zero   (ii)   exactly one root is zero

   (iii) Exactly one root at infinity   (iv) both roots at infinity

Sol.

(i)   If c = 0 → equation become $ax^2 + bx = 0 \to x(ax + b) = 0$ which have atleast on root zero.

(ii)   for exactly one root zero, $c = 0$ $a \neq 0$, $b \neq 0$

(iii)   for exactly one root to be at infinity → $cx^2 + bx + a = 0$ has exactly one root at origin

      $\to a = 0, b \neq 0, c \neq 0$

(iv)   for both roots at infinity → $cx^2 + bx + a = 0$ has both roots at origin → $a = 0, b = 0, c \neq 0$.

Ex19. The equation $2ax^2 - 2x - a = 0$, $a \in N$ has rational roots, then 'a' may take values

   (A) 2   (B) 12   (C)   70   (D) 408

Sol. $D = 4 + 8a^2$ must be a perfect square

i.e. $1 + 2a^2$ must be a perfect square. All are correct options.

Ex20. If $x = 3 + \sqrt{5}$ then find the value of $x^4 - 12x^3 + 44x^2 - 48x + 17$

Sol. You may be thinking of direct substitution but it may be very lengthy calculation. So we take different approach.

$x = 3 + \sqrt{5} \to (x - 3)^2 = 5 \to x^2 - 6x + 4 = 0$.

Now divide $x^4 - 12x^3 + 44x^2 - 48x + 17$ by $x^2 - 6x + 4$

$$\begin{array}{r}
x^2-6x+4 \\
x^2-6x+4\overline{)x^4-12x^3+44x^2-48x+17} \\
\underline{x^4-6x^3+4x^2} \\
-6x^3+40x^2-48x \\
\underline{-6x^3+36x^2-24x} \\
4x^2-24x+17 \\
\underline{4x^2-24x+16} \\
1
\end{array}$$

$\to x^4 - 12x^3 + 44x^2 - 48x + 17 = (x^2 - 6x + 4)(x^2 - 6x + 4) + 1 \forall x \in R$ ....(1)

For $x = 3 + \sqrt{5}$, $x^2 - 6x + 4 = 0$

So, if we put x = 3 + √5 in (1) we get

$$\left(3+\sqrt{5}\right)^4 - 12\left(3+\sqrt{5}\right)^3 + 44\left(3+\sqrt{5}\right)^2 - 48\left(3+\sqrt{5}\right) + 17 = 0 \times 0 + 1$$

So, value of $x^4 - 12x^3 + 44x^2 - 48x + 17$ at $x = 3 + \sqrt{5}$ is 1.

WRITE YOUR NOTES HERE:

EXERCISE - 2

1. α, β are the roots of $x^2 - 6x + m = 0$. If $2\alpha + \beta = 8$ then value of m is _____

2. Find a, if $ax^2 - 4x + 9 = 0$ has integral roots.

3. If the sum of roots of the equation $(a + 1)x^2 + (2a + 3)x + (3a + 4) = 0$ is -1 then product of roots is _____

4. If the difference of the roots of $ax^2 + 2hx + b = 0$ is 1, prove that $4(h^2 - ab) = a^2$.

5. The students while solving a quadratic in x, one copied the constant term incorrectly and got the roots 3 and 2. The other copied constant term and coefficient of x^2 as -6 and 1 respectively. The correct roots are _____

6. If $a + b + c = 0$ and $a, b, c \in Q$. Prove that roots of equation

   $(b + c - a)x^2 + (c + a - b) x + (a + b - c) = 0$ are rational

7. If roots of $\dfrac{x^2 - bx}{ax - c} = \dfrac{k-1}{k+1}$ are equal and opposite then value of k is _____

8. If α, β are roots of $ax^2 + bx + c = 0$, $a \neq 0$ then find the quadratic having roots

   (i) α^2, β^2 (ii) $\dfrac{\alpha}{\beta}, \dfrac{\beta}{\alpha}$

9. If a quadratic equation in (x or y) is formed from $y^2 = 4ax$ and $y = mx + c$. If the resultant quadratic have equal roots then prove that $c = a/m$.

10. The values of k for which one root of the equation $(k^2 - 5k + 3)x^2 + (3k - 1)x + 2 = 0$ is twice as large as other, is/are _____

11. If α, β are roots of $x^2 - 2x + 4 = 0$. The value of $\alpha^5 + \beta^5$ is _____

12. If $p, q \in \{1, 2, 3, 4\}$, the number of equations of the form $px^2 + qx + 1 = 0$ having real roots is/are

13. If $x^2 - x - 1 = 0$, then value of $x^3 - 2x + 1$ is _____

14. Find all integers n such that $(n^2 - n - 1)^{n+2} = 1$

15. If $f(x) = ax^7 + bx^5 + cx^3 - 6$, and $f(-9) = 3$, find $f(9)$.

16. Find the remainder when $x^{81} + x^{49} + x^{25} + x^9 + x$ is divided by $x^3 - x$.

17. The quadratic polynomials $p(x) = a (x - 3)^2 + bx + 1$ and $q(x) = 2x^2 + c(x - 2) + 13$ are equal for all values of x. Find the values of a, b and c.

18. Find the value of k for which the graph of $y = x^2 + (2x + 3) k + 4(x + 2) + 3k - 5$ intersects the axis of x at two distinct points.

SYMMETRIC EXPRESSION

Every expression $f(x, y)$ is said to be symmetric if $f(x, y) = f(y, x)$. In general, A polynomial $P(x_1, x_2, ..., x_n)$ is called symmetric if it is left unchanged by any permutation of the variables $x_1, x_2, ..., x_n$.

Ex.   Let $f(x, y) = x^2 y + xy^2$ then you can see $f(x, y) = f(y, x)$.

Note: Every symmetric expression of roots of $ax^2 + bx + c = 0$ can be express in terms of sum of roots and product of roots.

Ex.   Let $f(\alpha, \beta) = \alpha\beta^3 + \beta\alpha^3$ is a symmetric expression of roots.

$\rightarrow f(\alpha, \beta) = \alpha\beta(\beta^2 + \alpha^2) = \alpha\beta [(\alpha + \beta)^2 - 2\alpha\beta]$ which is written in the form of $\alpha + \beta$, and $\alpha\beta$.

Some examples of symmetric expressions are:

(i) $\alpha^2 + \beta^2$ (ii) $\alpha^2 + \alpha\beta + \beta^2$ (iii) $\dfrac{1}{\alpha} + \dfrac{1}{\beta}$ (iv) $\dfrac{\alpha}{\beta} + \dfrac{\beta}{\alpha}$

(v) $\alpha\beta^2 + \beta\alpha^2$ (vi) $\left(\dfrac{\alpha}{\beta}\right)^2 + \left(\dfrac{\beta}{\alpha}\right)^2$ (vii) $\alpha^3 + \beta^3$ (viii) $\alpha^7 + \beta^7$

EQUATION TRANSFORMATIONS

Ex1.   If α and β are roots of equation $x^2 - 3x + 2 = 0$ then find the equation having roots

(i) $\dfrac{1}{\alpha}$ & $\dfrac{1}{\beta}$ (ii) $-\alpha, -\beta$ (iii) $k\alpha, k\beta$ (iv) α^n, β^n

(v) $k + \alpha, k + \beta$ (vi) $\alpha^{1/n}$ & $\beta^{1/n}$

Sol.   You can observe that roots are 1, 2 so let $\alpha = 2$, $\beta = 1$.

Sum of roots $\alpha + \beta = 3$, product of roots $\alpha\beta = 2$

(i)   The quadratic equation having roots $1/\alpha, 1/\beta$ is $x^2 - \left(\dfrac{1}{\alpha} + \dfrac{1}{\beta}\right)x + \dfrac{1}{\alpha\beta} = 0 \rightarrow x^2 - \left(\dfrac{\alpha + \beta}{\alpha\beta}\right)x + \dfrac{1}{\alpha\beta} = 0$

$\rightarrow x^2 - (3/2)x + (1/2) = 0 \rightarrow 2x^2 - 3x + 1 = 0$.

Note: *you can obtain the same equation by replacing* $x \rightarrow \dfrac{1}{x}$.

The equation: $x^2 - 3x + 2 = 0$ becomes $\left(\dfrac{1}{x}\right)^2 - 3\left(\dfrac{1}{x}\right) + 2 = 0 \rightarrow 2x^2 - 3x + 1 = 0$.

(ii)   *Replace: x by – x:*

The required equation is: $(-x)^2 - 3(-x) + 2 = 0 \rightarrow x^2 + 3x + 2 = 0$ having roots – 1 and – 2.

(iii)   *Replace $x \rightarrow x/k$*

The required equation is: $\left(\dfrac{x}{k}\right)^2 - 3\left(\dfrac{x}{k}\right) + 2 = 0 \rightarrow x^2 - 3kx + 2k^2 = 0$

(iv)   *Replace $x \rightarrow x^{1/n}$*

The required equation is: $\left(x^{1/n}\right)^2 - 3\left(x^{1/n}\right) + 2 = 0 \rightarrow x^{2/n} - 3x^{1/n} + 2 = 0$

(v)   *Replace $x \to x - k$*

The required equation is: $(x - k)^2 - 3(x - k) + 2 = 0 \to x^2 - (2k + 3)x + k^2 + 3k + 2 = 0$.

(vi)   *Replace $x \to x^n$*

The required equation is: $(x^n)^2 - 3(x^n) + 2 = 0 \to x^{2n} - 3x^n + 2 = 0$.

Ex2.   If α, β are roots of $x^2 - 2x + 5 = 0$ then find quadratic equation having roots $\alpha^3 + \alpha^2 - \alpha + 20$ and $\beta^3 + 4\beta^2 - 7\beta + 30$

Sol. α, β are roots of $x^2 - 2x + 5 = 0 \to \alpha^2 - 2\alpha + 5 = 0$ and $\beta^2 - 2\beta + 5 = 0$

$\alpha^3 + \alpha^2 - \alpha + 20 = (\alpha + 3)(\alpha^2 - 2\alpha + 5) + 5 = 5$

$\beta^3 + 4\beta^2 - 7\beta + 30 = (\beta + 6)(\beta^2 - 2\beta + 5) = 0$

So, the equation having roots 5 and 0 is $x^2 - 5x = 0$.

IDENTITIES Vs EQUATION

Let's have some examples of identities.

(i) $(a + x)^2 = a^2 + 2ax + x^2 \; \forall \; a, x \in R$ (ii) $\sin^2 x + \cos^2 x = 1, \forall \; x \in R$

(iii) $e^{i\pi} = -1$ (iv) $2 + 3 = 6 - 1$

(v) $1 + \tan^2 x = \sec^2 x \; \forall \; x \in R - \left\{(2n+1)\dfrac{\pi}{2}\right\}, n \in I$ (vi) $\sin^{-1} x + \cos^{-1} x = \pi/2 \; \forall \; x \in [-1, 1]$

Here are some examples of equations

(i) $3x - 2 = 0$ (ii) $x^2 - 3x + 2 = 0$

(iii) $\sin(x) = 1$ (iv) $e^x = x + \sin x$

What is the difference between identities and equation?

You can see. The identity is an equation which is true for all it's domain whereas equation is becoming true for few numbers in its domain.

Note: If a polynomial equation of n degree has more than 'n' roots then it becomes an identity.

❖   If $ax^2 + bx + c = 0$ has more than 2 roots then it becomes an identity.

Proof: Let $ax^2 + bx + c = 0$ has three roots α, β and γ $(\alpha \neq \beta \neq \gamma)$ then

$a\alpha^2 + b\alpha + c = 0$ ....(1)

$a\beta^2 + b\beta + c = 0$ ....(2)

$a\gamma^2 + b\gamma + c = 0$ ....(3)

from (1) and (2) we get: $a(\alpha^2 - \beta^2) + b(\alpha - \beta) = 0$

$\to (\alpha - \beta)[a(\alpha + \beta) + b] = 0$.

$\therefore \alpha \neq \beta$ so, $a(\alpha + \beta) + b = 0$ ....(4)

Similarly, $a(\alpha + \gamma) + b = 0$ .... (5) and $a(\beta + \gamma) + b = 0$ ....(6)

By subtracting (5) from (4) we get $a(\beta - \gamma) = 0. \; \therefore \beta \neq \gamma \to a = 0$

From (4) If $a = 0 \to b = 0$. From (1) we get $a = b = c = 0$.

Hence $ax^2 + bx + c = 0$ become $0x^2 + 0x + 0 = 0$ is an identity and true $\forall \; x \in R$.

Ex3. If $(a^2 - 1)x^2 + (a - 1)x + a^2 - 4a + 3 = 0$ is an identity in x, then find the value of a.

Sol. For quadratic $ax^2 + bx + c = 0$ to be an identity. Necessary condition is $a = 0$, $b = 0$ and $c = 0$.

So, $a^2 - 1 = 0 \rightarrow a = \pm 1$ ....(1)

$a - 1 = 0 \rightarrow a = 1$ ....(2)

$a^2 - 4a + 3 = 0 \rightarrow a = 1, 3$ ....(3)

from (1), (2) and (3) $a = 1$ only.

Ex4. Find the value of 'k' for which equation $(k^2 - 3k + 2)x^2 - (k^2 - 5k + 4)x + k - k^2 = 0$ posses more than two roots.

Sol. $k^2 - 3k + 2 = 0 \rightarrow k = 1, 2$ ....(1)

$k^2 - 5k + 4 = 0 \rightarrow k = 1, 4$ ....(2)

$k - k^2 = 0 \rightarrow k = 0, 1$ ....(3)

from (1), (2) and (3) $k = 1$. This makes, every coefficient zero, simultaneously

Ex5. Show that $\dfrac{(x+b)(x+c)}{(b-a)(c-a)} + \dfrac{(x+c)(x+a)}{(c-b)(a-b)} + \dfrac{(x+a)(x+b)}{(a-c)(b-c)} = 1$ is an identity.

Sol. If we put $x = - a$ then above equation become $1 + 0 + 0 = 1$ (true)

Similarly, $x = - b$ and $- c$ are roots. Hence its an identity.

EQUATIONS AND ALGEBRAIC MANIPULATIONS

Here we are going to discuss about the ways of approach and methods to solve equations. At initial stage we perform lot of mistakes, so here are some ways to reduce the mistakes.

❖ **CANCELLATION MISTAKE**

Let's have some basic examples of cancellation

Ex1. Solve the following equations

(i) $x^2 - 3x + 2 = 3(x - 1)$ (ii) $x^2 - 2x = 5x$

Sol. (i) $x^2 - 3x + 2 = 3(x - 1) \rightarrow (x - 1)(x - 2) = 3 (x - 1)$

$\rightarrow \cancel{(x-1)}(x-2) = 3\cancel{(x-1)} \rightarrow (x-2) = 3 \rightarrow x = 5.$

Here you can see, you are getting one root only for given quadratic.

Can you recognise your mistake?

Such type of cancellation leads to *loss of genuine roots.*

CORRECT WAY:

$x^2 - 3x + 2 = 3(x - 1) \rightarrow (x - 1)(x - 2) = 3 (x - 1)$

$\rightarrow (x - 1)(x - 2) - 3 (x - 1) = 0 \rightarrow (x - 1)[(x - 2) - 3] = 0 \rightarrow (x - 1)(x - 5) = 0$

So, $x = 1, 5$ are genuine solutions.

Rule: *Avoid cancellation as far as possible.*

(ii) **Wrong way:** $x^2 - 2x = 5x \rightarrow x^2 = 7x \rightarrow x \cdot \cancel{x} = 7\cancel{x} \rightarrow x = 7$

Here, you will lose a genuine solution x = 0 due to cancellation.

Correct Way:

x² - 2x = 5x → x(x - 2) = 5x → x(x - 2) – 5x = 0 → x (x - 2 - 5) = 0 → x(x - 7) = 0 → x = 0, 7

Ex2. Select the correct cancellation(s) with reason

(A) $\cancel{6}(x-1)=\cancel{6}(x^2-6x+5) \to$ (x - 1) = x² – 6x + 5 as 6 is a constant and 6 ≠ 0.

(B) $\cancel{(x^2+x+1)}(x-2)=\cancel{(x^2+x+1)} \to (x-2)=1$ as x² + x + 1 ≠ 0 ∀ x ∈ R

(C) $0=0 \to 3\times\cancel{0}=2\times\cancel{0} \to 3=2$

(D) $\dfrac{x}{\cancel{x-1}}=\dfrac{1}{\cancel{x-1}} \to x=1$

Sol. (A) $\cancel{6}(x-1)=\cancel{6}(x^2-6x+5) \to$ (x - 1) = x² – 6x + 5 is a correct cancellation

In this case we will not lose any root.

(B) $\cancel{(x^2+x+1)}(x-2)=\cancel{(x^2+x+1)} \to (x-2)=1$ as x² + x + 1 ≠ 0 ∀ x ∈ R

The above cancellation will be good if we were looking for real roots only but it will also cause loss of genuine imaginary roots

As (x² + x + 1)(x - 2) = (x² + x + 1) → (x² + x + 1)(x - 2 - 1) = 0

→ (x² + x + 1)(x - 3) = 0 → $x=3, \dfrac{-1+i\sqrt3}{2}, \dfrac{-1-i\sqrt3}{2}$. We were losing roots $x = \dfrac{-1+i\sqrt3}{2}, \dfrac{-1-i\sqrt3}{2}$.

(C) In an equation, if we cancel the term becoming zero then its leads us to absurd results.

(D) $\dfrac{x}{\cancel{x-1}}=\dfrac{1}{\cancel{x-1}} \to x=1$

In this cancellation we are having domain related problem.

Here we are getting x = 1 as root but in actual the equation is not define for x = 1.

Hence, x = ϕ (No, solutions)

Hence, only option (A) is correct.

Ex3. The equation $x-\dfrac{2}{x-1}=1-\dfrac{2}{x-1}$ has

(A) no root   (B) one root   (C) two equal roots   (D) infinitely many root

Sol. $x-\dfrac{2}{x-1}=1-\dfrac{2}{x-1}$ as x ≠ 1 (since, equation is not define for x = 1)

→ $x-\dfrac{\cancel{2}}{\cancel{x-1}}=1-\dfrac{\cancel{2}}{\cancel{x-1}} \to$ x = 1 but x ≠ 1. So, no root of the equation

❖ **SQUARING MISTAKE**

Let's has an equation x = 1. Its root is x = 1 (only one root)

Now if we square both side then we will get the equation like x² – 1 = 0

$\rightarrow (x - 1)(x + 1) = 0 \rightarrow x = -1, 1$

Here you can observe that we are increasing the roots. Here $x = -1$ is an extra solution arises due to squaring both side. So, we need to avoid squaring.

What we need to do if squaring is necessary?

You just solve the equation after squaring both sides. It retain the genuine solution but ad some extraneous solution, so at the end we need to check for extraneous solutions by satisfying the roots in original equation. Let's have some cases for crystal clear picture.

Ex1. The equation $\sqrt{x+1} - \sqrt{x-1} = \sqrt{4x-1}$ has

(A) No solution

(B) One solution

(C) two solution

(D) more than two solution

Sol. $\sqrt{x+1} - \sqrt{x-1} = \sqrt{4x-1}$

On squaring both side we get

$x + 1 + x - 1 - 2\sqrt{x+1}\sqrt{x-1} = 4x - 1 \rightarrow 2x - 2\sqrt{x^2 - 1} = 4x - 1$

$\rightarrow 1 - 2x = 2\sqrt{x^2 - 1}$. (Now, again square both side)

$\rightarrow 1 + 4x^2 - 4x = 4 (x^2 - 1) \rightarrow 1 - 4x = -4 \rightarrow x = \dfrac{5}{4}$.

At the moment we need to be alert that we have done squaring so we may get extraneous roots.

Cross Check:

Let check the arriving root $x = \dfrac{5}{4}$ by satisfying it in original equation $\sqrt{x+1} - \sqrt{x-1} = \sqrt{4x-1}$.

$\sqrt{\dfrac{5}{4}+1} - \sqrt{\dfrac{5}{4}-1} = \sqrt{4\left(\dfrac{5}{4}\right)-1} \rightarrow \sqrt{\dfrac{9}{4}} - \sqrt{\dfrac{1}{4}} = \sqrt{4} \rightarrow \dfrac{3}{2} - \dfrac{1}{2} = 2 \rightarrow 1 = 2$ (which is not possible)

So $x = 5/4$ is not a solution. Hence option (A) is correct.

Ex2. Find the values of x, if $x = \sqrt{30 + \sqrt{30 + \sqrt{30 + \sqrt{30 + ...\infty}}}}$

Sol. You can see $x = \sqrt{30 + \underbrace{\left(\sqrt{30 + \sqrt{30 + \sqrt{30 + ...\infty}}}\right)}_{x}}$

Observe $\underbrace{\sqrt{30 + \sqrt{30 + \sqrt{30 + ...\infty}}}}_{x}$

The equation becomes $x = \sqrt{30 + x}$.

Now on squaring both sides we get: $x^2 = 30 + x$

$\rightarrow x^2 - x - 30 = 0 \rightarrow (x - 6)(x + 5) = 0 \rightarrow x = 6$ or -5

Still here we did squaring we need to check our solution

As $\sqrt{R} \geq 0$ so $x \geq 0$ hence $x \neq -5$ so $x = 6$ is the only answer.

Ex3. Solve the equation: $2x - 3 = \sqrt{2x-1}$

Sol. On squaring both side we get

$4x^2 + 9 - 12x = 2x - 1$

$\rightarrow 4x^2 - 14x + 10 = 0 \rightarrow 2x^2 - 7x + 5 = 0 \rightarrow 2x^2 - 5x - 2x + 5 = 0$

$\rightarrow x(2x - 5) - 1(2x - 5) = 0 \rightarrow (x - 1)(2x - 5) = 0 \rightarrow x = 1$ or $\dfrac{5}{2}$.

Here you can see $x = 1$ is not satisfying the original equation so $x = \dfrac{5}{2}$ is the only root.

Ex4. In the real number system, the equation $\sqrt{x+3-4\sqrt{x-1}} + \sqrt{x+8-6\sqrt{x-1}} = 1$ has

(A) No solution   (B) Exactly two distinct solutions

(C) Exactly four distinct solutions   (D) Infinitely many solutions

Sol. In such a case, take one root other side and then square both side.

$\rightarrow \sqrt{x+3-4\sqrt{x-1}} = 1 - \sqrt{x+8-6\sqrt{x-1}}$

On squaring both side we get: $x+3-4\sqrt{x-1} = 1 + x + 8 - 6\sqrt{x-1} - 2\sqrt{x+8-6\sqrt{x-1}}$

$\rightarrow 2\sqrt{x+8-6\sqrt{x-1}} = 6 - 2\sqrt{x-1} \rightarrow \sqrt{x+8-6\sqrt{x-1}} = 3 - \sqrt{x-1}$

Now, square both side we get: $x+8-6\sqrt{x-1} = 9 + x - 1 - 6\sqrt{x-1}$

$\rightarrow x+8-6\sqrt{x-1} = x+8-6\sqrt{x-1}$ This is true always. Hence option (D) is correct.

Ex5. What is the product of the real roots of the equation: $x^2 + 18x + 30 = 2\sqrt{x^2+18x+45}$?

Sol. Let $x^2 + 18x + 30 = t$ then $x^2 + 18x + 30 = 2\sqrt{x^2+18x+45} \rightarrow t = 2\sqrt{(t+15)}$.

Now on squaring both side we get: $t^2 = 4t + 60 \rightarrow t^2 - 4t - 60 = 0 \rightarrow (t-10)(t+6) = 0 \rightarrow t = -6, 10$

$t = -6 \rightarrow x^2 + 18x + 30 = -6$ but $x^2 + 18x + 30 \geq 0$ (observe in equation $\sqrt{R} \geq 0$)

so, $x^2 + 18x + 30 \neq -6$. Hence $x^2 + 18x + 30 = 10 \rightarrow x^2 + 18x + 20 = 0$

so, product of real roots = 20.

METHOD – II

Let $\sqrt{x^2+18x+45} = t \geq 0 \rightarrow$ the given equation will be: $x^2 + 18x + 45 - 2\sqrt{x^2+18x+45} - 15 = 0$

$\rightarrow t^2 - 2t - 15 = 0 \rightarrow t = 5, -3$ but $t \neq -3$ so $t = 5$

$\rightarrow \sqrt{x^2+18x+45} = 5 \rightarrow x^2 + 18x + 45 = 25 \rightarrow x^2 + 18x + 20 = 0$

So, product of real roots = 20.

❖ **DOMAIN RELATED MISTAKE**

Ex1. Solve the equation $\sin x = 0$ where $x \in [0, 2\pi)$

Sol. Obviously we know that $\sin x = 0 \rightarrow x = 0, \pi$ are the answers.

But we also know the formula: $\sin(x) = \dfrac{2\tan\left(\dfrac{x}{2}\right)}{1+\tan^2\left(\dfrac{x}{2}\right)} = 0 \to \tan\left(\dfrac{x}{2}\right) = 0$

As $x \in [0, 2\pi)$ then $\dfrac{x}{2} \in [0, \pi)$ hence $\tan\left(\dfrac{x}{2}\right) = 0 \to \dfrac{x}{2} = 0 \to x = 0$

You can see the conversion of equation lost the solution $x = \pi$.

So, avoid the formula or be careful while using the formula

$$\sin(x) = \frac{2\tan\left(\dfrac{x}{2}\right)}{1+\tan^2\left(\dfrac{x}{2}\right)} \quad \text{and} \quad \cos(x) = \frac{1-\tan^2\left(\dfrac{x}{2}\right)}{1+\tan^2\left(\dfrac{x}{2}\right)}.$$

Ex2. Solve the equation: $\tan x - \sec x = 0 \ \forall \ x \in [0, 2\pi)$

Sol. $\tan x - \sec x = 0 \to \dfrac{\sin x}{\cos x} - \dfrac{1}{\cos x} = 0 \to \dfrac{\sin x - 1}{\cos x} = 0$

$\text{Sin}x = 1 \to x = \pi/2$ but at this point $\cos x$ is becoming zero so $x \in \emptyset$

Here you can see that splitting $\tan x$ in $\sin x$ and $\cos x$ may create problem. So careful, while applying the same

Let's generate problem solving ability related to different equations

Ex1. Solve the following equations: (*First solve your own*)

(i) $5x^2 + \sqrt{5}\,x - 1 = 0$

(ii) $ab\,x^2 + (a^2 - b^2)x - ab = 0$

(iii) $\dfrac{a}{ax-1} + \dfrac{b}{bx-1} = a + b, \ x \neq \dfrac{1}{a}, \dfrac{1}{b}$

(iv) $x^2 - 2ax + a^2 - b^2 - c^2 + 2bc = 0$

(v) $5^{x+1} + 5^{2-x} = 5^3 + 1$

(vi) $4^x - 3^{x-\frac{1}{2}} = 3^{x+\frac{1}{2}} - 2^{2x-1}$

(vii) $\sqrt{\dfrac{x}{1-x}} + \sqrt{\dfrac{1-x}{x}} = \dfrac{13}{6}$

(viii) $\sqrt{x^2 - 7x + 17} + \sqrt{x^2 - 7x + 8} = 9$

(ix) $\left(2+\sqrt{3}\right)^{x^2-3x+3} + \left(2-\sqrt{3}\right)^{x^2-3x+3} = 4$

(x) $\left(x^2+2\right)^2 + 8x^2 = 6x\left(x^2+2\right)$

(xi) $x^2 + a\,|x+a| - 3a^2 = 0, \ a > 0$

(xii) $\sqrt{x+1} - \sqrt{x-1} = 1$

Sol. (i) $5x^2 + \sqrt{5}\,x - 1 = 0$

$D = b^2 - 4ac = 5 + 20 = 25$ So as per formula $x = \dfrac{-b \pm \sqrt{b^2 - 4ac}}{2a}$, $x = \dfrac{-\sqrt{5} \pm 5}{10}$

(ii) $ab\,x^2 + (a^2 - b^2)x - ab = 0 \to ab\,x^2 + a^2x - b^2x - ab = 0$

$$\to ax(bx + a) - b(bx + a) = 0 \to (ax - b)(bx + a) = 0 \to x = \frac{b}{a}, -\frac{a}{b}$$

(iii) $\dfrac{a}{ax-1} + \dfrac{b}{bx-1} = a + b, \; x \neq \dfrac{1}{a}, \dfrac{1}{b}$

$$\to \frac{a}{ax-1} - b + \frac{b}{bx-1} - a = 0 \to \frac{a-abx+b}{ax-1} + \frac{b-abx+a}{bx-1} = 0 \to (a+b-abx)\left(\frac{1}{ax-1} + \frac{1}{bx-1}\right) = 0$$

$$\to (a+b-abx)\frac{(bx-1+ax-1)}{(ax-1)(bx-1)} = 0 \to x = \frac{a+b}{ab} \; \& \; \frac{2}{a+b}$$

(iv) $x^2 - 2ax + a^2 - b^2 - c^2 + 2bc = 0 \to (x - a)^2 - (b - c)^2 = 0$

$\to (x - a + b - c)(x - a - b + c) = 0 \to x = a + c - b, a + b - c$

(v) $5^{x+1} + 5^{2-x} = 5^3 + 1 \to 5 \cdot 5^x + \dfrac{25}{5^x} = 126 \;$ let $5^x = t$

$$\to 5t + \frac{25}{t} = 126 \to 5t^2 - 126\,t + 25 = 0 \to 5t^2 - 125\,t - t + 25 = 0$$

$\to 5t(t - 25) - 1(t - 25) = 0 \to (5t - 1)(t - 25) = 0 \to t = 25 \; \& \; 1/5$

So, if t = 5^x = 25 = $5^2 \to$ x = 2 & If t = 5^x = 1/5 $\to$ x = - 1, So, x = - 1, 2 are solutions.

METHOD – II

I know you may be thinking of comparing as shown

$5^{x+1} + 5^{2-x} = 5^3 + 5^0 \to$ x + 1 = 3 and 2- x = 0 $\to$ x = 2 in both case and

x + 1 = 0 and 2 – x = 3 $\to$ x = - 1 in another case.

Though the answers are correct but sometimes it create problem. It's a hit and trial way. Good for objective approach.

(vi) $4^x - 3^{x-\frac{1}{2}} = 3^{x+\frac{1}{2}} - 2^{2x-1} \to 4^x - \dfrac{3^x}{\sqrt{3}} = 3^x\sqrt{3} - \dfrac{2^{2x}}{2}$

$$\to 4^x + \frac{2^{2x}}{2} = 3^x\sqrt{3} + \frac{3^x}{\sqrt{3}} \to 4^x\left(1 + \frac{1}{2}\right) = 3^x\left(\sqrt{3} + \frac{1}{\sqrt{3}}\right) \to 4^x \times \frac{3}{2} = 3^x \times \frac{4}{\sqrt{3}}$$

$$\to \frac{2^{2x}}{2^3} = \frac{3^x}{3^{3/2}} \to 2^{2x-3} = \left(\sqrt{3}\right)^{2x-3} \to \left(\frac{2}{\sqrt{3}}\right)^{2x-3} = 1 \to 2x - 3 = 0 \to x = 3/2$$

(vii) $\sqrt{\dfrac{x}{1-x}} + \sqrt{\dfrac{1-x}{x}} = \dfrac{13}{6}$

Let $\sqrt{\dfrac{x}{1-x}} = t \to t + \dfrac{1}{t} = \dfrac{13}{6} \to 6t^2 - 13t + 6 = 0$

$6t^2 - 9t - 4t + 6 = 0 \to 3t(2t - 3) - 2(2t - 3) = 0 \to (3t - 2)\,(2t - 3) = 0 \to t = 2/3 \; or \; 3/2$

Case – I (t = 2/3): $\sqrt{\dfrac{x}{1-x}} = \dfrac{2}{3} \rightarrow \dfrac{x}{1-x} = \dfrac{4}{9} \rightarrow \dfrac{x}{(1-x)+x} = \dfrac{4}{9+4} \rightarrow x = 4/13$ (apply dividendo)

Case – II (t = 3/2): $\sqrt{\dfrac{x}{1-x}} = \dfrac{3}{2} \rightarrow \dfrac{x}{1-x} = \dfrac{9}{4} \rightarrow \dfrac{x}{(1-x)+x} = \dfrac{9}{4+9} \rightarrow x = 9/13$ (apply dividendo)

(viii) $\sqrt{x^2 - 7x + 17} + \sqrt{x^2 - 7x + 8} = 9$

Observe that $x^2 - 7x$ is repeating so, let $x^2 - 7x + 8 = t$

$\rightarrow \sqrt{t+9} + \sqrt{t} = 9 \rightarrow \sqrt{t+9} = 9 - \sqrt{t}$ ....(1)

On squaring equation (1) both side we get

$t + 9 = 81 + t - 18\sqrt{t} \rightarrow 18\sqrt{t} = 72 \rightarrow \sqrt{t} = 4 \rightarrow t = 16$

$\rightarrow x^2 - 7x + 8 = 16 \rightarrow x^2 - 7x - 8 = 0 \rightarrow x = -1, 8$

So, x = - 1 and 8 are the solutions.

(ix) $\left(2+\sqrt{3}\right)^{x^2-3x+3} + \left(2-\sqrt{3}\right)^{x^2-3x+3} = 4$

In this question, first you need to observe $\left(2+\sqrt{3}\right)\left(2-\sqrt{3}\right) = 1 \rightarrow 2-\sqrt{3} = \dfrac{1}{2+\sqrt{3}}$

The given equation becomes: $\left(2+\sqrt{3}\right)^{x^2-3x+3} + \dfrac{1}{\left(2+\sqrt{3}\right)^{x^2-3x+3}} = 4$

Let $\left(2+\sqrt{3}\right)^{x^2-3x+3} = t \rightarrow t + \dfrac{1}{t} = 4 \rightarrow t^2 - 4t + 1 = 0 \rightarrow t = 2+\sqrt{3}, 2-\sqrt{3} = \dfrac{1}{2+\sqrt{3}}$

Case – I $\left(2+\sqrt{3}\right)^{x^2-3x+3} = 2+\sqrt{3}$

$x^2 - 3x + 3 = 1 \rightarrow x^2 - 3x + 2 = 0 \rightarrow x = 1, 2$

Case – II $\left(2+\sqrt{3}\right)^{x^2-3x+3} = 2-\sqrt{3} = \left(2+\sqrt{3}\right)^{-1}$

$x^2 - 3x + 3 = -1 \rightarrow x^2 - 3x + 4 = 0 \rightarrow \dfrac{3\pm i\sqrt{7}}{2}$

so, x = 1, 2 and $\dfrac{3\pm i\sqrt{7}}{2}$ are the solutions for given equation.

(x) $\left(x^2+2\right)^2 + 8x^2 = 6x\left(x^2+2\right)$

Let $x^2 + 2 = t \rightarrow t^2 - 6xt + 8x^2 = 0 \rightarrow (t - 2x)(t - 4x) = 0$

$\rightarrow (x^2 + 2 - 2x)(x^2 + 2 - 4x) = 0 \rightarrow (x^2 - 2x + 2)(x^2 - 4x + 2) = 0$

$\rightarrow x = 2\pm\sqrt{2}, 1\pm i$

(xi) $x^2 + a|x + a| - 3a^2 = 0$, $a > 0$

Case – I $(x \geq -a)$

The equation becomes: $x^2 + a(x + a) - 3a^2 = 0 \rightarrow x^2 + ax - 2a^2 = 0 \rightarrow x = a, -2a$ but $x \neq -2a$ as $x \geq -a$

Case – II $(x < -a)$

The equation becomes: $x^2 - a(x + a) - 3a^2 = 0 \rightarrow x^2 - ax - 4a^2 = 0 \rightarrow x = \left(\dfrac{1 \pm \sqrt{17}}{2}\right)a$ but $x \neq \dfrac{1 + \sqrt{17}}{2}a$

So, $x = \left(\dfrac{1 - \sqrt{17}}{2}\right)a$ is the answer.

Hence, $x = a$ and $\left(\dfrac{1 + \sqrt{17}}{2}\right)a$ are the roots of given equation.

(xii) $\sqrt{x+1} - \sqrt{x-1} = 1 \rightarrow \sqrt{x+1} = 1 + \sqrt{x-1}$ ......(1)

On squaring (1) both side we get: $x + 1 = 1 + x - 1 + 2\sqrt{x-1} \rightarrow 1 = 2\sqrt{x-1}$

Now square again: $1 = 4(x - 1) \rightarrow x = 5/4$

But it's not satisfying the original equation hence $x \in \phi$

Ex2. Solve for integers x, y, z:

$x + y = 1 - z$, $x^3 + y^3 = 1 - z^2$.

Sol. Eliminating z from the given set of equations, we get $x^3 + y^3 + \{1 - (x + y)\}^2 = 1$

$\rightarrow (x + y)(x^2 - xy + y^2) + 1 + (x + y)^2 - 2(x + y) = 1$

$\rightarrow (x + y)(x^2 - xy + y^2 + x + y - 2) = 0$

Case I: x + y = 0

Then $z = 1$ and $(x, y, z) = (m, -m, 1)$, where m is an integer give one family of solutions.

Case II: x + y ≠ 0

Then $x^2 - xy + y^2 + x + y - 2 = 0$ (Must be)

This can be written in the form $(2x - y + 1)^2 + 3(y + 1)^2 = 12$

Here there are two possibilities:

$2x - y + 1 = 0$, $y + 1 = \pm 2$; $2x - y + 1 = \pm 3$, $y + 1 = \pm 1$.

Analysing all these cases we get

$(x, y, z) = (0, 1, 0), (-2, -3, 6), (1, 0, 0), (0, -2, 3), (-2, 0, 3), (-3, -2, 6)$

Ex3. Solve the system of equations: $\dfrac{2x_1^2}{1 + x_1^2} = x_2$, $\dfrac{2x_2^2}{1 + x_2^2} = x_3$ and $\dfrac{2x_3^2}{1 + x_3^2} = x_1$

Sol. Obviously $(x_1, x_2, x_3) = (0, 0, 0)$ is a solution to the given system.

From the given system, you can observe that equations are cyclic in nature so is their solution too.

Let $(x_1, x_2, x_3) \neq (0, 0, 0)$ then we can assume $x_1 = \dfrac{1}{y_1}$, $x_2 = \dfrac{1}{y_2}$ and $x_3 = \dfrac{1}{y_3}$

$\rightarrow \dfrac{2x_1{}^2}{1+x_1{}^2} = x_2 \rightarrow \dfrac{2}{y_1{}^2+1} = \dfrac{1}{y_2} \rightarrow 2y_2 = 1+y_1{}^2$ ......(1)

Similarly, other equation can be written as

$2y_3 = 1+y_2{}^2$ ....(2) and $2y_1 = 1+y_3{}^2$ ....(3)

Now, on adding equation (1), (2) and (3) we get

$(y_1 - 1)^2 + (y_2 - 1)^2 + (y_3 - 1)^2 = 0 \rightarrow y_1 = y_2 = y_3 = 1 = x_1 = x_2 = x_3$

So, $(x_1, x_2, x_3) = (0, 0, 0)$ and $(1, 1, 1)$ are the solution to given system of equations.

Ex4. Solve the system of equations: $x + y - z = 7$, $x^2 + y^2 - z^2 = 37$, $x^3 + y^3 - z^3 = 1$

Sol. $x + y = 7 + z$ .... (1)

$x^2 + y^2 - z^2 = 37 \rightarrow (x + y)^2 - 2xy = 37 + z^2$

$\rightarrow (7 + z)^2 - 2xy = 37 + z^2 \rightarrow 2xy = 49 + 14z + z^2 - 37 - z^2 \rightarrow 2xy = 12 + 14z$

$\rightarrow xy = 6 + 7z$ .... (2)

Again, $x^3 + y^3 = 1 + z^3 \rightarrow (x + y)(x^2 + y^2 - xy) = (1 + z)(1 - z + z^2)$ ..... (3)

Using (1) and (2) we get equation (3) as

$(7 + z)(37 + z^2 - 6 - 7z) = (1 + z)(1 - z + z^2)$

$\rightarrow (7 + z)(31 + z^2 - 7z) = (1 + z)(1 - z + z^2)$

$\rightarrow 217 + 7z^2 - 49z + 31z + z^3 - 7z^2 = 1 - z + z^2 + z - z^2 + z^3$

$\rightarrow 217 - 18z = 1 \rightarrow 18z = 216 \rightarrow z = 12$ ....(4)

From (1), (2) & (4) we get $x + y = 19$ and $xy = 90$ $\rightarrow x = 10, y = 9$ or $x = 9, y = 10$

So, the solution set $(x, y, z) = (10, 9, 12)$ & $(9, 10, 12)$.

Ex5. Let x, y, z be non – zero real numbers such that $\dfrac{x}{y}+\dfrac{y}{z}+\dfrac{z}{x}=7$ and $\dfrac{y}{x}+\dfrac{z}{y}+\dfrac{x}{z}=9$, then

$\dfrac{x^3}{y^3}+\dfrac{y^3}{z^3}+\dfrac{z^3}{x^3} - 3$ is equal to

(A) 152   (B) 153   (C) 154   (D) 155

Sol. Let $\dfrac{x}{y} = a, \dfrac{y}{z} = b$ & $\dfrac{z}{x} = c \rightarrow abc = 1$ ....(1)

Given: $a + b + c = 7$ ....(2) and $\dfrac{1}{a}+\dfrac{1}{b}+\dfrac{1}{c}=9$ .... (3)

From (1) and (3) we get $ab + bc + ca = 9abc = 9$ ......(4)

Now $a^3 + b^3 + c^3 - 3abc = (a + b + c)(a^2 + b^2 + c^2 - [ab + bc + ca])$

$\rightarrow a^3 + b^3 + c^3 - 3abc = (a + b + c)((a + b + c)^2 - 3[ab + bc + ca])$

$\rightarrow a^3 + b^3 + c^3 - 3 = 7(7^2 - 3\times9) = 154$. Option (C) is correct.

Ex6. Let a, b, x, y be real numbers such that $a^2 + b^2 = 81$, $x^2 + y^2 = 121$ and $ax + by = 99$. Then the set of all possible values of ay – bx is

(A) $\left(0, \dfrac{9}{11}\right]$ (B) $\left(0, \dfrac{9}{11}\right)$ (C) $\{0\}$ (D) $\left[\dfrac{9}{11}, \infty\right)$

Sol. $a^2 + b^2 = 81$, $x^2 + y^2 = 121$ and $ax + by = 99$

$(a^2 + b^2)(x^2 + y^2) = 81 \times 121$ ....(1)

and $(ax + by)^2 = 99^2$ ....(2)

on subtraction of (2) from (1) we get $(ay - bx)^2 = 0 \rightarrow ay - bx = 0$.

So, option (C) is correct.

Ex7. Let a, b, c, d be numbers in the set {1, 2, 3, 4, 5, 6} such that curve $y = 2x^3 + ax + b$ & $y = 2x^3 + cx + d$ have no point in common. The maximum possible value of $(a - c)^2 + b - d$ is

(A) 0   (B) 5   (C) 30   (D) 36

Sol. The curve has no common point $\rightarrow$ Equation $2x^3 + ax + b = 2x^3 + cx + d$ has no real roots

$\rightarrow (a - c)x = d - b \rightarrow x = \dfrac{d-b}{a-c}$ should be Not Define else it has one common point.

$x = \dfrac{d-b}{a-c} = $ Not Define if $a = c \rightarrow (a - c)^2 + b - d \,|_{MAX} = 0 + 6 - 1 = 5$. Hence option (B) is correct.

Ex8. If p & q are prime numbers such that $x^2 - px + q = 0$ has distinct positive integral roots, find p & q.

Sol. Let $\alpha, \beta \in N$ such that $\alpha < \beta$ and roots of the equation $x^2 - px + q = 0$

$\rightarrow \alpha + \beta = p$ and $\alpha\beta = q$ (Prime) $\rightarrow \alpha = 1$ and $\beta = q$

But $\alpha + \beta = p \rightarrow p = q + 1$. So, q and p are two consecutive primes

$\rightarrow q = 2$ and $p = 3$.

☺ ☺ *Let's have a mind refreshing problem. Just solve it and enjoy learning.* ☺ ☺

MIND REFRESHMENT # 2

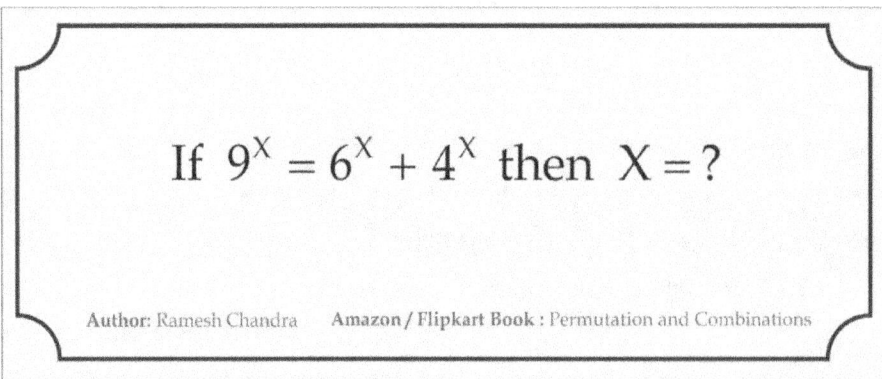

If $9^X = 6^X + 4^X$ then $X = ?$

Author: Ramesh Chandra   Amazon / Flipkart Book : Permutation and Combinations

EXERCISE - 3

1. The number of real values of x satisfying the equation

$$\sqrt{x^2-6x+9}+\sqrt{x^2-6x+6}=1 \text{ is/are } \underline{\qquad\qquad}$$

2. The number of irrational solutions of the equation

$$\sqrt{x^2+\sqrt{x^2+11}}+\sqrt{x^2-\sqrt{x^2+11}}=4 \text{ is/are } \underline{\qquad\qquad}$$

3. Solve the following equations:

(i) $\sqrt{2x+9}+x=13$

(ii) $\sqrt{x^2+x-6}+2-x=\sqrt{x^2-7x+10}$

(iii) $\sqrt{1+4x-x^2}=x-1$

(iv) $(16-x^2)\sqrt{3-x}=0$

(v) $\sqrt[3]{16-x^3}=4-x$

(vi) $\sqrt{11-x}+\sqrt{x-5}=2\sqrt{3}$

4. Solve the following equations

(i) $\dfrac{1}{x+4}-\dfrac{1}{x+5}=\dfrac{1}{20}$

(ii) $x+\dfrac{1}{x}=\dfrac{17}{4}$

(iii) $\dfrac{x}{x+1}+\dfrac{x+1}{x}=\dfrac{169}{60}$

(iv) $2\left(x^2+\dfrac{1}{x^2}\right)-9\left(x+\dfrac{1}{x}\right)+14=0$

(v) $6\left(x^2+\dfrac{1}{x^2}\right)-25\left(x-\dfrac{1}{x}\right)+12=0$

(vi) $3^{x+2}+3^{-x}=10$

(vii) $\dfrac{x+3}{2x-7}-\dfrac{2x-1}{x-3}=0$

(viii) $3x^4-20x^3-94x^2-20x+3=0$

(ix) $5\sqrt{\dfrac{3}{x}}+7\sqrt{\dfrac{x}{3}}=\dfrac{68}{3}$

(x) $(x+1)(x+2)(x+3)(x+4)=24$

(xi) $x^2-5x+2\sqrt{x^2-5x+3}=12$

(xii) $12x^4-56x^3+89x^2-56x+12=0$

5. Solve the following equations:

(i) $\sqrt{4x^2+5x+1}-2\sqrt{x^2-x+1}=9x-3$

(ii) $\left(x+3\sqrt{x}+2\right)\left(x+9\sqrt{x}+18\right)=168x$

6. If α, β are roots of equation $8x^2-3x+27=0$, then value of $4\left[\left(\dfrac{\alpha^2}{\beta}\right)^{\frac{1}{3}}+\left(\dfrac{\beta^2}{\alpha}\right)^{\frac{1}{3}}\right]$ is___

7. Solve the following equations:

(i) $3^m-2^n=1$

(ii) $x^3=2y^3+4z^3$, $x, y, z \in I$

(iii) $|x-3|^{(x^2-8x+15)/(x-2)}=1$

(iv) $(x-5)(x-7)(x+6)(x+4)=504$

(v) $\sqrt[3]{14+x}+\sqrt[3]{14-x}=4$

(vi) $9^x-3^{x+1}-4=0$

8. Solve the system of equations

$x+y+u=4$, $y+u+v=-5$, $u+v+x=0$, $v+x+y=-8$

9. Solve the system

$(x+y)(x+z)=30$, $(y+z)(y+x)=15$, $(z+x)(z+y)=18$.

CROSS MULTIPLICATION METHOD

The topic is of class 10th. You may be aware of it but still learn it again.

We have system of linear equations in two variables as

$$a_1 x + b_1 y = c_1$$
$$a_2 x + b_2 y = c_2$$

The system can be solve by Graphical Method as well as Algebraic Method

The algebraic method consisting three sub method as

    (1) Elimination Method   (2) By Substitution Method

    (3) Cross Multiplication Method

Here we are discussing the third method "Cross Multiplication Method" by taking an example and then generalised the proceeding.

Ex.   Solve the system of equation:

$$2x - y = 3$$
$$x - 2y = 0$$ by Cross Multiplication Method.

Sol.   as you can see x = 2 and y = 1 satisfy the system so it's the required solution. You can check it by elimination or substitution method.

Let's rearrange the system as shown: $2x - y - 3 = 0$
$x - 2y - 0 = 0$

Now observe the way of representation as shown.

$$2x - y - 3 = 0$$
$$x - 2y - 0 = 0$$

$$\frac{x}{\begin{vmatrix} -1 & -3 \\ -2 & 0 \end{vmatrix}} = \frac{y}{\begin{vmatrix} -3 & 2 \\ 0 & 1 \end{vmatrix}} = \frac{1}{\begin{vmatrix} 2 & -1 \\ 1 & -2 \end{vmatrix}}$$

Now question is, How to solve $\begin{vmatrix} -1 & -3 \\ -2 & 0 \end{vmatrix}, \begin{vmatrix} -3 & 2 \\ 0 & 1 \end{vmatrix}$ or $\begin{vmatrix} 2 & -1 \\ 1 & -2 \end{vmatrix}$?

These are the determinants of order 2 x 2.

Let's have a general 2 x 2 determinant as shown in figure: $\begin{vmatrix} a & b \\ c & d \end{vmatrix} = ad - bc$

So, $\begin{vmatrix} -1 & -3 \\ -2 & 0 \end{vmatrix} = (-1)(0) - (-2)(-3) = -6$.

$\begin{vmatrix} -3 & 2 \\ 0 & 1 \end{vmatrix} = (-3)(1) - (0)(2) = -3$ and $\begin{vmatrix} 2 & -1 \\ 1 & -2 \end{vmatrix} = (2)(-2) - (1)(-1) = -3$

So, $\frac{x}{\begin{vmatrix} -1 & -3 \\ -2 & 0 \end{vmatrix}} = \frac{y}{\begin{vmatrix} -3 & 2 \\ 0 & 1 \end{vmatrix}} = \frac{1}{\begin{vmatrix} 2 & -1 \\ 1 & -2 \end{vmatrix}} \rightarrow \frac{x}{-6} = \frac{y}{-3} = \frac{1}{-3} \rightarrow x = \frac{-6}{-3} = 2$ and $y = \frac{-3}{-3} = 1$

x = 2 and y = 1 is solution that we already be calculated.

Now, let's move on generalised way:

Que. Solve the system of equation

$a_1x + b_1y + c_1 = 0$
$a_2x + b_2y + c_2 = 0$ by Cross multiplication method

Sol. The system can be expressed as

$$\frac{x}{\begin{vmatrix} b_1 & c_1 \\ b_2 & c_2 \end{vmatrix}} = \frac{y}{\begin{vmatrix} c_1 & a_1 \\ c_2 & a_2 \end{vmatrix}} = \frac{1}{\begin{vmatrix} a_1 & b_1 \\ a_2 & b_2 \end{vmatrix}} \rightarrow x = \frac{\begin{vmatrix} b_1 & c_1 \\ b_2 & c_2 \end{vmatrix}}{\begin{vmatrix} a_1 & b_1 \\ a_2 & b_2 \end{vmatrix}} \text{ and } y = \frac{\begin{vmatrix} c_1 & a_1 \\ c_2 & a_2 \end{vmatrix}}{\begin{vmatrix} a_1 & b_1 \\ a_2 & b_2 \end{vmatrix}} \text{ only if } \begin{vmatrix} a_1 & b_1 \\ a_2 & b_2 \end{vmatrix} \neq 0$$

The system has solution based on certain situations

(I)   The system has unique solution if $\begin{vmatrix} a_1 & b_1 \\ a_2 & b_2 \end{vmatrix} \neq 0$ or $\frac{a_1}{a_2} \neq \frac{b_1}{b_2}$

In this case lines $a_1x + b_1y + c_1 = 0$ and $a_2x + b_2y + c_2 = 0$ intersect only at single point, if solved by graphical method.

(II) The system has infinite solution if every determinant is zero or $\frac{a_1}{a_2} = \frac{b_1}{b_2} = \frac{c_1}{c_2}$

In this case lines $a_1x + b_1y + c_1 = 0$ and $a_2x + b_2y + c_2 = 0$ intersect at infinite points, as they overlap each other, if solved by graphical method.

(III) The system has no solution if $\frac{a_1}{a_2} = \frac{b_1}{b_2} \neq \frac{c_1}{c_2}$.

In this case lines $a_1x + b_1y + c_1 = 0$ and $a_2x + b_2y + c_2 = 0$ becomes parallel without overlapping. So no any intersecting point is there, if solved by graphical method.

PRACTICE PROBLEM

Ex.   Solve the following system of equations using Cross Multiplication Method

(i) $3x - y = 2$
     $2x + 3y = 5$

(ii) $x - y + 1 = 0$
     $2x + 5y - 5 = 0$

(iii) $2x - 3y + 1 = 0$
      $4x - 6y - 5 = 0$

(iv) $5x + y - 2 = 0$
     $10x + 2y - 4 = 0$

Ans.   (i) x = 1, y = 1   (ii) x = 0, y = 1   (iii) No solution   (iv) Infinite solutions

COMMON ROOTS CONDITION

Consider two distinct quadratic equations $a_1x^2 + b_1x + c_1 = 0$ and $a_1x^2 + b_1x + c_1 = 0$ having exactly one root common. Then, if α, β are roots of first in which α is common root then other quadratic has roots α, γ.

Since, α is root of both so it will satisfy both the equations:

$$a_1\alpha^2 + b_1\alpha + c_1 = 0$$

$$a_2\alpha^2 + b_2\alpha + c_2 = 0$$

Now, by cross multiplication method we can have $\dfrac{\alpha^2}{\begin{vmatrix} b_1 & c_1 \\ b_2 & c_2 \end{vmatrix}} = \dfrac{\alpha}{\begin{vmatrix} c_1 & a_1 \\ c_2 & a_2 \end{vmatrix}} = \dfrac{1}{\begin{vmatrix} a_1 & b_1 \\ a_2 & b_2 \end{vmatrix}}$

Now, $\dfrac{\alpha^2}{\begin{vmatrix} b_1 & c_1 \\ b_2 & c_2 \end{vmatrix}} = \dfrac{\alpha}{\begin{vmatrix} c_1 & a_1 \\ c_2 & a_2 \end{vmatrix}} \rightarrow \alpha = \dfrac{\begin{vmatrix} b_1 & c_1 \\ b_2 & c_2 \end{vmatrix}}{\begin{vmatrix} c_1 & a_1 \\ c_2 & a_2 \end{vmatrix}}$ ....(1) and $\dfrac{\alpha}{\begin{vmatrix} c_1 & a_1 \\ c_2 & a_2 \end{vmatrix}} = \dfrac{1}{\begin{vmatrix} a_1 & b_1 \\ a_2 & b_2 \end{vmatrix}} \rightarrow \alpha = \dfrac{\begin{vmatrix} c_1 & a_1 \\ c_2 & a_2 \end{vmatrix}}{\begin{vmatrix} a_1 & b_1 \\ a_2 & b_2 \end{vmatrix}}$ ....(2)

From (1) and (2) we can have $\alpha = \dfrac{\begin{vmatrix} b_1 & c_1 \\ b_2 & c_2 \end{vmatrix}}{\begin{vmatrix} c_1 & a_1 \\ c_2 & a_2 \end{vmatrix}} = \dfrac{\begin{vmatrix} c_1 & a_1 \\ c_2 & a_2 \end{vmatrix}}{\begin{vmatrix} a_1 & b_1 \\ a_2 & b_2 \end{vmatrix}} \rightarrow \left(\begin{vmatrix} c_1 & a_1 \\ c_2 & a_2 \end{vmatrix}\right)^2 = \left(\begin{vmatrix} b_1 & c_1 \\ b_2 & c_2 \end{vmatrix}\right)\left(\begin{vmatrix} a_1 & b_1 \\ a_2 & b_2 \end{vmatrix}\right)$

$\left(\begin{vmatrix} c_1 & a_1 \\ c_2 & a_2 \end{vmatrix}\right)^2 = \left(\begin{vmatrix} b_1 & c_1 \\ b_2 & c_2 \end{vmatrix}\right)\left(\begin{vmatrix} a_1 & b_1 \\ a_2 & b_2 \end{vmatrix}\right)$ is the required condition of exactly one root common.

Now, question is how to remember the formula?

You can observe "$B^2 = A \times C$" where $B = \begin{vmatrix} c_1 & a_1 \\ c_2 & a_2 \end{vmatrix}$, $A = \begin{vmatrix} b_1 & c_1 \\ b_2 & c_2 \end{vmatrix}$, $C = \begin{vmatrix} a_1 & b_1 \\ a_2 & b_2 \end{vmatrix}$

Think about A, B and C in system of equations and have your mind set to get them for future.

Ex1. Find the value of k for which $x^2 - 5x + 6 = 0$ & $2x^2 - kx + 6 = 0$ has exactly one root common.

Sol. As per condition of common roots for system $\begin{matrix} x^2 - 5x + 6 = 0 \\ 2x^2 - kx + 6 = 0 \end{matrix}$: i.e. $B^2 = AC$

$\rightarrow \begin{vmatrix} 6 & 1 \\ 6 & 2 \end{vmatrix}^2 = \begin{vmatrix} -5 & 6 \\ -k & 6 \end{vmatrix} \times \begin{vmatrix} 1 & -5 \\ 2 & -k \end{vmatrix} \rightarrow (12 - 6)^2 = (-30 + 6k)(-k + 10)$

$\rightarrow 36 = 30k - 6k^2 - 300 + 60k \rightarrow 6k^2 - 90k + 336 = 0 \rightarrow k^2 - 15k + 56 = 0 \rightarrow k = 7, 8$.

Now, question arises that why are we getting two values of 'k' instead having only one root common?

ANALYSIS

$x^2 - 5x + 6 = 0 \rightarrow x = 2$ or 3. Now, if

(i) $x = 2$ is a common root then it will satisfy the equation $2x^2 - kx + 6 = 0 \rightarrow 2(2)^2 - 2k + 6 = 0$
$\rightarrow k = 7$.

(ii) $x = 3$ is a common root then it will satisfy the equation $2x^2 - kx + 6 = 0 \rightarrow 2(3)^2 - 3k + 6 = 0$
$\rightarrow k = 8$.

Ex2. If the equations $ax^2 + 2cx + b = 0$ and $ax^2 + 2bx + c = 0$ $(b \neq c)$ have a common root, then $a + 4b + 4c$ is equal to

(A) -2 (B) -1 (C) 0 (D) 1

Sol. Let α be common root $\rightarrow a\alpha^2 + 2c\alpha + b = 0$ ..... (1) and $a\alpha^2 + 2b\alpha + c = 0$ ....(2)

Subtracting (2) from (1) $\rightarrow 2(c - b)\alpha + (b - c) = 0 \rightarrow \alpha = 1/2$ as $(b \neq c)$

If we put $\alpha = 1/2$ in equation (1) then we get $a + 4b + 4c = 0$. Hence, option (C) is correct.

Ex3. If the quadratic equations $x^2 + ax + b = 0$ and $x^2 + bx + a = 0$ ($a \neq b$) have a common root then the value of $a + b$ is _____

Sol. Let α is the common root then

$\alpha^2 + a\alpha + b = 0$ .....(1) and $\alpha^2 + b\alpha + a = 0$ ....(2)

Now on subtraction (2) from (1) we get

$\alpha(a - b) + (b - a) = 0 \rightarrow \alpha = 1$ which is the common root so it will satisfy both the equations

$\rightarrow 1 + a + b = 0 \rightarrow a + b = -1$.

CONDITION FOR BOTH ROOTS COMMON

First, let's observe the quadratics having both roots common. Here you can see

(i) $x^2 - 3x + 2 = 0$ has roots $x = 1, 2$.

(ii) $2x^2 - 6x + 4 = 0$ has roots $x = 1, 2$.

(iii) $kx^2 - 3kx + 2k = 0$ has roots $x = 1, 2$. Where $k \in R - \{0\}$

So, what are you observing here? You can see, coefficients are in proportion

Hence, the system

$$a_1 \alpha^2 + b_1 \alpha + c_1 = 0$$
$$a_2 \alpha^2 + b_2 \alpha + c_2 = 0$$

has both roots common if $\dfrac{a_1}{a_2} = \dfrac{b_1}{b_2} = \dfrac{c_1}{c_2}$.

So, condition for both roots to be common is: $\dfrac{a_1}{a_2} = \dfrac{b_1}{b_2} = \dfrac{c_1}{c_2}$.

Ex4. Find the value of a & b if the equations $ax^2 + 2x + 3 = 0$ and $x^2 - 4x + b = 0$ have both roots common.

Sol. As per condition of both roots common: $\dfrac{a}{1} = \dfrac{2}{-4} = \dfrac{3}{b}$

$\rightarrow a = -1/2$ and $b = -6$.

Ex5. If the equation $x^2 - 4x + 5 = 0$ & $x^2 + ax + b = 0$ where $a, b \in R$ have common root(s). Find a & b

Sol. as we observe: $x^2 - 4x + 5 = 0$ having Discriminant $D = (-4)^2 - 4 \times 5 = -4 < 0$

$\rightarrow$ Roots are imaginary and both equations have, real coefficients so roots of $x^2 - 4x + 5 = 0$ occur in conjugate pairs.

$\rightarrow$ both roots are common $\rightarrow \dfrac{1}{1} = \dfrac{-4}{a} = \dfrac{5}{b} \rightarrow a = -4, b = 5$.

Ex6. If $a, b, c \in I^+$ and $ax^2 + bx + c = 0$ and $x^2 + 2x + 9 = 0$ have a common root. Find the minimum value of $a + b + c$.

Sol. Discriminant of $x^2 + 2x + 9 = 0$ is $D = 4 - 4 \times 9 = -32 < 0 \rightarrow$ roots are imaginary.

Hence, if one root is common $\rightarrow$ both roots are common due to imaginary roots & real coefficients.

$\rightarrow \dfrac{a}{1} = \dfrac{b}{2} = \dfrac{c}{9} = k$, where $k \in I^+$ (Let)

$\rightarrow a = k, b = 2k, c = 9k$. So, $a + b + c = 12k$. For being minimum, k must be equal to 1

Hence, $a + b + c \mid_{MIN} = 12$.

Ex7. If one root of the equation $x^2 - x + 3a = 0$ is double the root of the equation $x^2 - x + a = 0$ then find the value of 'a' $(a \neq 0)$

Sol. let α, β are roots of equation $x^2 - x + a = 0$ then

2α, γ are the roots of equation $x^2 - x + 3a = 0$

$$\alpha^2 - \alpha + a = 0 \quad ....(1)$$
$$(2\alpha)^2 - (2\alpha) + 3a = 0 \quad ....(2)$$

$\begin{aligned} \alpha^2 - \alpha + a = 0 \\ 4\alpha^2 - 2\alpha + 3a = 0 \end{aligned} \Bigg\} \rightarrow \begin{aligned} x^2 - x + a = 0 \\ 4x^2 - 2x + 3a = 0 \end{aligned} \Bigg\}$ have exactly one root in common.

So, $B^2 = AC \rightarrow \begin{vmatrix} a & 1 \\ 3a & 4 \end{vmatrix}^2 = \begin{vmatrix} -1 & a \\ -2 & 3a \end{vmatrix} \times \begin{vmatrix} 1 & -1 \\ 4 & -2 \end{vmatrix} \rightarrow a^2 = (-a)(2)$

$\rightarrow a(a + 2) = 0$, since $a \neq 0$ so $a + 2 = 0 \rightarrow a = -2$.

Ex8. The equations $ax^2 + bx + a = 0$, $x^3 - 2x^2 + 2x - 1 = 0$ have two roots common where $a, b \in R$ then $a + b$ must be equal to

(A) 1   (B) -1   (C) 0   (D) None

Sol. $x^3 - 2x^2 + 2x - 1 = 0 \rightarrow (x - 1)(x^2 - x + 1) = 0$.

The equation $x^2 - x + 1 = 0$ has imaginary roots so $x = 1$ can't be common root.

$\rightarrow x^2 - x + 1 = 0$ and $ax^2 + bx + a = 0$ has both roots common

$\rightarrow \dfrac{a}{1} = \dfrac{b}{-1} = \dfrac{a}{1} \rightarrow a = -b \rightarrow a + b = 0$

We can also think of like.

$x^2 - x + 1 = 0$ and $x^2 + \dfrac{b}{a}x + 1 = 0$ are same $\rightarrow \dfrac{b}{a} = -1 \rightarrow a + b = 0$.

Hence, option (C) is correct.

Ex9. If $ax^2 + bx + c = 0$ & $bx^2 + cx + a = 0$ have a common root & a, b, c are non – zero real number then find the value of $\dfrac{a^3 + b^3 + c^3}{abc}$.

Sol. if you observe at $x = 1$, both are becoming $a + b + c$ but it doesn't mean that $x = 1$ is a common root.

So, have traditional way, Apply $B^2 = AC$

$\rightarrow \begin{vmatrix} c & a \\ a & b \end{vmatrix}^2 = \begin{vmatrix} b & c \\ c & a \end{vmatrix} \times \begin{vmatrix} a & b \\ b & c \end{vmatrix} \rightarrow (bc - a^2)^2 = (ab - c^2)(ac - b^2) \rightarrow b^2 c^2 + a^4 - 2a^2 bc = a^2 bc - c^3 a - ab^3 + b^2 c^2$

$\rightarrow a^4 + ab^3 + c^3 a - 3a^2 bc = 0 \rightarrow a(a^3 + b^3 + c^3 - 3abc) = 0$.

As a $\neq$ 0 so $a^3 + b^3 + c^3 - 3abc = 0 \rightarrow \dfrac{a^3 + b^3 + c^3}{abc} = 3$.

Ex10. If the equation $4x^2 \sin^2\theta - 4x \sin\theta + 1 = 0$ and $a(b - c) x^2 + b(c - a)x + c(a - b) = 0$ have a common root and 2nd equation has equal roots. Find the general solution of θ.

Sol. As you can see x = 1 satisfying the 2nd equation so as per question x = 1 is the common root, so it will satisfy the first equation as well.

$\rightarrow 4 \sin^2\theta - 4 \sin\theta + 1 = 0 \rightarrow (2 \sin\theta - 1)^2 = 0 \rightarrow \sin\theta = 1/2 = \sin(\pi/6)$

$\rightarrow \theta = n\pi + (- 1)^n(\pi/6)$ where $n \in I$.

FACTORIZATION OF QUADRATIC EXPRESSIONS ($ax^2 + 2hxy + by^2 + 2gx + 2fy + c = 0$)

The topic is based on conic section (pair of straight lines) but still we are studying in algebra to learn factorization.

The general expression ax + by + c = 0, where a, b, c $\in$ R (Not all zero simultaneously) represents a straight line in 2D Cartesian Co-ordinate system (X – Y Plane).

Consider two lines L_1: x + y – 2 = 0 and L_2: 3x – 2y + 5 = 0. If we multiply them then the resulting equation represents pair of straight lines. We can have proof of statement too.

The pair of straight line = (x + y – 2)(3x – 2y + 5) = 0 $\rightarrow 3x^2 + xy - 2y^2 - x + 9 y - 10 = 0$ ......(1)

PROOF:

Observe the figure. You can see Points P, Q, R and S.

Point Q is a general point lie on line L_2, hence it will satisfy $L_2 = 0$

$\rightarrow L_2 |_{At\,Q} = 0$

Similarly, $L_1 |_{At\,R} = 0$, $L_1 |_{At\,P} = 0$ and $L_2 |_{At\,P} = 0$

But $L_1 |_{At\,S} \neq 0$ and $L_2 |_{At\,S} \neq 0$ where S is a general point (Outside, not on pair of Straight Lines).

Now Think of equation $L_1 L_2 = 0$. ....(2)

$(L_1 |_{At\,Q}) \times (L_2 |_{At\,Q}) = 0 \rightarrow (\neq 0) \times (0) = 0 \rightarrow 0 = 0$, True

$\rightarrow$ Point Q satisfy equation (2) so it will lies on the curve $L_1 L_2 = 0$

Similarly, Points P and R lies on the curve $L_1 L_2 = 0$ But Point S will not lies on this curve.

As, $(L_1 |_{At\,S}) \times (L_2 |_{At\,S}) = 0 \rightarrow (\neq 0) \times (\neq 0) = 0$ So, Point S is not on the curve.

$\rightarrow L_1 L_2 = 0$ represents pair of straight lines $L_1 = 0$ & $L_2 = 0$.

□

Now, move on to current topic, factorization of quadratic expressions

The pair of straight line = (x + y – 2)(3x – 2y + 5) = 0 $\rightarrow 3x^2 + xy - 2y^2 - x + 9 y - 10 = 0$ ......(1)

You can see, the expansion is easy but in reverse (Factorisation), it may be difficult task to you.

So, what should be our strategy?

Look at: $3x^2 + xy - 2y^2 - x + 9 y - 10 = 0$ and focus only at $3x^2 + xy - 2y^2$. You can factorize this part easily.

Like: $3x^2 + xy - 2y^2 = 3x^2 + 3xy - 2xy - 2y^2$ (Same as you perform in quadratic expression)

$\rightarrow 3x^2 + xy - 2y^2 = 3x (x + y) - 2y (x + y) \rightarrow (3x - 2y)(x + y)$

$\rightarrow 3x^2 + xy - 2y^2 = (3x - 2y)(x + y)$.

Now, you have done half part of factorization.

Can we write $3x^2 + xy - 2y^2 - x + 9y - 10 = (3x - 2y)(x + y)$? Why? What is the problem?

What about $- x + 9y - 10$? How we get this?

Can we write: $3x^2 + xy - 2y^2 - x + 9y - 10 = (3x - 2y + k_1)(x + y + k_2)$ where k_1 & k_2 are real constants

Yes, we can write. Now task is to find k_1 & k_2.

On comparing the coefficients of x we get: $k_1 + 3k_2 = -1$ .... (2)

On comparing the coefficients of y we get: $k_1 - 2k_2 = 9$ ....(3)

On solving equations (2) & (3) we get $k_1 = 5$, $k_2 = -2$

$\rightarrow 3x^2 + xy - 2y^2 - x + 9y - 10 = (3x - 2y + k_1)(x + y + k_2)$

$\rightarrow 3x^2 + xy - 2y^2 - x + 9y - 10 = (3x - 2y + 5)(x + y - 2)$

Ex1. Factorize the following pair of straight lines

(i) $4x^2 - 7xy - 2y^2 + 6x - 3y + 2 = 0$

(ii) $8x^2 + 24xy + 18y^2 + 10x + 15y + 3 = 0$

(iii) $2x^2 - 3xy - 2y^2 + 4x - 3y + 2 = 0$

(iv) $x^2 + 4xy + 4y^2 + 8x + 16y + 16 = 0$

Sol.

(i) $4x^2 - 7xy - 2y^2 + 6x - 3y + 2 = 0$

Look at: $4x^2 - 7xy - 2y^2 = 4x^2 - 8xy + xy - 2y^2 = 0$

$= 4x(x - 2y) + y(x - 2y)$

$= (4x + y)(x - 2y)$

$\rightarrow 4x^2 - 7xy - 2y^2 = (4x + y)(x - 2y)$

So, $4x^2 - 7xy - 2y^2 + 6x - 3y + 2 = (4x + y + k_1)(x - 2y + k_2)$

Now, on comparing coefficients of x, y both side, we get

$k_1 + 4k_2 = 6$ ..... (1) and $-2k_1 + k_2 = -3$ .... (2)

Now, from (1) and (2) we get $k_1 = 2$, $k_2 = 1$

So, $4x^2 - 7xy - 2y^2 + 6x - 3y + 2 = 0 \rightarrow (4x + y + 2)(x - 2y + 1)$

Now, solve remaining your own. ☺ ☺

(ii) $8x^2 + 24xy + 18y^2 + 10x + 15y + 3 = 0 \rightarrow (2x + 3y + 1)(4x + 6y + 3) = 0$

(iii) $2x^2 - 3xy - 2y^2 + 4x - 3y + 2 = 0 \rightarrow (x - 2y + 1)(2x + y + 2) = 0$

(iv) $x^2 + 4xy + 4y^2 + 8x + 16y + 16 = 0 \rightarrow (x + 2y + 4)^2 = 0$

Now, at the moment, we have question here that "can we factorize any second degree equation of type $ax^2 + 2hxy + by^2 + 2gx + 2fy + c = 0$?"

The answer is NO. There are certain conditions in which we can express as product of two lines.

We can factorize $ax^2 + 2hxy + by^2 + 2gx + 2fy + c = 0$ only if it represent a pair of straight lines.

CONDITIONS TO FACTORIZE ($ax^2 + 2hxy + by^2 + 2gx + 2fy + c = 0$)

(i) $\Delta = \begin{vmatrix} a & h & g \\ h & b & f \\ g & f & c \end{vmatrix} = 0$ or $\Delta = abc + 2fgh - af^2 - bg^2 - ch^2 = 0$

(ii) $h^2 - ab \geq 0$

Where determinant $\begin{vmatrix} a & h & g \\ h & b & f \\ g & f & c \end{vmatrix}$ can be solved as shown in figure (Observe the pattern)

$$\begin{vmatrix} a & h & g \\ h & b & f \\ g & f & c \end{vmatrix} = \begin{matrix} a & h & g & a & h \\ h & b & f & h & b \\ g & f & c & g & f \end{matrix}$$

$$= (abc + hfg + hfg) - (bg^2 + af^2 + ch^2)$$

Or

$$\begin{vmatrix} a & h & g \\ h & b & f \\ g & f & c \end{vmatrix} = a\begin{vmatrix} b & f \\ f & c \end{vmatrix} - h\begin{vmatrix} h & f \\ g & c \end{vmatrix} + g\begin{vmatrix} h & b \\ g & f \end{vmatrix}$$

$$= a(bc - f^2) - h(hc - fg) + g(hf - gb)$$

$$= abc + 2fgh - af^2 - bg^2 - ch^2$$

Ex2. If $3x^2 + 2\alpha xy + 2y^2 + 2ax - 4y + 1$ can be resolve into two linear factors, then prove that α is a root of the equation $x^2 + 4ax + 2a^2 + 6 = 0$

Sol. The equation $3x^2 + 2\alpha xy + 2y^2 + 2ax - 4y + 1 = 0$ represent pair of straight lines so

(i) $\Delta = 0$ and (ii) $h^2 - ab \geq 0$

$$\Delta = 0 \rightarrow \begin{vmatrix} 3 & \alpha & a \\ \alpha & 2 & -2 \\ a & -2 & 1 \end{vmatrix} = 0 \rightarrow 3\begin{vmatrix} 2 & -2 \\ -2 & 1 \end{vmatrix} - \alpha\begin{vmatrix} \alpha & -2 \\ a & 1 \end{vmatrix} + a\begin{vmatrix} \alpha & 2 \\ a & -2 \end{vmatrix} = 0$$

$\rightarrow -6 - \alpha(\alpha + 2a) + a(-2\alpha - 2a) = 0$

$\rightarrow -6 - \alpha^2 - 4a\alpha - 2a^2 = 0 \rightarrow \alpha^2 + 4a\alpha + 2a^2 + 6 = 0$

$\rightarrow \alpha$ is a root of equation $x^2 + 4ax + 2a^2 + 6 = 0$

Ex3. Show that $\Delta = 0$ for $(x - y)^2 + (x + y)^2 = 0$ but can't be resolve into pair of straight lines (Linear Factors) and represent a point $(0, 0)$.

Sol. $(x - y)^2 + (x + y)^2 = 0 \rightarrow x^2 + y^2 = 0$

Here you can observe that

(i) $\Delta = \begin{vmatrix} 1 & 0 & 0 \\ 0 & 1 & 0 \\ 0 & 0 & 0 \end{vmatrix} = 0$ (ii) $h^2 - ab = 0^2 - 1 \times 1 = -1 < 0$

Second condition failed to be a pair of straight lines.

But you can see $(x - y)^2 + (x + y)^2 = 0$ is possible only if $x - y = 0$ and $x + y = 0$

i.e. intersection point of the lines $x - y = 0$ and $x + y = 0$

That is $(0, 0)$

THEORY OF EQUATIONS

In this topic we will study about polynomial functions and various methods to find and analyze the roots of polynomial equations. We already discussed some important theorems related to the topic, like Fundamental Theorem of Algebra, Gauss alternative formulation, Remainder Theorem, Factor Theorem etc.

THEOREM – 1

Every polynomial of degree n has n and only n zeroes.

1) DESCARTE'S RULE OF SIGNS

To determine the nature of some of the roots of a polynomial equation it is not always necessary to solve it; for instance, the truth of the will be readily admitted.

STATEMENT:

In any single variable real polynomial equation $f(x) = 0$ (ordered by descending variable exponent), the number of real positive roots cannot exceed the number of changes in the signs of the coefficients of the terms in $f(x)$, and the number of real negative roots cannot exceed the number of changes in the signs of the coefficients of $f(-x)$.

Rule # 1: The maximum (+) ve real roots of real polynomial equation $f(x) = 0$ are number of sign change. (*Ordered by descending variable exponent*)

Ex1. The equation $f(x) = (x - 1)(x - 2)(x + 1) = 0$ has three roots 1, 2 and – 1.

If we expand it then above equation will be written as: $x^3 - 2x^2 - x + 2 = 0$

The sign changes are as shown: $(+) x^3 (-) 2 x^2 (-) x (+) 2 = 0$, i.e. $(+) (-) (-) (+)$

As per rule # 1, there are 2 sign change from left to right, hence maximum number of (+) ve roots are 2.

Ex2. $f(x) = 3x^{10} - 4x^5 + 2x^4 - 6 x + 5 = 0$

Here you can observe, the sign: $(+) 3x^{10} (-) 4x^5 (+) 2x^4 (-) 6 x (+) 5 = 0$

There are 4 sign change so, Maximum numbers of positive roots are 4.

Note: *The number of positive roots may be 0, 2 and 4 but not 1 & 3 in this case.*

Rule#2: The maximum (-) ve real roots of real polynomial equation $f(x) = 0$ are the number of sign change in $f(- x) = 0$, (*ordered by descending variable exponent*)

Ex3. Let $f(x) = x^3 - 2 x^2 - x + 2 = 0$ then $f (- x) = - x^3 - 2 x^2 + x + 2 = 0$.

The number of sign change in $f(- x) = (-) x^3 (-) 2 x^2 (+) x (+) 2 = 0$ is 1 so maximum number of negative real roots = 1.

Ex4. Let $f(x) = x^7 - 2x^6 + x^5 + x^3 - x - 1 = 0$

$\rightarrow f(- x) = - x^7 - 2x^6 - x^5 - x^3 + x - 1 = 0$

Total number of sign change = 2, so Maximum number of (-)ve real roots of $f(x) = 0$ are 2.

Rule#3: If a = Maximum number of (+)ve real roots,

b = Maximum number of (-)ve real roots of real polynomial equation p(x) = 0 of degree 'n' then, Minimum number of imaginary roots are n – (a + b).

Ex5. Discuss the nature of roots of equation $x^{10} - 2x^9 - x^7 + x^6 + 2x^5 - 3x^2 - 2x + 1 = 0$

Sol. Let $f(x) = x^{10} - 2x^9 - x^7 + x^6 + 2x^5 - 3x^2 - 2x + 1 = 0$

There are 4 sign change in f(x), so f(x) = 0 has at most 4 real roots.

$f(-x) = x^{10} + 2x^9 + x^7 + x^6 - 2x^5 - 3x^2 + 2x + 1 = 0$ has 2 sign change so, the equation has at most 2 negative real roots

Since degree of polynomial is 10, hence Minimum number of imaginary roots = 10 – (4 + 2) = 4.

THEOREM · 2

If the equation $a_n x^n + a_{n-1} x^{n-1} + a_{n-2} x^{n-2} + \ldots\ldots + a_1 x + a_0 = 0$, *where* $a_n \neq 0$ *and* $a_0, a_1, a_2, a_3, \ldots, a_n \in R$ *has an Imaginary root* $\alpha + i\beta$ *then it also has another root* $\alpha - i\beta$ *where* $i = \sqrt{-1}$ *and* $\alpha, \beta \in R$.

PROOF:

Let $f(x) = a_n x^n + a_{n-1} x^{n-1} + a_{n-2} x^{n-2} + \ldots\ldots + a_1 x + a_0 = 0$ has root $\alpha + i\beta \to f(\alpha + i\beta) = 0$. .....(1)

Consider $(x - (\alpha + i\beta))(x - (\alpha - i\beta)) = (x - \alpha - i\beta)(x - \alpha + i\beta) = (x - \alpha)^2 + \beta^2$ this is a quadratic.

Now, as per remainder theorem we can have

$f(x) = [(x - \alpha)^2 + \beta^2]Q(x) + Ax + B \ \forall \ x \in C$ ....(2)

where, Q(x) = quotient, Remainder = Ax + B

From, (1) and (2) we get

$f(\alpha + i\beta) = [(\alpha + i\beta - \alpha)^2 + \beta^2]Q(\alpha + i\beta) + A(\alpha + i\beta) + B = 0$

$\to A(\alpha + i\beta) + B = 0 \to A\alpha + B + iA\beta = 0$

$\to A\alpha + B = 0$ and $A\beta = 0$

Since, $(\alpha + i\beta)$ is imaginary so, $\beta \neq 0$, hence $A = 0 \to B = 0 \to$ Remainder = 0

$\to f(x) = [(x - \alpha)^2 + \beta^2]Q(x) + 0$

$\to f(x) = [(x - (\alpha + i\beta))(x - (\alpha - i\beta))]Q(x)$

$\to f(x) = 0$ has another root $\alpha - i\beta$.

THEOREM · 3

If $p + \sqrt{q}$ *is the root of equation* $a_n x^n + a_{n-1} x^{n-1} + a_{n-2} x^{n-2} + \ldots\ldots + a_1 x + a_0 = 0$, *where* $a_n \neq 0$ *and* $a_0, a_1, a_2, \ldots, a_n \in Q$ *then* $p - \sqrt{q}$ *must be another root, where* $p, q \in Q$.

PROOF:

Let $f(x) = a_n x^n + a_{n-1} x^{n-1} + a_{n-2} x^{n-2} + \ldots\ldots + a_1 x + a_0$ is an 'nth' degree polynomial with rational coefficients having root $p + \sqrt{q}$, where $p, q \in Q$.

$\to f(x) = [(x - (p + \sqrt{q}))(x - (p - \sqrt{q}))]Q(x) + Ax + B \ \forall \ x \in R$.

$\to f(x) = [(x - p)^2 - q^2]Q(x) + Ax + B \ \forall \ x \in R$.

$\to f(p + \sqrt{q}) = 0 = A(p + \sqrt{q}) + B \ \to A = 0 \to B = 0$.

So, $f(x) = [(x - (p + \sqrt{q}))(x - (p - \sqrt{q}))]Q(x)$

Hence, f(x) = 0 has another root $x = p - \sqrt{q}$

THEOREM – 4

If the rational number p/q, a fraction in its lowest form (ie HCF ($|p|$, $|q|$) = 1, q ≠ 0) is a root of equation $a_n x^n + a_{n-1} x^{n-1} + a_{n-2} x^{n-2} + \dots\dots + a_1 x + a_0 = 0$ *where* $a_n \neq 0$ *and* $a_1, a_2, \dots, a_n \in I$ *then p is a divisor of* a_0 *and q is a divisor of* a_n.

PROOF

Since p/q is a root of given equation

$$\rightarrow a_n \left(\frac{p}{q}\right)^n + a_{n-1} \left(\frac{p}{q}\right)^{n-1} + a_{n-2} \left(\frac{p}{q}\right)^{n-2} + \dots\dots + a_1 \left(\frac{p}{q}\right) + a_0 = 0$$

Now, multiply both sides by q^n we get

$$a_n p^n + a_{n-1} p^{n-1} q + a_{n-2} p^{n-2} q^2 + \dots\dots + a_1 p q^{n-1} + a_0 q^n = 0 \ \dots\dots(1)$$

Now, divide both side by p, we get

$$a_n p^{n-1} + a_{n-1} p^{n-2} q + a_{n-2} p^{n-3} q^2 + \dots\dots + a_1 q^{n-1} = -\frac{a_0 q^n}{p}$$

Since, p & q are relatively prime and LHS is an Integer hence p must divide a_0

Similarly, from Equation (1) we get

$$a_{n-1} p^{n-1} q + a_{n-2} p^{n-2} q^2 + \dots\dots + a_1 p q^{n-1} + a_0 q^n = -a_n p^n$$

Divide this equation by q, we get

$$a_{n-1} p^{n-1} + a_{n-2} p^{n-2} q + \dots\dots + a_1 p q^{n-2} + a_0 q^{n-1} = -\frac{a_n p^n}{q}$$

LHS is an integer and p, q are relatively prime so, q must divide a_n.

Corollary:

Every rational root of the equation $x^n + a_{n-1} x^{n-1} + a_{n-2} x^{n-2} + \dots\dots + a_1 x + a_0 = 0$, where $a_i \in I$, must be integer. Moreover, every such root must be divisor of the constant a_0.

THEOREM - 5

If α is an r – times repeated root of polynomial equation f(x) = 0 then α is (r - 1) times repeated root of f'(x) = 0, where f'(x) is the derivative of f(x).

PROOF:

Given that α is r times repeated root of f(x) = 0.

Then f(x) = (x - α)^r Q(x) where Q(α) ≠ 0 (Using Factor Theorem)

Now, by applying product rule of differentiation, we obtain:

f'(x) = r(x - α)$^{r-1}$ Q(x) + (x - α)^r Q'(x)

→ f'(x) = (x - α)$^{r-1}$ [r Q(x) + (x - α)Q'(x)]

Observe that r Q(α) + (x - α)Q'(α) ≠ 0 and f'(x) = (x - α)$^{r-1}$[r Q(x) + (x - α)Q'(x)]

You can see, (x - α)$^{r-1}$ as factor of f'(x)

→ r - 1 times repeated root of f'(x) = 0.

Note: *The above theorems are very important to remember. Let's move on to next topic*

2) RELATION BETWEEN THE ROOTS AND COEFFICIENTS OF A POLYNOMIAL EQUATION

Let's have some experiments before actual start of the topic.

Experiment # 1

Consider the equation $(x - 1)(x - 2)(x - 3)(x - 4) = 0$

The equation has 4 roots as, 1, 2, 3 and 4. If we expand it then we get

$[(x - 1)(x - 2)](x - 3)(x - 4)$

$= [x^2 - (1+2)x + 1 \times 2](x^2 - (3 + 4)x + 3 \times 4)$

$= x^4 - (1+2+3+4) x^3 + (1 \times 2 + 1 \times 3 + 1 \times 4 + 2 \times 3 + 2 \times 4 + 3 \times 4) x^2 - (1 \times 2 \times 3 + 1 \times 2 \times 4 + 1 \times 3 \times 4 + 2 \times 3 \times 4) x + 1 \times 2 \times 3 \times 4$.

So, what are you observing here?

You can see the following

(i)   The coefficients of $x^3 = - (1 + 2 + 3 + 4) = (-)$ sum of roots taking one at a time.

(ii) The coefficients of $x^2 = + (1 \times 2 + 1 \times 3 + 1 \times 4 + 2 \times 3 + 2 \times 4 + 3 \times 4) = (+)$ sum of product of roots taking two roots at a time.

(iii) The coefficients of $x = - (1 \times 2 \times 3 + 1 \times 2 \times 4 + 1 \times 3 \times 4 + 2 \times 3 \times 4) = (-)$ Sum of product of roots taking three at a time.

(iv) The coefficients of $x^0 = + (1 \times 2 \times 3 \times 4) = (+)$ Sum of product of roots taking 4 at a time.

Are you observing (+), (-) sign along with x^r coefficient?

(a) we are getting (+) sign if r = even integer

(b) we are getting (-) sign if r = odd integer

Que1. Find the coefficient of x^{99} and x^{98} in the expansion of $(x - 1) (x - 2) (x - 3).....(x - 100) = 0$

Sol.   The coefficient of $x^{99} = (-) (1 + 2 + 3 + 4 + ...+ 100) = -\dfrac{(100)(101)}{2} = -5050$.

The coefficient of $x^{98} = (+)(1 \times 2 + 1 \times 3 + ...+ 1 \times 100 + 2 \times 3 + 2 \times 4 +....+ 2 \times 100 + 3 \times 4 + ...+ 99 \times 100)$

Now, let $X = (1 \times 2 + 1 \times 3 + ...+ 1 \times 100 + 2 \times 3 + 2 \times 4 +....+ 2 \times 100 + 3 \times 4 + ...+ 99 \times 100)$

We already discussed the concept in chapter 2 (Number Theory).

If we think of $(1 + 2 + 3 +.....+ 100)^2 = 1^2 + 2^2 + 3^2 + ...+ 100^2 + 2X$

$\rightarrow 2X = (5050)^2 - [1^2 + 2^2 + 3^2 + ...+ 100^2]$

$\rightarrow 2X = 5050^2 - \left(\dfrac{100 \times 101 \times 201}{6}\right) = 5050^2 - 5050 \times 67 = 5050 \times 4983 = 25164150$

$\rightarrow X = 12582075$

The coefficient of $x^{98} = (+) 12582075 = 12582075$.

Experiment # 2

Consider the equation $(x - \alpha) (x - \beta) (x - \gamma) = 0$

Here, you can see the equation have roots α, β and γ.

If we expand the expression $(x - \alpha) (x - \beta)(x - \gamma)$ then we get $x^3 - (\alpha + \beta + \gamma)x^2 - (\alpha\beta + \beta\gamma + \gamma\alpha)x + \alpha\beta\gamma$

$\rightarrow (x - \alpha)(x - \beta)(x - \gamma) = x^3 - (\alpha + \beta + \gamma)x^2 - (\alpha\beta + \beta\gamma + \gamma\alpha)x + \alpha\beta\gamma$ ....(1)

Now, If we have any cubic equation $ax^3 + bx^2 + cx + d = 0$, $(a \neq 0)$ having roots α, β and γ then

$ax^3 + bx^2 + cx + d = k(x - \alpha)(x - \beta)(x - \gamma)$

Why are we writing k and what should be its value?

To balance equation, k must be 'a'

$\rightarrow ax^3 + bx^2 + cx + d = a(x - \alpha)(x - \beta)(x - \gamma)$

$\rightarrow ax^3 + bx^2 + cx + d = a[x^3 - (\alpha + \beta + \gamma)x^2 - (\alpha\beta + \beta\gamma + \gamma\alpha)x + \alpha\beta\gamma]$

On comparing both sides we get following:

(i) $\alpha + \beta + \gamma = (-)\dfrac{b}{a}$

(ii) $\alpha\beta + \beta\gamma + \gamma\alpha = (+)\dfrac{c}{a}$

(iii) $\alpha\beta\gamma = (-)\dfrac{d}{a}$

STATEMENT:

If α, β and γ are roots of the equation $ax^3 + bx^2 + cx + d = 0$, $(a \neq 0)$ then we have

(i) $\alpha + \beta + \gamma = -\dfrac{b}{a}$ (ii) $\alpha\beta + \beta\gamma + \gamma\alpha = \dfrac{c}{a}$ (iii) $\alpha\beta\gamma = -\dfrac{d}{a}$

Ex1. If roots of equation $x^3 + px^2 + qx + r = 0$ are in arithmetic progression, show that $2p^3 - 9pq + 27r = 0$

Sol. Consider the roots $a - d$, a, $a + d$

$\rightarrow (a - d) + a + (a + d) = -p/1$

$\rightarrow 3a = -p \rightarrow a = -p/3$, this is a root of equation

$\rightarrow \left(-\dfrac{p}{3}\right)^3 + p\left(-\dfrac{p}{3}\right)^2 + q\left(-\dfrac{p}{3}\right) + r = 0$

$\rightarrow -\dfrac{p^3}{27} + \dfrac{p^3}{9} - \dfrac{pq}{3} + r = 0$. Now, multiply both sides by -27

$\rightarrow 2p^3 - 9pq + 27r = 0$.

Ex2. If 2 & 3 are roots of the equation $2x^3 + mx^2 - 13x + n = 0$, find third root

Sol. Let, third root be α

$\rightarrow$ Sum of product of roots taking two at a time $= 2\alpha + 3\alpha + 6 = (-13/2)$

$\rightarrow 5\alpha = -25 / 2 \rightarrow \alpha = -5/2$.

Ex3. Find the roots α, β & γ of the equation $x^3 - 11x^2 + 36x - 36 = 0$ if $\dfrac{1}{\alpha} + \dfrac{1}{\beta} = \dfrac{2}{\gamma}$.

Sol. *Can you guess the nature of roots? (Rational roots must be integral)*

As α, β, γ are roots of the equation, so

(i) $\alpha + \beta + \gamma = 11$ (ii) $\alpha\beta + \beta\gamma + \gamma\alpha = 36$ (iii) $\alpha\beta\gamma = 36$

If we divide (ii) by (iii) we get $\dfrac{1}{\alpha} + \dfrac{1}{\beta} + \dfrac{1}{\gamma} = 1 \to \dfrac{2}{\gamma} + \dfrac{1}{\gamma} = 1 \to \gamma = 3$

From, (i) & (iii) we get $\alpha + \beta = 8$ and $\alpha\beta = 12$

$\to (\alpha, \beta) = (2, 6)$ or $(6, 2)$

$\to (\alpha, \beta, \gamma) = (2, 6, 3)$ or $(6, 2, 3)$

Ex4. Let a & b are two roots of the equation $x^3 + px^2 + qx + r = 0$ satisfying the relation $ab + 1 = 0$. Prove that $r^2 + pr + q + 1 = 0$, $(r \neq 0)$

Sol. Let third root be α then product of roots $ab\alpha = -r$, but $ab + 1 = 0 \to ab = -1$

$\to \alpha = r$, i.e. third root is r

$\to$ r will satisfy the equation $\to r^3 + pr^2 + qr + r = 0$

$\to r(r^2 + pr + q + 1) = 0$

As, $r \neq 0 \to r^2 + pr + q + 1 = 0$

Ex5. Solve the equation $x^3 - 9x^2 + 14x + 24 = 0$, given that two of whose roots are in the ratio 3: 2.

Sol. Let the roots are $3\alpha, 2\alpha, \beta$

(i) $3\alpha + 2\alpha + \beta = 9 \to 5\alpha + \beta = 9$

(ii) $6\alpha^2 + 2\alpha\beta + 3\alpha\beta = 14 \to 6\alpha^2 + 5\alpha\beta = 14$

(iii) $6\alpha^2\beta = -24$

From (i) & (ii) we get: $6\alpha^2 + 5\alpha (9 - 5\alpha) = 14 \to 19\alpha^2 - 45\alpha + 14 = 0$

$\to 19\alpha^2 - 38\alpha - 7\alpha + 14 = 0 \to (19\alpha - 7)(\alpha - 2) = 0$

$\alpha = 2$ or $7/19$

Case – (I) $\alpha = 2 \to \beta = -1$ which satisfy the third relation, so roots are 6, 4 & -1

Case – (II) $\alpha = 7/19 \to \beta = 9 - 35/19 = 136/19$ but it's not satisfying the third relation $6\alpha^2\beta = -24$.

So, roots are 6, 4 and – 1.

Ex6. If the roots of $x^3 + px^2 + qx + r = 0$ are in geometric progression. Find the relation between p, q, r.

Sol. Let the roots be $\dfrac{a}{r}, a, ar$ then

(i) $\dfrac{a}{r} + a + ar = -p \to a\left(\dfrac{1}{r} + 1 + r\right) = -p$ (ii) $\dfrac{a^2}{r} + a^2 + a^2 r = q \to a^2\left(\dfrac{1}{r} + 1 + r\right) = q$

(iii) $a^3 = -r$

If we divide (ii) from (i), we get $a = -q/p$

Now, from (iii), we have $(-q/p)^3 = -r \to q^3 = p^3 r$

So, required relation is $q^3 = p^3 r$

Ex7. Let $f(x) = P_0(x) + P_1(x) e^x + P_2(x) e^{2x} + ... + P_n(x) e^{nx}$, where n be an integer ≥ 1 and $P_0(x), P_1(x), ..., P_n(x)$ are polynomials. If $f(x) = 0$ for any arbitrary large number x, then

(A) $P_n(x)$ is positive for atleast one $x \in R$

(B) $P_{n-1}(x) \neq P_{n-2}(x)$ for at least one negative x

(C) $P_{n-1}(-2) = 0$

(D) none of these

Sol. $\lim_{x \to \infty} \dfrac{f(x)}{e^{nx}} \to P_n(x) = 0$

Similarly, $P_0(x) = P_1(x) = ... , P_{n-1}(x) = 0$

Hence, option (C) is correct.

GENERAL THEORY OF EQUATIONS

Let $\alpha_1, \alpha_2, \alpha_3, \alpha_4, ..., \alpha_n$ are roots of equation $a_n x^n + a_{n-1} x^{n-1} + a_{n-2} x^{n-2} + ... + a_1 x + a_0 = 0$, where $a_n \neq 0$ and $a_0, a_1, a_2, a_3, ..., a_n \in C$ then

$$a_n x^n + a_{n-1} x^{n-1} + a_{n-2} x^{n-2} + ... + a_1 x + a_0 = a_n (x - \alpha_1)(x - \alpha_2)(x - \alpha_3).....(x - \alpha_n) = 0$$

(1) $\displaystyle\sum_{i=1}^{n} \alpha_i = \alpha_1 + \alpha_2 + \alpha_3 + ... + \alpha_n = (-1)^1 \left(\dfrac{a_{n-1}}{a_n}\right) = $ Sum of roots taken one at a time

(2) $\displaystyle\sum_{1 \leq i < j \leq n} \alpha_i \alpha_j = \alpha_1 \alpha_2 + \alpha_1 \alpha_3 + ... + \alpha_1 \alpha_n + \alpha_2 \alpha_3 + \alpha_2 \alpha_4 + ... + \alpha_{n-1} \alpha_n = (-1)^2 \left(\dfrac{a_{n-2}}{a_n}\right)$

(3) $\displaystyle\sum_{1 \leq i < j < k \leq n} \alpha_i \alpha_j \alpha_k = (-1)^3 \left(\dfrac{a_{n-3}}{a_n}\right)$

....

....

....

(r) $\displaystyle\sum \alpha_1 \alpha_2 \alpha_3 .....\alpha_r = (-1)^r \left(\dfrac{a_{n-r}}{a_n}\right)$

....

(n) $\alpha_1 \alpha_2 \alpha_3 .....\alpha_n = (-1)^n \left(\dfrac{a_0}{a_n}\right)$

Ex1. Consider the biquadratic equation $x^4 + px^3 + qx^2 + rx + s = 0$ having roots α, β, γ and δ.

Statement-1: The product of two roots of equation is equal to product of the other two then $r^2 = p^2 s$.

Statement-2: If $\alpha\beta = \gamma\delta$ then equation can be written as $x^4 + px^3 + qx^2 + rx + s = (x^2 + \ell x + \sqrt{s})(x^2 + mx + \sqrt{s})$

(A) Statement-1 is True, Statement-2 is a correct explanation for Statement-1.

(B) Statement-1 is True, Statement-2 is True, & Statement-2 is not a correct explanation for Statement-1

(C) Statement-1 is True, Statement-2 is False.

(D) Statement-1 is False, Statement-2 is True.

Sol. Given α, β, γ and δ as roots and Let $\alpha\beta = \gamma\delta$

(i) $\alpha + \beta + \gamma + \delta = -p$

(ii) $\sum \alpha\beta\gamma = -r \rightarrow \alpha\beta\gamma + \beta\gamma\delta + \gamma\delta\alpha + \delta\alpha\beta = -r \rightarrow \alpha\beta(\gamma + \beta + \alpha + \delta) = -r$ (Assuming $\alpha\beta = \gamma\delta$)

(iii) $\alpha\beta\gamma\delta = s \rightarrow \alpha\beta = \sqrt{s}$

Now $\alpha\beta(\alpha + \beta + \gamma + \delta) = -r \rightarrow p\sqrt{s} = r \rightarrow r^2 = p^2 s$

Hence, statement – 1 is correct and statement – 2 is also correct explanation so option (A) is correct.

Ex2. Let $f(x) = x^4 + ax^3 + bx^2 + cx + d$ be a polynomial with real coefficients and real zeroes. If $|f(i)| = 1$, (where $i = \sqrt{-1}$ and $|a + ib| = \sqrt{a^2 + b^2}$) then $a + b + c + d$ is equal to

(A) -1　　　　　　　(B) 1　　　　　　　(C) 0　　　　　　　(D) can't be determine

Sol. Let $f(x) = (x - x_1)(x - x_2)(x - x_3)(x - x_4)$

$\rightarrow |f(i)| = \sqrt{1 + x_1^2}\sqrt{1 + x_2^2}\sqrt{1 + x_3^2}\sqrt{1 + x_4^2} = 1 \rightarrow x_1 = x_2 = x_3 = x_4 = 0$

$\rightarrow$ all four roots are zero. So, $f(x) = x^4$

$\therefore a + b + c + d = 0 \rightarrow$ Option (C) is correct.

Ex3. Find the set of values of λ for which there is atleast one triplet (x, y, z) of R - {0} such that

$\dfrac{x}{z} + \dfrac{y}{x} + \dfrac{z}{y} = \dfrac{y}{z} + \dfrac{z}{x} + \dfrac{x}{y} = \lambda$.

Sol. Let $a = \dfrac{x}{y}$, $b = \dfrac{y}{z}$ and $c = \dfrac{z}{x}$.

Then as per question: $abc = 1$, $a + b + c = \lambda$ and $\dfrac{1}{a} + \dfrac{1}{b} + \dfrac{1}{c} = \lambda \rightarrow ab + bc + ca = \lambda abc = \lambda$

$\rightarrow a, b, c$ are roots of equation: $t^3 - \lambda t^2 + \lambda t - 1 = 0$

$\rightarrow (t^3 - 1) - \lambda t(t - 1) \rightarrow (t - 1)(t^2 + (1 - \lambda)t + 1) = 0$

Since, it has three real solution so, Discriminant of $t^2 + (1 - \lambda)t + 1 = 0$ must be ≥ 0

$\rightarrow (1 - \lambda)^2 - 4 \geq 0 \rightarrow \lambda \in (-\infty, -1] \cup [3, \infty)$.

Ex4. Consider the sequence of quadratic expressions $Q_r(x) = a_r x^2 + b_r x - 1$ where $a_r = (-1)^r$, $b_r \in (2, \infty)$ $\forall$ $r \in \{0, 1, 2, 3, \ldots., n\}$

Statement-1: If $b_{r+1}^2 - b_r^2 = 4$ and $b_{r+2} - b_r = k$ where k is positive real constant, then roots $\in (0, 1)$ of all the quadratic equations $Q_r(x) = 0$, will be in Harmonic Progression.

Statement-2: $x = 1$, lies between and outside the roots of the equation $Q_r(x) = 0$ if r is odd and even respectively.

(A) Statement-1 is True, Statement-2 is a correct explanation for Statement-1.

(B) Statement-1 is True, Statement-2 is True, & Statement-2 is not a correct explanation for Statement-1

(C) Statement-1 is True, Statement-2 is False.

(D) Statement-1 is False, Statement-2 is True.

Sol. (I) clearly, $b_{r+1}^2 - b_r^2 = 4$ govern the condition of exactly one root common between $Q_{r+1}(x) = 0$ and $Q_r(x) = 0$.

Now, for $\left.\begin{array}{l} Q_0(x)=x^2+b_0 x-1=0 \\ Q_1(x)=-x^2+b_1 x-1=0 \end{array}\right\}$ Let α_1 be it's common root then

$$\alpha_1=\frac{2}{b_0+b_1}, \alpha_2=\frac{2}{b_1+b_2}, \alpha_3=\frac{2}{b_2+b_3},\dots\dots\dots, \text{ and so on}$$

$$\therefore \quad b_{r+2}-b_r=k \rightarrow b_{r+2}-b_r=k=\frac{2}{\alpha_{r+2}}-\frac{2}{\alpha_{r+1}}=\text{constant} \rightarrow \alpha_1, \alpha_2, \alpha_3 \text{ are in HP}$$

(II) $a_r f(1)=a_r (a_r+b_r-1)$

   If r = even then $a_r f(1) > 0$ & if r = odd then $a_r f(1) < 0$

Ex5.   If $x^5 - x^3 + x = a$, when $x > 0$, then the maximum value of $2a - x^6$ is equal to _____

Sol.   We have $a=x(x^4-x^2+1)=\frac{x(x^6+1)}{x^2+1} \rightarrow x^6+1=a\left(\frac{x^2+1}{x}\right) \geq 2a$ (As AM ≥ GM $\rightarrow x+\frac{1}{x} \geq 2 \, \forall x \in R^+$)

$\rightarrow 2a - x^6 \leq 1$, so maximum value of $2a - x^6$ is 1.

THEOREM · 6

The sum of rth powers of roots of the polynomial equation $f(x) = a_n x^n + a_{n-1}x^{n-1} + a_{n-2}x^{n-2}+\dots\dots+a_1 x+a_0=0$,

where $a_n \neq 0$ and $a_0, a_1, a_2, a_3,\dots, a_n \in R$ is the coefficient of x^{-r} in expansion of $\frac{xf'(x)}{f(x)}$.

PROOF

Let $\alpha_1, \alpha_2, \alpha_3, \alpha_4,\dots\dots, \alpha_n$ are roots of equation $a_n x^n + a_{n-1}x^{n-1}+a_{n-2}x^{n-2}+\dots\dots+a_1 x+a_0=0$

Then $f(x)=a_n x^n+a_{n-1}x^{n-1}+a_{n-2}x^{n-2}+\dots\dots+a_1 x+a_0=a_n(x-\alpha_1)(x-\alpha_2)(x-\alpha_3)\dots\dots(x-\alpha_n)=0$

Now, take log both sides, we get

   $\ln f(x) = \ln(a_n) + \ln(x - \alpha_1) + \ln(x - \alpha_2) + \ln(x - \alpha_3) + \dots\dots\dots+ \ln(x - \alpha_n)$

   Now Differentiate both side, we get

$$\frac{f'(x)}{f(x)}=0+\frac{1}{x-\alpha_1}+\frac{1}{x-\alpha_2}+\dots\dots+\frac{1}{x-\alpha_n}$$

$$\rightarrow \frac{xf'(x)}{f(x)}=\frac{x}{x-\alpha_1}+\frac{x}{x-\alpha_2}+\dots\dots+\frac{x}{x-\alpha_n}=\frac{1}{1-\left(\frac{\alpha_1}{x}\right)}+\frac{1}{1-\left(\frac{\alpha_2}{x}\right)}+\dots\dots+\frac{1}{1-\left(\frac{\alpha_n}{x}\right)}$$

As we know that: $1 + x + x^2 + x^3 +\dots+ \infty = \frac{1}{1-x} \, \forall |x|<1$. So,

$$\frac{xf'(x)}{f(x)}=\left[1+\left(\frac{\alpha_1}{x}\right)^1+\left(\frac{\alpha_1}{x}\right)^2+\dots+\infty\right]+\left[1+\left(\frac{\alpha_2}{x}\right)^1+\left(\frac{\alpha_2}{x}\right)^2+\dots+\infty\right]+\dots+\left[1+\left(\frac{\alpha_n}{x}\right)^1+\left(\frac{\alpha_n}{x}\right)^2+\dots+\infty\right]$$

$$\rightarrow \alpha_1^r+\alpha_2^r+\alpha_3^r+\dots+\alpha_n^r=\text{Coefficient of } x^{-r} \text{ in the expansion of} \frac{xf'(x)}{f(x)}.$$

EXERCISE - 4

1. Solve: $x^4 - 4x^2 + 8x + 35 = 0$, given $2 + i\sqrt{3}$ is a root.

2. Solve: $x^4 - 5x^3 + 4x^2 + 8x - 8 = 0$, given that one of root is $1 - \sqrt{5}$.

3. Find a polynomial equation of the lowest degree with rational coefficients having $\sqrt{3}$ and $1 - 2i$ as two of its roots.

4. Solve: $4x^5 + x^3 + x^2 - 3x + 1 = 0$, given that it has rational roots.

5. Show that the equation $x^3 + qx + r = 0$ has two equal roots if $27r^2 + 4q^3 = 0$.

6. Solve the equation $15x^3 - 23x^2 + 9x - 1 = 0$ whose roots are in harmonic progression.

7. Solve the equation $x^3 - 11x^2 + 38x - 40 = 0$, given that the ratio of two of its root is $2 : 1$.

8. Find the roots of $4x^3 + 20x^2 - 23x + 6 = 0$. If it's given that two roots are equal.

9. Find the greatest possible number of real roots of the equation $x^5 - 6x^2 - 4x + 5 = 0$

10. Show that $x^5 - 3x^2 + 4 = 0$ has at least two imaginary roots.

11. If α, β, γ are roots of the equation, $x^3 - x - 1 = 0$ then $\dfrac{1+\alpha}{1-\alpha} + \dfrac{1+\beta}{1-\beta} + \dfrac{1+\gamma}{1-\gamma}$ has the value equals to___

12. If α, β, γ are the roots of $x^3 - 7x + 7 = 0$, find the value of $\alpha^{-4} + \beta^{-4} + \gamma^{-4}$.

13. If a, b, c are roots of equation $x^3 + px^2 + qx + r = 0, (r \neq 0)$ then value of $\left| \dfrac{a}{b} + \dfrac{b}{c} + \dfrac{c}{a} + \dfrac{b}{a} + \dfrac{c}{b} + \dfrac{a}{c} - \dfrac{pq}{r} \right|$ is_

14. **Statement –1:** Coefficient of x^{51} in the expansion of $(x - 1)(x^2 - 2)(x^3 - 3) \ldots (x^{10} - 10)$ is $- 1$

    Statement –2: Coefficient of $x^{\frac{n(n+1)}{2} - 4}$, $n \geq 4$ in the expansion of $(x - 1)(x^2 - 2)(x^3 - 3) \ldots (x^n - n)$ is $- 4 + (-1)(-3) = - 1$.

    (A) Statement - 1 is true, Statement - 2 is true & Statement – 2 is a correct explanation for Statement - 1.

    (B) Statement-1 is true, Statement -2 is true & Statement -2 is not a correct explanation for Statement - 1.

    (C) Statement -1 is true, Statement - 2 is false.

    (D) Statement -1 is false, Statement - 2 is true.

15. Solve $x^4 - 2x^3 - 12x^2 + 10x + 3 = 0$.

16. If α is an imaginary root of equation $x^7 - 1 = 0$. Form a cubic equation whose roots are $\alpha + \alpha^6$, $\alpha^2 + \alpha^5$ and $\alpha^3 + \alpha^4$.

17. If α and β are roots of the equation $px^2 + qx + r = 0$, $p \neq 0$. If p, q, r in arithmetic progression and $\dfrac{1}{\alpha} + \dfrac{1}{\beta} = 4$, then the value of $|\alpha - \beta|$ is

    (A) $\dfrac{\sqrt{34}}{9}$ (B) $\dfrac{2\sqrt{13}}{9}$ (C) $\dfrac{\sqrt{61}}{9}$ (D) $\dfrac{2\sqrt{17}}{9}$

18. Find the value of k for which the equations $3x^2 + 4kx + 2 = 0$ and $2x^2 + 3x - 2 = 0$ have a common root.

19. Find the value of 'α' for which the system of inequality $x^2 + 2x + \alpha \leq 0$ and $x^2 - 4x - 6\alpha \leq 0$ has a unique solution.

20. For what value of 'a' do the curves $y = 1 + \dfrac{x^2}{a^3}$ & $y = 4\sqrt{x}$ possesses only one point in common.

GRAPH OF QUADRATIC EXPRESSIONS

To draw the curve of $y = ax^2 + bx + c$, $a \neq 0$. $a, b, c \in R$. Let's start with very basic curve $y = x^2$.

You can see in table that values of $y = x^2$ at $x = 0, 1, -1, 2, -2, 3 \ldots$ are $0, 1, 1, 4, 4, 9, \ldots$, respectively. So, following points lies on the curve $y = x^2$

$(0, 0), (1, 1), (-1, 1), (2, 4), (-2, 4), (3, 9), \ldots$ etc.

x	0	1	-1	2	-2	3
y	0	1	1	4	4	9

Let's plot the curve.

SCALE:

X: One box = 1 unit.

Y: One box = 1 unit.

Observe the curve of $y = x^2$ as shown in figure.

The curve is symmetric about y – axis. The curve is known as parabolic opening upwards.

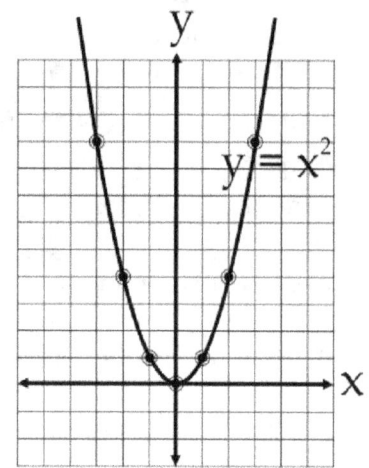

If you plot the curves of $y = 2x^2$, $y = 3x^2$, $y = 5x^2$ then you will observe the similar nature of these curves and all will be opening upward.

Now, If you will plot the curve of $y = -x^2$, $y = -2x^2$, $y = -5x^2$, etc. then you will see the parabola opening downward.

Let's have comparison curves of $y = k\, x^2$, $k \in$ Real constant.

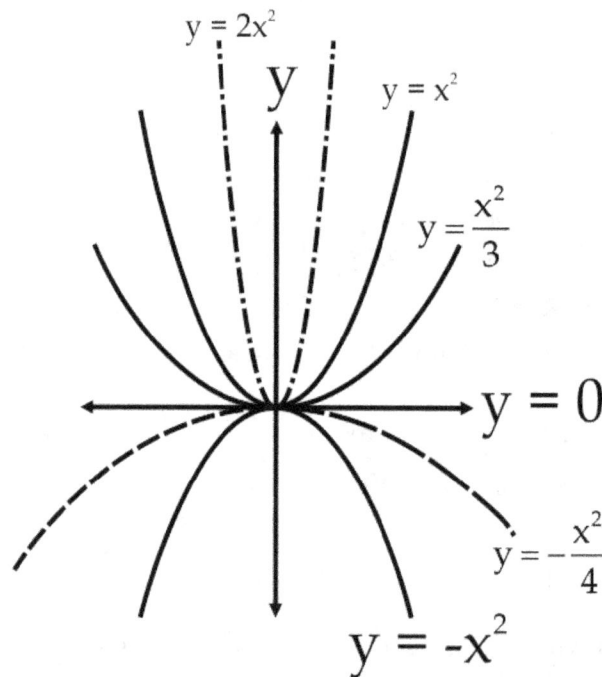

Observe the pattern of $y = k\, x^2$ with different values of k. Draw some curve your own to have actual feeling of learning.

OBSERVATION:

The curve of $y = ax^2$ is parabolic opening upward if $a > 0$ and opening downward if $a < 0$.

UPWARD & DOWNWARD SHIFTING OF $Y = X^2$.

Let's plot the following curves

    (i) $y = x^2$ (ii) $y = x^2 + 2$ (iii) $y = x^2 - 3$

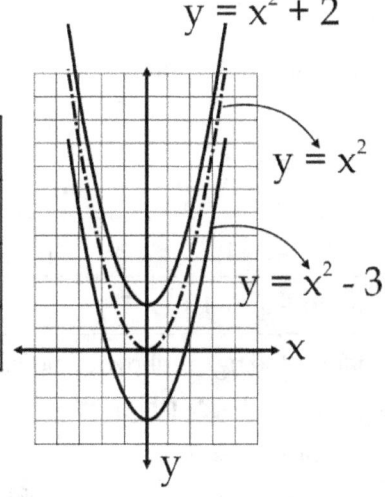

x	0	1	2	3	-2
$y = x^2$	0	1	4	9	4
$y = x^2 + 2$	2	3	6	11	6
$y = x^2 - 3$	-3	-2	1	6	1

Observe the table and curve closely and analyse the situation. You may find that if we have the curve of $y = x^2$ then $y = x^2 + k$ will be obtain by shifting k unit upward and $y = x^2 - k$ will be obtain by shifting the curve $y = x^2$ by k unit downward where k > 0.

NOTE: *If we have $y = f(x)$ curve then we can draw the curve of $y = f(x) + k$ & $y = f(x) - k$ by shifting $y = f(x)$ curve k unit, upward & downward respectively, where k > 0.*

LEFTWARD & RIGHTWARD SHIFTING OF CURVE $Y = X^2$

Let's plot the following curves

    (i) $y = x^2$ (ii) $y = (x - 1)^2$ (iii) $y = (x + 4)^2$

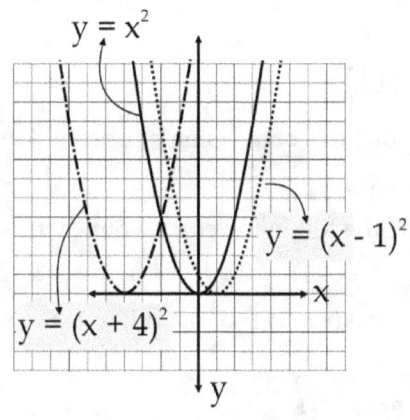

x	0	1	2	3	4
$y = x^2$	0	1	4	9	16
$y = (x - 1)^2$	1	0	1	4	9
$y = (x + 2)^2$	4	9	16	25	36

SCALE: X: One box = 1 unit.   Y: One box = 1 unit.

Observe the shifting of curve $y = x^2$.

In, $y = x^2$, at $x = x_0$ we will get the value as $y = x_0^2$ so, (x_0, x_0^2) lie on the curve $y = x^2$

Same height x_0^2 will be achieve at $x = x_0 + 1$ in $y = (x - 1)^2$ ($\rightarrow$ curve shifted rightward by '1' unit) and at $x = x_0 - 4$ in $y = (x + 4)^2$ ($\rightarrow$ curve shifted leftward by '4' unit).

NOTE: *If we have y = f(x) curve then we can draw the curve of y = f(x - k) & y = f(x + k) by shifting y = f(x) curve k unit rightward & leftward respectively, where k > 0.*

Ex1. Draw the following curves in single graph paper and observe the transformation.

(i) $y = 2x^2$ (ii) $y = 2(x - 3)^2$ (iii) $y = 2(x - 3)^2 + 1 = 2x^2 - 12x + 19$

Sol. Consider the table for $y = 2x^2$, $y = 2(x - 3)^2$ & $y = 2(x - 3)^2 + 1 = 2x^2 - 12x + 19$.

x	- 2	- 1	0	1	2	3
$y = 2x^2$	8	2	0	2	8	18
$y = 2(x - 3)^2$	50	32	18	8	2	0
$y = 2(x - 3)^2 + 1$	51	33	19	9	3	1

Analyze the figure shown below.

GRAPHICAL TRANSFORMATION

Step (i) Draw $y = 2x^2$

Step (ii) Sift the curve of $y = 2x^2$ rightward by 3 unit to get the curve $y = 2(x - 3)^2$

Step (iii) Now, shift the curve $y = 2(x - 3)^2$ upward by '1' unit to get $y = 2(x - 3)^2 + 1$.

Here you can see that $y = 2(x - 3)^2 + 1$

$\rightarrow y = 2x^2 - 12x + 19$

This is a quadratic expression of the form

$y = ax^2 + bx + c$, $a \neq 0$. $a, b, c \in R$.

So, to draw the curve of $y = ax^2 + bx + c$, $a \neq 0$, $a, b, c \in R$, *first we need to convert it in the form*

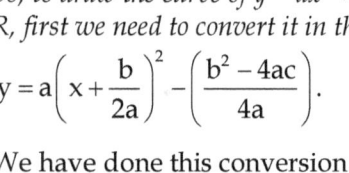
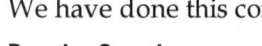

$$y = a\left(x + \frac{b}{2a}\right)^2 - \left(\frac{b^2 - 4ac}{4a}\right).$$

We have done this conversion at start of the topic.

Practice Question:

Ex2. Draw the following curve your own.

(i) $y = (x - 1)^2 + 2$ (ii) $y = 5(x - 2)^2$ (iii) $y = 2(x - 3)^2 + 5$ (iv) $y = - 2(x + 1)^2 + 6$

(v) $y = - (x + 1)^2 - 3$ (vi) $y = x^2 - x - 6$ (vii) $y = - 3x^2 + x - 1$ (viii) $y = - x^2 + x + 1$

VERTEX OF PARABOLA

Point of intersection of parabola and its axis is known as vertex of parabola. The co-ordinates of vertex of

parabola $y = ax^2 + bx + c$, $a \neq 0$, $a, b, c \in R$ is $V\left(-\dfrac{b}{2a}, -\dfrac{D}{4a}\right)$ where $D = b^2 - 4ac$ i.e. Discriminant.

Note: you may have observed the opening of parabola $y = ax^2 + bx + c$, it's opening upward if $a > 0$ and opening downward if $a < 0$.

Look at the figure shown below.

Parabola $y = ax^2 + bx + c$

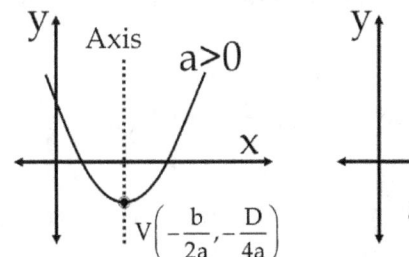

SHORT CUT:

To have a rough idea about the curve, we need to know about opening and vertex co-ordinates of parabola so, do the following steps to plot the curve of parabola $y = ax^2 + bx + c$

Step 1:   know the opening from a. Opening upward if a > 0 and downward if a < 0

Step2:   Find the vertex co –ordinates $x = -\dfrac{b}{2a}$ & $y = -\dfrac{D}{4a}$, i.e.$V\left(-\dfrac{b}{2a}, -\dfrac{D}{4a}\right)$.

Step3:   Locate the vertex on graph paper and plot the curve.

Step 4:   write the important point on the curve like intersection of curve with x – axis and y- axis etc.

Ex 3:   Plot the curve of $y = x^2 - x - 6$.

Sol.   Step 1: a = 1 > 0 → parabola opening upward.

Step 2: $x = -\dfrac{b}{2a} = -\dfrac{(-1)}{2 \times 1} = \dfrac{1}{2}$, $y = -\dfrac{D}{4a} = -\dfrac{\left((-1)^2 - 4 \times 1 \times (-6)\right)}{4 \times 1} = -\dfrac{25}{4} \to V\left(\dfrac{1}{2}, -\dfrac{25}{4}\right)$.

Step 3 and 4 are shown in figure.

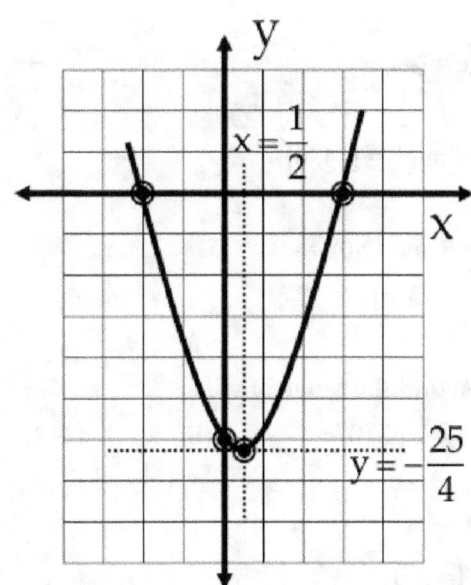

Scale: X: One box = 1 unit.   Y: One box = 1 unit.

Ex4. Find the maximum value of $Q(x) = -2x^2 + x + 1$ and vertex coordinates.

Sol. Since, $a = -2 < 0$ so, $Q(x)$ will be a parabola opening downward, Hence maximum value of $Q(x)$ will be at vertex.

$$\to x = -\frac{b}{2a} \to x = -\frac{1}{2(-2)} = \frac{1}{4}. \text{ So, } Q(x)\big|_{MAX} = Q\left(\frac{1}{4}\right) = -2\left(\frac{1}{4}\right)^2 + \left(\frac{1}{4}\right) + 1 = \frac{9}{8}$$

Alternate:

$$Q(x)\big|_{MAX} = -\frac{D}{4a} = -\frac{(1 + 4 \times 2 \times 1)}{4(-2)} = \frac{9}{8}$$

Vertex co-ordinates: $V\left(\frac{1}{4}, \frac{9}{8}\right)$

Ex5. Find the minimum value of $Q(x) = x^2 + 2x + 5$

Sol. as $a = 1 > 0 \to$ the graph of $y = Q(x)$ is parabola opening upward. Hence, minimum value will be at it's vertex.

So, $Q(x)\,|_{MIN} = -\dfrac{D}{4a} = -\dfrac{(4 - 4 \times 5)}{4} = 4$.

Ex6. Consider $y = x^2 - 2x - 3$ then find range of y if

(i) $x \in R$ (ii) $x \in (0, 3]$ (iii) $x \in [-2, 0)$

Sol. $a = 1 > 0 \to$ graph of $y = x^2 - 2x - 3$ is parabola opening upward.

The vertex of the parabola is at V (1, - 4)

Look at the figure shown:

Scale: X: One box = 1 unit.   Y: One box = 1 unit.

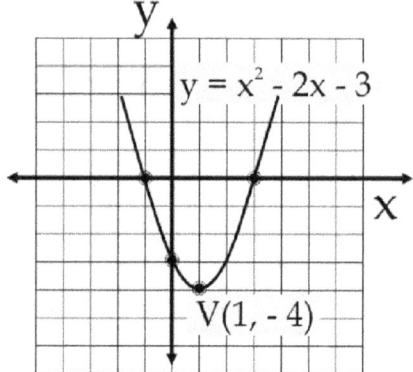

(i)   If $x \in R$

   $x^2 - 2x - 3\,|_{MIN} = -4 \to$ Range = $[-4, \infty)$

(ii)   If $x \in (0, 3]$

   Then values of $y \in [-4, 0] \to$ Range = $[-4, 0]$

(iii)   If $x \in [-2, 0)$

   As $Q(-2) = (-2)^2 - 2(-2) - 3 = 5$ and $Q(0) = -3$.

   $\to y \in (-3, 5]$. Hence, Range = $(-3, 5]$.

Ex7. Consider $y = x^4 - 4x^2 + 1$ then find the range of y if

(i) $x \in R$ (ii) $x \in [0, 3)$ (iii) $x \in [-4, -2]$

Sol. Let $x^2 = t \geq 0$

   $\to y = t^2 - 4t + 1$ where $t \geq 0$

   Vertex of $y = t^2 - 4t + 1$ is V (2, - 3)

Consider the figure of $y = t^2 - 4t + 1$ where $t \geq 0$

$y = t^2 - 4t + 1 = (t - 2)^2 - 3$

$(t - 2)^2 - 3 = 0 \rightarrow t = 2 \pm \sqrt{3}$

(i) $x \in R \rightarrow t \geq 0 \rightarrow y \in [-3, \infty)$

     Range of $y = [-3, \infty)$

(ii) $x \in [0, 3) \rightarrow t \in [0, 9)$

     $\rightarrow y \in [-3, 46)$

(iii) $x \in [-4, -2) \rightarrow t \in (4, 16]$

     $\rightarrow y(4) = (4 - 2)^2 - 3 = 1$ & $y(16) = (16 - 2)^2 - 3 = 193$

     $\rightarrow y \in (1, 193]$ so, Range is $= (1, 193]$

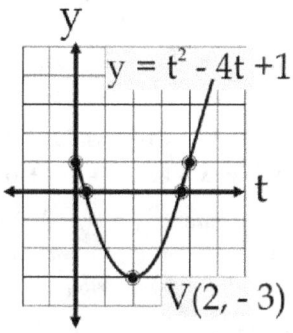

SIGN OF QUADRATIC EXPRESSIONS

In this topic, we will study about the inequality part of quadratic.

Consider $y = f(x) = ax^2 + bx + c$ where $a \neq 0$, $a, b, c \in R$. There are two possibilities of opening of parabola

So, let's make two cases (i) $a > 0$ and (ii) $a < 0$ and analyse the situation.

CASE (i) a > 0

Consider the graphs of $y = ax^2 + bx + c$ where $a \neq 0$, $a, b, c \in R$, as shown below.

The vertex of parabola is $V\left(-\dfrac{b}{2a}, -\dfrac{D}{4a}\right)$. Where $D = b^2 - 4ac$

$x = -\dfrac{b}{2a}$ is the line having vertex of parabola.

(a) If $D < 0$, $a > 0 \rightarrow -\dfrac{D}{4a} > 0$ hence vertex will be above x – axis.

❖ **Note:** If $a > 0$ and $D < 0 \rightarrow y = ax^2 + bx + c > 0 \, \forall \, x \in R$.

(b) If $D = 0$ then $-\dfrac{D}{4a} = 0$ hence vertex will be on the x- axis.

$\rightarrow ax^2 + bx + c = 0$ has repeated root $x = -\dfrac{b}{2a}$.

❖ **Note:** If $a > 0$ and $D = 0$ then $ax^2 + bx + c \geq 0 \, \forall \, x \in R$.

(c) If $D > 0$ and $a > 0$ then $ax^2 + bx + c = 0$ has real roots $x = \alpha$ & β.

Here you can observe that $ax^2 + bx + c > 0 \, \forall \, x \in (-\infty, \alpha) \cup (\beta, \infty)$ and $ax^2 + bx + c < 0 \, \forall \, x \in (\alpha, \beta)$.

❖ **Note:** If $a > 0$, $D > 0$ then $ax^2 + bx + c > 0 \, \forall \, x \in (-\infty, \alpha) \cup (\beta, \infty)$ and $ax^2 + bx + c < 0 \, \forall \, x \in (\alpha, \beta)$.

CASE (ii) a < 0

Consider the graphs of $y = ax^2 + bx + c$ where $a \neq 0$, $a, b, c \in R$, as shown below.

If $a < 0 \rightarrow$ parabola is opening downward.

(a) If $a < 0$ and $D > 0 \rightarrow -\dfrac{D}{4a} > 0 \rightarrow$ Vertex of parabola will be above x – axis.

$\rightarrow$ the curve $y = ax^2 + bx + c$ will intersect the x-axis at two real points $x = \alpha$ & β.

❖ **Note:** $y = ax^2 + bx + c > 0 \ \forall \ x \in (\alpha, \beta)$ & $y = ax^2 + bx + c < 0 \ \forall \ x \in (-\infty, \alpha) \cup (\beta, \infty)$

(b)   If $a < 0$ & $D = 0$ then $y = ax^2 + bx + c = 0$ has repeated real roots

$x = -\dfrac{b}{2a}$ and vertex will lie on x – axis.

Case: a < 0

❖ **Note:** If $a < 0$ and $D = 0$ then $ax^2 + bx + c \leq 0 \ \forall \ x \in R$.

(C)   If $a < 0$ and $D < 0$ then you can see that the curve is entirely below x – axis $\rightarrow ax^2 + bx + c < 0 \ \forall \ x \in R$

❖ **Note:** If $a < 0$ and $D < 0$ then $ax^2 + bx + c < 0 \ \forall \ x \in R$.

To know the sign of quadratic expression, above six notes are very important to understand. So, have crystal clear concepts regard the same.

I am having example to apply the above concepts.

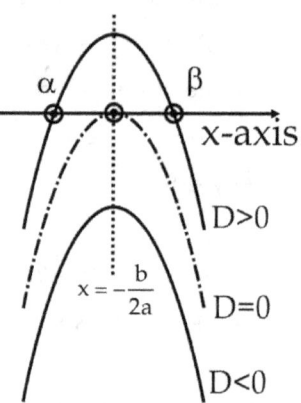

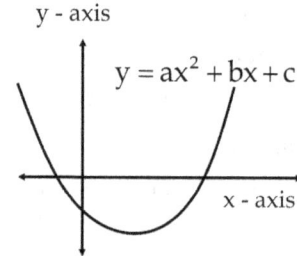

Ex1.   The figure shows the exact graph of $y = ax^2 + bx + c$ then select the wrong statement.

(A) $a > 0$

(B) $c > 0$

(C) $ab < 0$

(D) $D > 0$, D = Discriminant

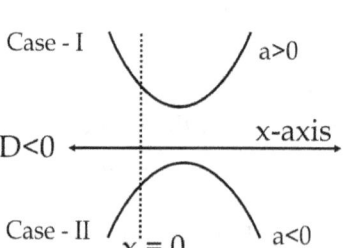

Sol.   Here you can see the parabola opening upward $\rightarrow a > 0$.

The curve $y = ax^2 + bx + c$ intersect y – axis below x – axis $\rightarrow y(0) = c < 0$. Option (B) is wrong.

The curve $y = ax^2 + bx + c$ intersect x – axis at two distinct points $\rightarrow D > 0$

The vertex is right side of y – axis $\rightarrow x = -\dfrac{b}{2a} > 0$

Here, $a > 0$ and $x = -\dfrac{b}{2a} > 0 \rightarrow b < 0 \rightarrow ab < 0$

Hence, option (B) is only wrong option.

Ex2.   If $c < 0$ & $ax^2 + bx + c = 0$ does not have any real roots then prove that

(i) $a - b + c < 0$

(ii) $9a + 3b + c < 0$

Sol.   Let $f(x) = ax^2 + bx + c$ and $ax^2 + bx + c = 0$ does not have any real roots $\rightarrow D < 0$.

Now, there are two possibilities of parabola as shown in figure.

But you can observe that

$f(0) = c < 0 \rightarrow$ the curve cut the y – axis below x – axis

$\rightarrow$ Case – II is our questions case.

$\rightarrow a < 0$ and $D < 0 \rightarrow f(x) = ax^2 + bx + c < 0 \ \forall \ x \in R$.

So, it's true for $x = -1$ and 3 also.

$\rightarrow f(-1) = a - b + c < 0$ and $f(3) = 9a + 3b + c < 0$

Ex3. For what least integral value of k, $(k - 2)x^2 + 8x + k + 4 > 0 \ \forall \ x \in R$?

Sol. $ax^2 + bx + c > 0 \ \forall \ x \in R$ only if $a > 0$ and $D < 0$

(i) $a > 0 \rightarrow (k - 2) > 0 \rightarrow k \in (2, \infty)$ ....(1)

(ii) $D < 0 \rightarrow 64 - 4(k - 2)(k + 4) < 0 \rightarrow 16 - (k - 2)(k + 4) < 0$

$\rightarrow 16 - [k^2 + 2k - 8] < 0 \rightarrow k^2 + 2k - 24 > 0 \rightarrow (k - 4)(k + 6) > 0 \rightarrow k \in (-\infty, -6) \cup (4, \infty)$ ....(2)

From (1) & (2) we get $k \in (4, \infty)$

∴ Least integral value of k = 5.

Ex4. The equation $ax^2 + bx + c = 0$, $a, b, c \in R$ has no real roots. Prove that $c \ (a + b + c) > 0 \ \forall \ x \in R$.

Sol. Let $f(x) = ax^2 + bx + c = 0$ has no real roots $\rightarrow D < 0$

$\rightarrow$ there are two possible situations shown in figure

As we observe that $f(0) = c$ and $f(1) = a + b + c$.

And $c \ (a + b + c) = f(0) \times f(1)$

If we focus on both the situations then we can say

$f(0)$ and $f(1)$ are of same sign $\rightarrow f(0) \times f(1) > 0 \ \forall \ x \in R$

$\rightarrow c \ (a + b + c) > 0 \ \forall \ x \in R$.

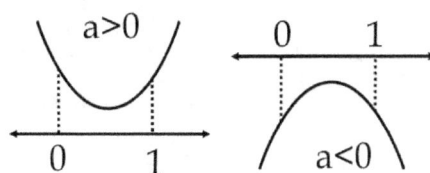

Ex5. Find the set of values of 'a' for which quadratic polynomials

(i) $(a + 4) x^2 - 2ax + 2a - 6 < 0 \ \forall \ x \in R$ (ii) $(a - 1) x^2 - (a + 1) x + (a + 1) > 0 \ \forall \ x \in R$

Sol. (i) $(a + 4) x^2 - 2ax + 2a - 6 < 0 \ \forall \ x \in R$ possible only if

(a) $a + 4 < 0 \rightarrow a \in (-\infty, -4)$ ......(1)

(b) $D < 0 \rightarrow 4a^2 - 4(a + 4)(2a - 6) < 0 \rightarrow a^2 - (a + 4)(2a - 6) < 0 \rightarrow a^2 + 2a - 24 > 0$

$\rightarrow (a - 4)(a + 6) > 0 \rightarrow a \in (-\infty, -6) \cup (4, \infty)$ ....(2)

From (1) and (2) we get $a \in (-\infty, -6)$.

(ii) $(a - 1) x^2 - (a + 1) x + (a + 1) > 0 \ \forall \ x \in R$ possible only if

(a) $a - 1 > 0 \rightarrow a \in (1, \infty)$ .... (1)

(b) $D < 0 \rightarrow (a + 1)^2 - 4(a - 1)(a + 1) < 0 \rightarrow (a + 1)^2 - 4 a^2 + 4 < 0$

$\rightarrow 3a^2 - 2a - 5 > 0 \rightarrow (3a - 5)(a + 1) > 0 \rightarrow a \in (-\infty, -1) \cup (5/3, \infty)$ .... (2)

From (1) and (2) we get $a \in (5/3, \infty)$.

Ex6. If $f(x) = ax^2 + bx + c > 0 \ \forall \ x \in R$. Prove that $f(x) + f'(x) + f''(x) > 0 \ \forall \ x \in R$.

Sol. $f(x) = ax^2 + bx + c$

$f'(x) = 2ax + b$ and $f''(x) = 2a$

So, Let $g(x) = f(x) + f'(x) + f''(x) = ax^2 + bx + c + (2ax + b) + (2a) = ax^2 + (b + 2a)x + 2a + b + c$

$\rightarrow g(x) = ax^2 + (b + 2a)x + 2a + b + c$.

Now, it's given that: $a > 0$ and $D = b^2 - 4ac < 0$ as $ax^2 + bx + c > 0 \ \forall \ x \in R$.

For, $g(x) = ax^2 + (b + 2a)x + 2a + b + c$.

(i) $a > 0$ given

(ii) $D = (b + 2a)^2 - 4a(2a + b + c)$

$= b^2 + 4ab + 4a^2 - 8a^2 - 4ab - 4ac$

$= (b^2 - 4ac) - 8a^2 < 0$ as $b^2 - 4ac < 0$

From (i) and (ii) we can say that $g(x) > 0 \; \forall \; x \in R$

$\to f(x) + f'(x) + f''(x) > 0 \; \forall \; x \in R$

Ex7. Consider the figure of real quadratic $y = ax^2 + bx + c$
as shown. Select **wrong** option (Where $D = b^2 - 4ac$, $i = \sqrt{-1}$)

(A) One root of $ax^2 + bx + c = 0$ is $x = \dfrac{-b + i\sqrt{-D}}{2a}$.

(B) $ax^2 + bx + c > 0 \; \forall \; x \in R, a \neq 0$

(C) $|a| + |b| + c = 0$ for at least one real triplet (a, b, c).

(D) $h = -\dfrac{b}{2a}$ & $k = -\dfrac{D}{4a}$

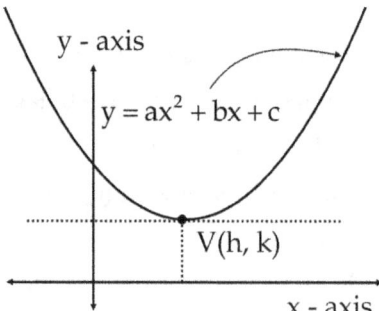

Sol. from the figure we can say (i) $a > 0$ and (ii) $D < 0 \to ax^2 + bx + c > 0 \; \forall \; x \in R$.

The roots of $ax^2 + bx + c = 0$ are $x = \dfrac{-b \pm \sqrt{D}}{2a} = \dfrac{-b \pm \sqrt{-(-D)}}{2a} = \dfrac{-b \pm i\sqrt{-D}}{2a}$

The vertex co-ordinates are $V(h, k) = \left(-\dfrac{b}{2a}, -\dfrac{D}{4a} \right)$.

As $y(0) = c > 0$ (from figure) so $|a| + |b| + c > 0$ and $|a| + |b| + c \neq 0$.

$\therefore$ there is no any real triplet (a, b, c) such that $|a| + |b| + c \neq 0$.

Hence, option (C) is wrong.

Ex8. If α & β ($\alpha < \beta$) are the roots of the equation $x^2 + bx + c = 0$, where $c < 0 < b$, then

(A) $0 < \alpha < \beta$ (B) $\alpha < 0 < \beta < |\alpha|$ (C) $\alpha < \beta < 0$ (D) $\alpha < 0 < |\alpha| < \beta$

Sol. $\alpha + \beta = -b < 0 \to$ Sum of roots is negative.

$\alpha\beta = c < 0 \to$ product of roots is also negative.

Now, $D = b^2 - 4c > 0 \to$ roots are real.

Hence, possible situation is as shown in figure.

From the figure we can have

$\alpha < 0 < \beta < |\alpha|$

$\therefore$ Option (B) is correct.

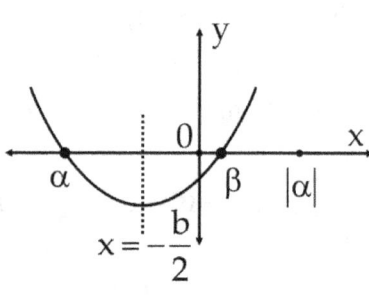

Ex9. If $ax^2 + bx + c = 0$ have real roots α, β such that $\alpha < \beta$, where $\alpha \neq 0$ and $\displaystyle\lim_{x \to \infty}\left(a + \dfrac{b}{x} + \dfrac{c}{x^2} \right) = 2$ then

$ax^2 + bx + c - |\alpha| = 0$ has

(A) both roots in (α, β) (B) both roots in $(-\infty, \alpha)$

(C) both roots in (β, ∞) (D) one root in $(-\infty, \alpha)$ and other in (β, ∞)

Sol. As $\lim\limits_{x \to \infty}\left(a + \dfrac{b}{x} + \dfrac{c}{x^2}\right) = 2 \to a = 2 > 0$.

$ax^2 + bx + c = 0$ have real roots α, β such that $\alpha < \beta \to D > 0$.

So, look at the figure of $y = ax^2 + bx + c$ and $y = ax^2 + bx + c - |\alpha|$ shown.

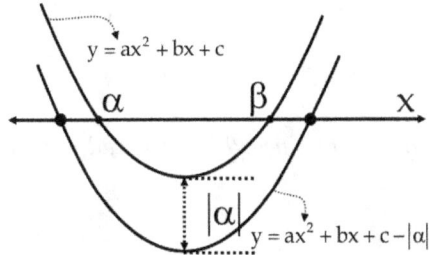

What are you observing in figure?

Shifting of parabola downward causes, one root in $(-\infty, \alpha)$ and other in (β, ∞).

Hence, Option (D) is correct.

Ex10. Let a, b, c, p, q be the real numbers. Suppose α, β are the roots of the equation $x^2 + 2px + q = 0$ & α, $1/\beta$ are the roots of the equation $ax^2 + 2bx + c = 0$, where $\beta \notin \{-1, 0, 1\}$

Statement 1: $(p^2 - q)(b^2 - ac) \geq 0$.

Statement 2: $b \notin pa$ or $c \notin qa$.

(A) Statement-1 is True, Statement-2 is a correct explanation for Statement-1.

(B) Statement-1 is True, Statement-2 is True, & Statement-2 is not a correct explanation for Statement-1

(C) Statement-1 is True, Statement-2 is False.

(D) Statement-1 is False, Statement-2 is True.

Sol. Observe the roots of equations

$x^2 + 2px + q = 0$ have roots α, β.

$ax^2 + 2bx + c = 0$ have roots α, $1/\beta$.

$\to$ exactly on root is common in both equations

Now, common root α may be real or imaginary.

(i) If $\alpha \in R$ then due to $p, q \in R$, β and $1/\beta$ are also a real number $\to D_1 = 4p^2 - 4q \geq 0$

Similarly, $D_2 = 4b^2 - 4ac \geq 0$.

$\to$ both equations have real roots $\to D_1 \times D_2 \geq 0$.

$\to (p^2 - q)(b^2 - ac) \geq 0$

(ii) If α is an imaginary root then β and $1/\beta$ must be imaginary, as a, b, c, p, q $\in R$.

$\to D_1 \leq 0$ and $D_2 \leq 0 \to D_1 \times D_2 \geq 0 \to (p^2 - q)(b^2 - ac) \geq 0$.

$\therefore$ Statement 1 is true

As $\beta \notin \{-1, 0, 1\} \rightarrow \beta^2 \neq 1, 0$

Now $\alpha\beta = q$ ....(1) and $\alpha/\beta = c/a$ .....(2)

From (1) and (2) we get $\dfrac{\alpha\beta}{\alpha/\beta} = \dfrac{qa}{c} = \beta^2 \neq 1 \rightarrow qa \neq c$

As $\beta \neq 1 \rightarrow \beta \neq 1/\beta \rightarrow \alpha + \beta \neq \alpha + 1/\beta$

$\rightarrow -2p \neq -2b/a \rightarrow ap \neq b$

∴ Statement 2 is also correct but not explaining first statement hence (B) is the correct option.

RANGE OF RATIONAL EXPRESSIONS

In this topic, I am going to discuss range of various form of expression related to Linear/Linear, Quadratic/Linear, Linear/Quadratic, Quadratic/Quadratic.

TYPE – I: $f(x) = \dfrac{\text{Linear}}{\text{Linear}}$

In general, the expression $y = \dfrac{ax + b}{cx + d}$ represents a rectangular hyperbola.

Ex1. Find the range of $y = \dfrac{4x - 1}{2x - 3} \ \forall \ x \in R - \{3/2\}$.

Sol. Observe the following limits

(i) $\displaystyle\lim_{x \to \infty} \dfrac{4x-1}{2x-3} = \lim_{x \to \infty} \dfrac{4 - \dfrac{1}{x}}{2 - \dfrac{3}{x}} = 2$ (ii) $\displaystyle\lim_{x \to -\infty} \dfrac{4x-1}{2x-3} = 2$

(iii) $\displaystyle\lim_{x \to \frac{3}{2}^+} \dfrac{4x-1}{2x-3} = \infty$ (iv) $\displaystyle\lim_{x \to \frac{3}{2}^-} \dfrac{4x-1}{2x-3} = -\infty$

(v) $y(3/2) = \text{N.D.}$

Look at the curve of the given expression.

$y = \dfrac{4x-1}{2x-3}$ Represents the rectangular hyperbola with asymptotes $x = 3/2$ and $y = 2$

Range : $y \in R - \{2\}$.

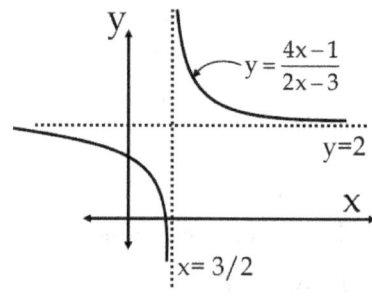

ALTERNATE - METHOD

$y = \dfrac{4x-1}{2x-3} \rightarrow 2xy - 3y = 4x - 1 \rightarrow x(2y - 4) = 3y - 1$.

$\rightarrow x = \dfrac{3y-1}{2y-4} \rightarrow y \in R - \{2\}$ as $x \in R - \{3/2\}$.

Ex2. If $a > b > c > d > 0$ then find the range of $y = \dfrac{ax - c}{bx - d}$

Sol. $y = \dfrac{ax-c}{bx-d} \rightarrow y(bx - d) = ax - c$

$\rightarrow$ byx – yd = ax – c $\rightarrow$ x(by – a) = dy – c

$\rightarrow x = \dfrac{dy - c}{by - a} \rightarrow y \neq a/b$ so range is y $\in$ R – {a/b}.

Look at the figure.

There is possibility of another rectangular hyperbola if (d/b) < c/a. Think!

TYPE – II: $f(x) = \dfrac{\text{Quadratic}}{\text{Linear}}$

Ex3. Find the range of $y = \dfrac{x^2 + 2x - 11}{x - 3}$.

Sol. $y = \dfrac{x^2 + 2x - 11}{x - 3} \rightarrow xy - 3y = x^2 + 2x - 11 \rightarrow x^2 + x(2 - y) + 3y - 11 = 0$

As x $\in$ R – {3}, D $\geq$ 0 $\rightarrow$ (2 - y)^2 – 4(3y - 11) $\geq$ 0 $\rightarrow$ y^2 – 16y + 48 $\geq$ 0

$\rightarrow$ (y - 4)(y - 12) $\geq$ 0 $\rightarrow$ y $\in$ (- ∞, 4]$\cup$[12, ∞).

Ex4. Find the range of following

   (i) $y = \dfrac{(x-1)(x-3)}{x-2}$ (ii) $y = \dfrac{(x-1)(x-2)}{x-3}$

Sol. (i) $y = \dfrac{(x-1)(x-3)}{x-2} \rightarrow y(x - 2) = x^2 - 4x + 3$

$\rightarrow$ x^2 – (4 + y)x + 3 + 2y = 0

as x $\in$ R – {2}, D $\geq$ 0 $\rightarrow$ (y + 4)^2 – 4(3 + 2y) $\geq$ 0

$\rightarrow$ y^2 + 4 $\geq$ 0 which is true $\forall$ y $\in$ R, so range is R

Look at the figure to analyse.

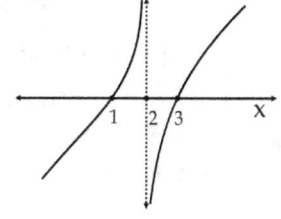

(ii) $y = f(x) = \dfrac{(x-1)(x-2)}{x-3} \rightarrow f'(x) = \dfrac{dy}{dx} = \dfrac{x^2 - 6x + 7}{(x-3)^2}$

f'(x) = 0 $\rightarrow$ x = 3$\pm$ $\sqrt{2}$

$\rightarrow$ f(x) has extremum at x = 3$\pm$ $\sqrt{2}$

f(3+ $\sqrt{2}$) = 3 + 2$\sqrt{2}$ and f(3 - $\sqrt{2}$) = 3 - 2$\sqrt{2}$.

Look at the figure plotted as shown.

$\rightarrow y \in \left(-\infty, 3 - 2\sqrt{2}\right] \cup \left[3 + 2\sqrt{2}, \infty\right)$

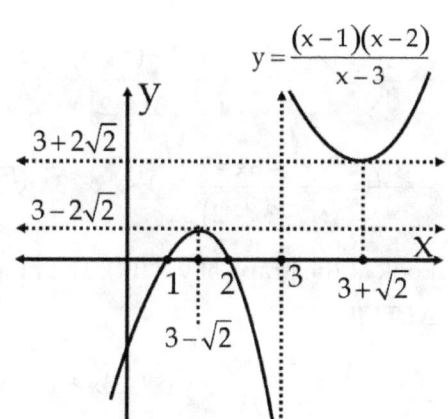

ALTERNATE

$y = \dfrac{(x-1)(x-2)}{x-3} \rightarrow y(x - 3) = x^2 - 3x + 2$

$\rightarrow$ x^2 – (3 + y)x + 3y + 2 = 0

As x $\in$ R – {3}, D $\geq$ 0 $\rightarrow$ (3 + y)^2 – 4(3y + 2) $\geq$ 0

$\rightarrow$ y^2 – 6y + 1 $\geq$ 0 $\rightarrow$ $y \in \left(-\infty, 3 - 2\sqrt{2}\right] \cup \left[3 + 2\sqrt{2}, \infty\right)$

TYPE – III: $f(x) = \dfrac{\text{Linear}}{\text{Quadratic}}$

Ex5. Find the range of $y = f(x) = \dfrac{x-2}{(x-1)(x-3)}$

Sol. Lets observe the curve.

$\lim\limits_{x \to 1^+} f(x) = \infty,\ \lim\limits_{x \to 1^-} f(x) = -\infty,\ \lim\limits_{x \to 3^+} f(x) = \infty$

$\lim\limits_{x \to 3^-} f(x) = -\infty,\ \lim\limits_{x \to \infty} f(x) = 0$ and $\lim\limits_{x \to -\infty} f(x) = 0$

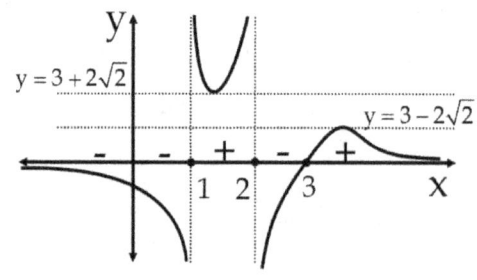

There is need of calculus to plot this curve. I am just plotting it for your observation.

As per graph, Range = R.

ALTERNATE

$y = \dfrac{x-2}{(x-1)(x-3)} \to y(x^2 - 4x + 3) = x - 2$

$\to yx^2 - (4y + 1)x + 3y + 2 = 0.$ As $x \in R - \{1, 3\}$

$D \geq 0 \to (4y + 1)^2 - 4y(3y + 2) \geq 0 \to 4y^2 + 1 \geq 0$ true $\forall\ y \in R.$ So, range = R.

Ex6. Find the range of $y = f(x) = \dfrac{(x-3)}{(x-1)(x-2)}.$

Sol. Lets observe the following limits to plot the curve.

$\lim\limits_{x \to \infty} f(x) = 0,\ \lim\limits_{x \to 2^+} f(x) = -\infty,\ \lim\limits_{x \to 2^-} f(x) = \infty$

$\lim\limits_{x \to 1^+} f(x) = \infty,\ \lim\limits_{x \to 1^-} f(x) = -\infty$

$y = f(x) = \dfrac{(x-3)}{(x-1)(x-2)} \to y = \dfrac{x-3}{x^2 - 3x + 2}$

$\to \dfrac{dy}{dx} = \dfrac{(x^2 - 3x + 2) \times 1 - (x-3) \times (2x-3)}{(x^2 - 3x + 2)^2}$

$\to \dfrac{dy}{dx} = \dfrac{-x^2 + 6x - 7}{(x^2 - 3x + 2)^2} = 0 \to x = 3 \pm \sqrt{2}.$ Here $f(3 + \sqrt{2}) = 3 - 2\sqrt{2}$ and $f(3 - \sqrt{2}) = 3 + 2\sqrt{2}$

Look at the graph of $y = f(x).$ The range of $f(x) = R - (3 - 2\sqrt{2}, 3 + 2\sqrt{2}).$

ALTERNATE:

$y = \dfrac{(x-3)}{(x-1)(x-2)} \to y(x^2 - 3x + 2) = x - 3 \to y x^2 - (3y + 1)x + 2y + 3 = 0.$

As $x \in R - \{1, 2\} \to D \geq 0$ hence $(3y + 1)^2 - 4y(2y + 3) \geq 0 \to y^2 - 6y + 1 \geq 0$

$\to (y - [3 - 2\sqrt{2}])(y - [3 + 2\sqrt{2}]) \geq 0 \to y \in (-\infty, 3 - 2\sqrt{2}] \cup [3 + 2\sqrt{2}, \infty).$

TYPE – IV: $f(x) = \dfrac{\text{Quadratic}}{\text{Quadratic}}$

Ex7. If x is real then prove that $\dfrac{x^2 - 3x + 4}{x^2 + 3x + 4} \in \left[\dfrac{1}{7}, 7\right]$.

Sol. Let $y = \dfrac{x^2 - 3x + 4}{x^2 + 3x + 4} \to y(x^2 + 3x + 4) = x^2 - 3x + 4$

$\to (y - 1) x^2 + (3y + 3) x + 4y - 4 = 0$

As $x \in R \to D \geq 0 \to (3y + 3)^2 - 4(y - 1)(4y - 4) \geq 0$

$\to 9 y^2 + 18y + 9 - 16(y^2 - 2y + 1) \geq 0$

$\to -7 y^2 + 50 y - 7 \geq 0 \to 7y^2 - 50y + 7 \leq 0$

$\to 7y^2 - 49y - y + 7 \leq 0 \to 7y (y - 7) - (y - 7) \leq 0$

$\to (7y - 1)(y - 7) \leq 0 \to y \in \left[\dfrac{1}{7}, 7\right]$.

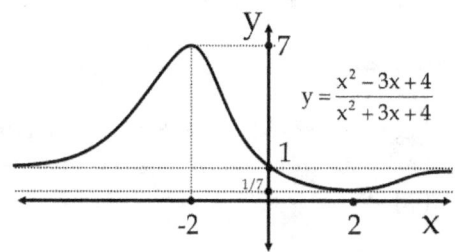

$y = \dfrac{x^2 - 3x + 4}{x^2 + 3x + 4}$

Look at the figure of f(x) as plotted using calculus.

☺ ☺ *Let's have a mind refreshing problem. Just solve it and divert your thinking little bit.* ☺ ☺

MIND REFRESHMENT # 3

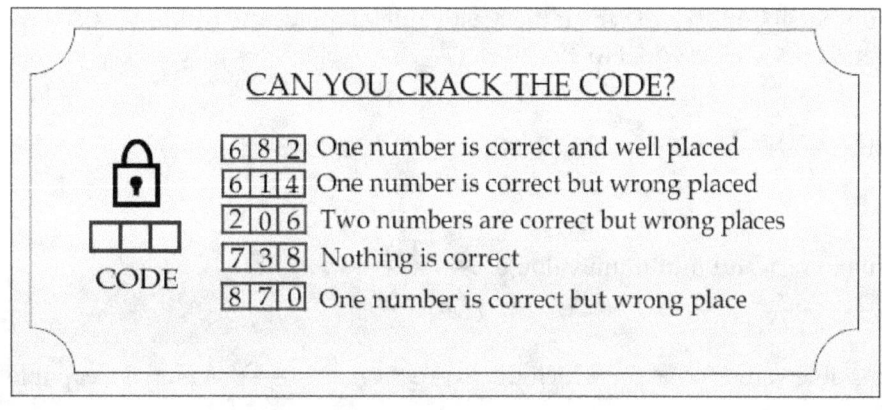

CAN YOU CRACK THE CODE?

6	8	2	One number is correct and well placed
6	1	4	One number is correct but wrong placed
2	0	6	Two numbers are correct but wrong places
7	3	8	Nothing is correct
8	7	0	One number is correct but wrong place

CODE

Now complete the given exercise and know your learning.

EXERCISE · 5

1. Plot the following curves

    (i) $y = -2x^2 + 4x$ (ii) $y = 5x^2 + x - 1$ (iii) $y = 3x^2 - x + 1$ (iv) $y = -x^2 - x - 2$

2. Consider $y = -2x^2 - 12x + 1$ then find the range of y if

    (i) $x \in R$ (ii) $x \in [-4, 1]$ (iii) $x \in (-2, 2]$

3. If $(1+x)(1+x^2)(1+x^4)(1+x^8)......(1+x^{128}) = \sum_{r=0}^{n} x^r$ then n is equal to

    (A) 255   (B) 127   (C) 63   (D) None of these

4. Let $f(x) = x^2 - (b+1)x + b$ and area of triangle formed by points $(\alpha, 0)$, $(\beta, 0)$ and$(0, f(0))$, where α and β are zero's of $f(x)$, is 3 units, then the value of b, is/are?

    (A) 3   (B) 1   (C) -2   (D) -1

5. Let $f(x) = \left(a + \dfrac{1}{a}\right)x^2 - 2x + 1$, where $a < 0$ and $m(a)$ be the maximum value of $f(x)$. As 'a' varies, then maximum value of $2 \cdot m(a)$, is?

6. For what value(s) of p the vertex of parabola $x^2 + 2px + 13$ lies at a distance of 5 units from the origin?

7. Discuss the least value of $f(x) = 2bx^2 - x^4 - 3b^2$ depending on the parameter b.

8. Find all numbers p for each of which the least value of quadratic trinomial $4x^2 - 4px + p^2 - 2p + 2$ on the interval $0 \le x \le 2$ is equal to 3.

9. Prove that $\dfrac{(x+1)(x-2)}{x(x+3)} \in R \ \forall x \in R - \{0, -3\}$.

10. Find the maximum and minimum value of $\dfrac{x^2 + 14x + 9}{x^2 + 2x + 3} \forall x \in R$.

11. Find all possible values of 'a' for which the expression $\dfrac{ax^2 - 7x + 5}{5x^2 - 7x + a}$ may be capable of taking all real values, x being any real quantity.

12. If $f(x) = \dfrac{x^3 + x - 2}{x^3 - 1}$ & $g(x) = \dfrac{x^2 + x + 2}{x^2 + x + 1}$ then find the set of values of x for which f(x), g(x) defined real (individually) and also find their range.

LOCATION OF ROOTS

Location of roots is a most important topic of QEE as IIT JEE Mains/Advanced exam is concern, so understand it clearly and have crystal clear view in the topic.

Ex1. Let $x^2 - (m - 3)x + m = 0$, $m \in R$. Find the set of values of m for which '2' lies between the roots of given equation.

Sol. The only condition for a real number 'k', lies between roots of $f(x) = ax^2 + bx + c = 0$, where $a \neq 0$ and a, b, c $\in$ R is $af(k) < 0$.

(i) Necessary Condition 'af(k) < 0'

For $x^2 - (m - 3)x + m = 0$, $m \in R$, a = 1 and k = 2, so as per condition

$1 \times f(2) < 0 \rightarrow 1 \times [2^2 - (m - 3) \times 2 + m] < 0 \rightarrow 10 - m < 0 \rightarrow m > 10$

$\rightarrow m \in (10, \infty)$ is the required answer.

Now question is, why are we applying af(k)<0 and not af(k) > 0 ? Let's try to know the answer

THEORY (OBSERVATION)

Let us sketches all possible curves of $f(x) = ax^2 + bx + c$, $a \neq 0$, a, b, c $\in$ R.

CASE – I (D > 0)

α, β are the real roots of the equation

$ax^2 + bx + c = 0 \rightarrow D > 0$.

Where $k_1 < \alpha < k_2 < \beta < k_3$.

The number k_2 lies between the roots of the equation $ax^2 + bx + c = 0$.

Observe and analyse the conditions as shown in figure.

$D = b^2 - 4ac > 0$

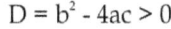

observation

(i) $a \cdot f(k_1) > 0$
(ii) $a \cdot f(k_2) < 0$
(iii) $a \cdot f(k_3) > 0$

CASE – II (D = 0)

Here $\alpha = \beta = -b/2a$ are equal roots of the equation $ax^2 + bx + c = 0 \rightarrow D = 0$.

(i) $a f(k_1) > 0$

(ii) $a f(k_2) > 0$

Where $k_1 < \alpha = \beta < k_2$

Observe the figure as shown and analyse in details. Extract information from curve.

$D = b^2 - 4ac = 0$

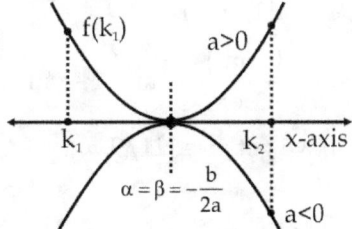

observation

(i) $a \cdot f(k_1) > 0$
(ii) $a \cdot f(k_2) > 0$

CASE – III (D < 0)

As $D < 0 \rightarrow$ there is no any real roots.

(i)   a f(k) > 0 , where $k \in R$.

Observe the condition as shown in figure.

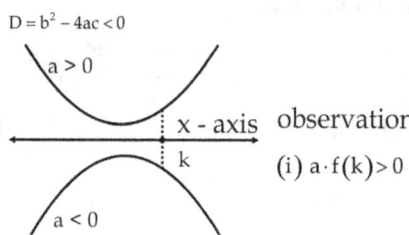

Note Results:

❖ a f(k) < 0 only when $\alpha < k < \beta$ or k lies between the roots of the Q.E.

❖ a f(k) > 0 → k lies outside the roots of $ax^2 + bx + c = 0$ or $k \in (-\infty, \alpha) \cup (\beta, \infty)$.

❖ a f(k) > 0 if $D = b^2 - 4ac < 0$.

❖ a f(k) < 0 is the sufficient condition for k lies between roots and need not to write D > 0.

■ CONDITIONS FOR BOTH ROOTS GREATER THAN x = k

Here are the following necessary required conditions.

(i) $D \geq 0$

(ii)   a f(k) > 0

(iii) $k < -\dfrac{b}{2a}$

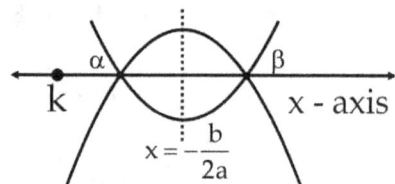

■ CONDITIONS FOR BOTH ROOTS LESS THAN x = k

Here are the following necessary required conditions.

(i) $D \geq 0$

(ii)   a f(k) > 0

(iii) $-\dfrac{b}{2a} < k$

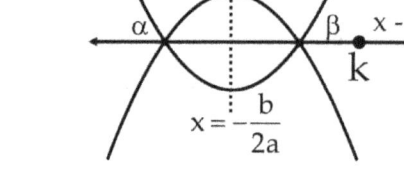

Hope you got the concepts. Let solve some examples to have better understanding.

Ex2.   Find all the values of the parameter 'm' for which both roots of $x^2 - 6mx + 2 - 2m + 9m^2 = 0$ exceed the number 3.

Sol. $a = 1 > 0$, k = 3 and $f(x) = x^2 - 6mx + 2 - 2m + 9m^2$.

The conditions for both roots greater than 3 are

(i) $D \geq 0 \rightarrow 36m^2 - 4(2 - 2m + 9m^2) \geq 0 \rightarrow m - 1 \geq 0 \rightarrow m \in [1, \infty)$.

(ii)   a f(k) > 0 → 1×f(3) > 0 → $3^2 - 6m×3 + 2 - 2m + 9m^2 > 0 \rightarrow 9m^2 - 20m + 11 > 0$

→ $9m^2 - 9m - 11m + 11 > 0 \rightarrow 9m(m - 1) - 11(m - 1) > 0 \rightarrow (9m - 11)(m - 1) > 0$

→ $m \in (-\infty, 1) \cup (11/9, \infty)$.

(iii) $k < -\dfrac{b}{2a} \rightarrow 3 < -\dfrac{(-6m)}{2×1} \rightarrow m > 1 \rightarrow m \in (1, \infty)$

From (i), (ii) & (iii) we get $m \in (11/9, \infty)$.

Ex3.   Find all the values of 'a' for which both roots of the equation $x^2 + x + a = 0$ exceeds the quantity 'a'.

Sol.   Let $f(x) = x^2 + x + a$, then conditions for both roots exceeds 'a' are as follow

(i) $D \geq 0 \rightarrow 1 - 4a \geq 0 \rightarrow a \in (-\infty, 1/4]$

(ii) $1 \times f(a) > 0 \rightarrow a^2 + 2a > 0 \rightarrow a(a + 2) > 0 \rightarrow a \in (-\infty, -2) \cup (0, \infty)$.

(iii) $a < -\dfrac{1}{2}$

From (i), (ii) & (iii) we get $a \in (-\infty, -2)$.

Ex4. Find the set of values of m for which $x^2 - (m - 3)x + m = 0$ have both roots less than 2.

Sol. Let $f(x) = x^2 - (m - 3)x + m$ then conditions for both roots less than 2, are as follow

(i) $D \geq 0 \rightarrow (m - 3)^2 - 4m \geq 0 \rightarrow m^2 - 10m + 9 \geq 0 \rightarrow (m - 9)(m - 1) \geq 0 \rightarrow m \in (-\infty, 1] \cup [9, \infty)$.

(ii) $1 \times f(2) > 0 \rightarrow 4 - 2(m - 3) + m > 0 \rightarrow 10 > m \rightarrow m \in (-\infty, 10)$

(iii) $-\dfrac{b}{2a} < k \rightarrow -\dfrac{-(m-3)}{2} < 2 \rightarrow m - 3 < 4 \rightarrow m \in (-\infty, 7)$

From (i), (ii) & (iii) we get $m \in (-\infty, 1]$.

Ex5. Find the value of m for which 2 lies between the roots of $x^2 - (m + 1)x + m^2 + m - 8 = 0$.

Sol. the required condition is $af(k) < 0$, where $a = 1$, $k = 2$ and $f(x) = x^2 - (m + 1)x + m^2 + m - 8$

$af(k) < 0 \rightarrow 4 - 2(m + 1) + m^2 + m - 8 < 0 \rightarrow m^2 - m - 6 < 0 \rightarrow (m - 3)(m + 2) < 0$

$\rightarrow m \in (-2, 3)$.

■ CONDITIONS FOR EXACTLY ONE ROOT LIE BETWEEN (K₁, K₂)

Let $f(x) = ax^2 + bx + c$, $a \neq 0$, a, b, c $\in$ R.

The necessary condition for exactly one root of $ax^2 + bx + c = 0$ lies between k_1 & k_2 is $f(k_1) \times f(k_2) < 0$, $a \neq 0$.

Some time it happens like $f(k_1) \times f(k_2) \leq 0$. Equality holds if one of root of equation $ax^2 + bx + c = 0$ is either k_1 or k_2. So in such case you need to be alert.

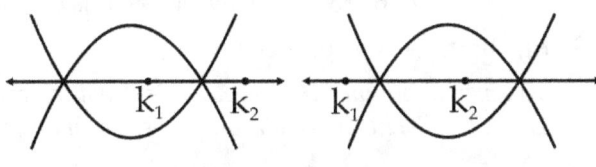

$f(k_1) \times f(k_2) < 0, a \neq 0$

Note. *$f(k_1) \times f(k_2) < 0$ include $D > 0$ so you need not to work on $D > 0$ while solving questions.*

In all other cases $f(k_1) \times f(k_2) > 0$.

Ex6. Find the set of values of m for which exactly one root of the equation $x^2 + mx + m^2 + 6m = 0$ lie in the interval (-2, 0).

Sol. Let $f(x) = x^2 + mx + m^2 + 6m$

As per condition, we have $f(-2) f(0) < 0 \rightarrow [(-2)^2 - 2m + m^2 + 6m] \times [m^2 + 6m] < 0$

$\rightarrow [m^2 + 4m + 4][m(m + 6)] < 0 \rightarrow (m+2)^2 m(m + 6) < 0 \rightarrow m \in (-6, 0) - \{-2\}$

Ex7. Find all possible values of 'a' for which exactly one root of $x^2 - (a + 1)x + 2a = 0$ lie in (0, 3).

Sol. Let $f(x) = x^2 - (a + 1)x + 2a$

As per condition, we have $f(0)f(3) < 0 \rightarrow 2a[9 - 3(a + 1) + 2a] < 0 \rightarrow 2a(6 - a) < 0$

$\rightarrow a(a - 6) > 0 \rightarrow a \in (-\infty, 0) \cup (6, \infty) \, ...(1)$

Observe: If a = 0

Then equation $x^2 - (a + 1)x + 2a = 0$ becomes $x^2 - x = 0 \rightarrow x = 0, 1$

Here root $x = 1$ lies between $(0, 3)$ so $a = 0$ should be in solution.

So, final answer is $a \in (-\infty, 0] \cup (6, \infty)$.

■ CONDITIONS FOR BOTH ROOT LIE BETWEEN (K_1, K_2)

Let $f(x) = ax^2 + bx + c$, $a \neq 0$, $a, b, c \in R$. Look at the graph of $y = f(x)$ in different situations.

The required conditions are

(i) $D \geq 0$

(ii) $af(k_1) > 0$

(iii) $af(k_2) > 0$

(iv) $k_1 < -b/2a < k_2$

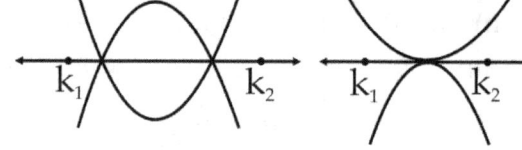

Ex8. α, β are roots of $x^2 + 2(k - 3)x + 9 = 0$, where $(\alpha \neq \beta)$ and $\alpha, \beta \in (-6, 1)$ then find the set of values of k.

Sol. Let $f(x) = x^2 + 2(k - 3)x + 9$. Then as per conditions

(i) $D > 0$ as $(\alpha \neq \beta) \rightarrow 4(k - 3)^2 - 4 \times 9 > 0 \rightarrow k(k - 6) > 0 \rightarrow k \in (-\infty, 0) \cup (6, \infty)$.

(ii) $af(k_1) > 0 \rightarrow 1 \times f(-6) > 0 \rightarrow 36 - 12(k - 3) + 9 > 0 \rightarrow 81 - 12k > 0 \rightarrow k \in (-\infty, 27/4)$

(iii) $af(k_2) > 0 \rightarrow 1 \times f(1) > 0 \rightarrow 1 + 2(k - 3) + 9 > 0 \rightarrow 2k + 4 > 0 \rightarrow k \in (-2, \infty)$

(iv) $k_1 < -b/2a < k_2 \rightarrow -6 < -(k - 3) < 1 \rightarrow 6 > k - 3 > -1 \rightarrow 9 > k > 2 \rightarrow k \in (2, 9)$

From all the above conditions (i), (ii), (iii) and (iv) we get $k \in (6, 27/4)$.

■ CONDITIONS FOR ATLEAST ONE ROOT LIE IN THE INTERVAL (K_1, K_2)

The current situation is union of two following conditions

(i)   Exactly one root lie between (k_1, k_2).

(ii) Both roots lie between (k_1, k_2).

You may also think like first take $D \geq 0$ and remove the conditions of no root lie between (k_1, k_2). You need to take care of boundary conditions. Do some analysis to get the concept correctly.

■ CONDITIONS FOR K_1, K_2 LIE BETWEEN ROOTS α and β

The situation is like k_1 and k_2 lies between roots

So, the necessary conditions are

(i) $af(k_1) < 0$

(ii) $af(k_2) < 0$

Where $f(x) = ax^2 + bx + c$, $a \neq 0$, $a, b, c \in R$.

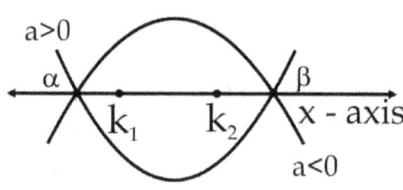

Ex9. Find the values of k for which one root of quadratic equation $(k - 5)x^2 - 2kx + k - 4 = 0$ is smaller than 1 and other root exceed 2.

Sol. Since, $(k - 5)x^2 - 2kx + k - 4 = 0$ is a quadratic so $k - 5 \neq 0 \rightarrow k \neq 5$.

Let $f(x) = (k - 5)x^2 - 2kx + k - 4$. As per conditions

(i) $(k - 5) f(1) < 0 \rightarrow (k - 5)(k - 5 - 2k + k - 4) < 0 \rightarrow -9(k - 5) < 0 \rightarrow k > 5$

(ii) $(k - 5) f(2) < 0 \rightarrow (k - 5)(4(k - 5) - 4k + k - 4) < 0 \rightarrow (k - 5)(k - 24) < 0 \rightarrow k \in (5, 24)$

From (i) and (ii) we get $k \in (5, 24)$.

Ex10. Let $a, b, c \in R$. If $ax^2 + bx + c = 0$ has two real roots α and β where $\alpha < -1$ and $\beta > 1$, then show that

$1 + \dfrac{c}{a} + \left|\dfrac{b}{a}\right| < 0.$

Sol.   As $ax^2 + bx + c = 0$ has two real roots α and β so $a \neq 0$.

Let $f(x) = x^2 + \dfrac{b}{a}x + \dfrac{c}{a}$ have zeroes α and β.

As per conditions $\alpha < -1$ and $\beta > 1 \rightarrow -1$ and 1 lies between roots so

(i) $af(k_1) < 0 \rightarrow f(-1) < 0 \rightarrow 1 - \dfrac{b}{a} + \dfrac{c}{a} < 0$

(ii) $af(k_2) < 0 \rightarrow f(1) < 0 \rightarrow 1 + \dfrac{b}{a} + \dfrac{c}{a} < 0$

So, from (i) and (ii) we can say that $1 + \left|\dfrac{b}{a}\right| + \dfrac{c}{a} < 0$

■   **CONDITIONS FOR AT LEAST ONE ROOT GREATER THAN K**

The situation is as shown in figure.

The same situation can be resolve by saying that

(i) $D \geq 0 \rightarrow$ both roots are real

(ii)   Subtract the conditions for both root less than k

You need to remove some more cases like

(a)   If $D = 0$ and $-b/2a = k = \alpha$ then remove α.

(b)   If $\alpha < \beta$ and $\beta = k$ then remove β.

Let's have an example for better understanding.

Ex11. Find set of values of 'a' for which the equation $x^2 + 2(a - 1) x + a + 5 = 0$ has atleast one positive root.

Sol.   Let $f(x) = x^2 + 2(a - 1) x + a + 5$. As per conditions

(i) $D \geq 0$: for real roots

$4(a - 1)^2 - 4(a + 5) \geq 0 \rightarrow 4(a^2 - 3a - 4) \geq 0 \rightarrow (a - 4)(a + 1) \geq 0 \rightarrow a \in (-\infty, -1] \cup [4, \infty)$ ......(1)

(ii)   Now we are getting conditions for both roots < 0

(a) $D \geq 0 \rightarrow a \in (-\infty, -1] \cup [4, \infty)$

(b) $af(0) > 0 \rightarrow a + 5 > 0 \rightarrow a \in (-5, \infty)$

(c) $-b/2a < 0 \rightarrow -(a - 1) < 0 \rightarrow a \in (1, \infty)$

From (a), (b) and (c) we get $a \in [4, \infty)$ .... (2)

Now remove (2) from (1) to get required condition $\rightarrow a \in (-\infty, -1]$

For $a = -5$, $f(x) = x^2 + 2(a - 1) x + a + 5 = 0 \rightarrow x^2 - 12 x = 0 \rightarrow x = 0, 12$.

$\rightarrow a = -5$ should be in the solution. This is already in the solution.

So, final answer is $a \in (-\infty, -1]$.

Solve the given exercise to get theory in depth and know about learning.

EXERCISE - 6

1. Find the set of values of 'p' for which the quadratic equation $(p - 5)x^2 - 2px - 4p = 0$ has atleast one positive root.

2. Find the set of values of k for which the roots of the equation $(k - 2)x^2 + 2kx + k + 3 = 0$ lies in the interval $(- 2, 1)$.

3. For what values of 'k' exactly one root of the equation $2^k x^2 - 4^k x + 2^k - 1 = 0$ lies between 1 and 2?

4. Find the values of x that maximize $f(x) = \left| \dfrac{3x+1}{9x^2 + 6x + 2} \right|$

5. Prove that for any real value of a the inequality, $(a^2 + 3)x^2 + (a + 2)x - 5 < 0$ is true for at least one negative x.

6. If $f(x) = 4x^2 + ax + (a - 3)$ is negative for atleast one x, find all possible values of a.

7. The value of parameter a for which $1 + \log_5(x^2 + 1) \le \log_5(ax^2 + 4x + a)$ is true $\forall\ x \in R$ is

   (A) [2, 9]　　　(B) [3, 7]　　　(C) [7, ∞)　　　(D) [5, ∞)

8. If a and b are two distinct non-zero real numbers such that $a - b = \dfrac{a}{b} = \dfrac{1}{a} - \dfrac{1}{b}$, then

   (A) $a > 0$　　　(B) $a < 0$　　　(C) $b > 0$　　　(D) $b < 0$

9. If the roots of the equation $x^2 - 2ax + a^2 + a - 3 = 0$ are real and less than 3, then

   (A) $a < 2$　　　(B) $2 \le a \le 3$　　　(C) $3 < a \le 4$　　　(D) $a > 4$

10. The value of a for which $2x^2 - 2(2a + 1)x + a(a + 1) = 0$ may have one root less than 'a' and another root greater than 'a' are given by

    (A) $1 > a > 0$　　　(B) $-1 < a < 0$　　　(C) $a \ge 0$　　　(D) $a > 0$ or $a < -1$

11. The equation $2^{2x} + (a - 1) 2^{x+1} + a = 0$ has roots of opposite sign then exhaustive set of values of 'a' is

    (A) $a < 0$　　　(B) $a \in (-1, 0)$　　　(C) $a \in (-\infty, 1/3)$　　　(D) $a \in (0, 1/3)$

12. If 2 lies between the roots of the equation $x^2 - mx + 2 = 0$, where $m \in R$ then value of $\left| \left[\left(\dfrac{3|t|}{9+t^2} \right)^m \right] \right|$ is

    (where [x] greatest integer $\le x$ and $|.|$ = Modulus)

    (A) 0　　　(B) 1　　　(C) 8　　　(D) 27

13. Find the sum of all the integral values of $a \in [1, 100]$ for which $x^2 - (a - 5)x + \left(a - \dfrac{15}{4} \right) = 0$ has atleast one root greater than zero.

14. If $b > a$, then the equation $(x - a)(x - b) = 1$ has -

    (A) Both roots in [a, b]　　　　　　(B) both roots in $(-\infty, a)$

    (C) Both roots in $(b, +\infty)$　　　　(D) one root in $(-\infty, a)$ and the other in $(b, +\infty)$

SOME SOLVED EXAMPLES

1. If the roots of the equation $x^3 - ax^2 + bx - c = 0$ are three consecutive integers, then what is the smallest possible value of b?

   (A) $-1/\sqrt{3}$ (B) -1 (C) 0   (D) 1

Solution:

Let three consecutive integers are $p - 1$, p & $p + 1$.

So, $p - 1 + p + p + 1 = a$ ..... (Sum of roots taken one at a time) $\rightarrow 3p = a$ ....(1)

$p(p - 1) + p(p + 1) + (p - 1)(p + 1) = b \rightarrow 3p^2 - 1 = b$ so the smallest value of b must be $= -1$.

Hence, (B) is correct option.

2. Let $f(x) = x^2 - (b + 1)x + b$ and area of triangle formed by points $(\alpha, 0)$, $(\beta, 0)$ and $(0, f(0))$, where α and β are zero's of $f(x)$, is 3 units, then the value of b, is/are?

   (A) 3   (B) 1   (C) -2 (D) -1

Solution:

$f(x) = x^2 - (b + 1)x + b$, $f(0) = b$. so, points are $(\alpha, 0)$, $(\beta, 0)$ and $(0, b)$

$$\text{Area of } \Delta = \frac{1}{2}\begin{Vmatrix}\alpha & 0 \\ \beta & 0 \\ 0 & b \\ \alpha & 0\end{Vmatrix} = 3 \rightarrow \left|\frac{b(\alpha - \beta)}{2}\right| = 3 \rightarrow |b(\alpha - \beta)| = 6 \rightarrow b^2(\alpha - \beta)^2 = 36$$

$\rightarrow b^2(b - 1)^2 = 36$ .... (1)

Hence A and C satisfying the relation (1)

3. Let $f(x) = \left(a + \dfrac{1}{a}\right)x^2 - 2x + 1$, where $a < 0$ and $m(a)$ be the maximum value of $f(x)$. As 'a' varies, then the greatest value of $2 \times m(a)$, is?

Solution:

$$f(x) = \left(a + \frac{1}{a}\right)x^2 - 2x + 1, a < 0 \rightarrow a + \frac{1}{a} \le -2 \forall a < 0$$

$$f(x)\big|_{MAX} = -\frac{D}{4a} = m(a) \rightarrow m(a) = -\frac{\left|4 - 4\left(a + \dfrac{1}{a}\right)\right|}{4\left(a + \dfrac{1}{a}\right)} = -\frac{1}{a + \dfrac{1}{a}} + 1$$

$$a + \frac{1}{a} \le -2 \forall a < 0 \rightarrow -\left(a + \frac{1}{a}\right) \ge 2 \rightarrow -\frac{1}{\left(a + \dfrac{1}{a}\right)} \in \left(0, \frac{1}{2}\right]. \text{ So, } m(a)\big|_{max} = \frac{1}{2} + 1 = 3/2$$

$\rightarrow 2m(a)\big|_{max} = 3.$

4. If $x = 2 + 2^{1/3} + 2^{2/3}$, then the value of $x^3 - 6x^2 + 6x + 3$ is equals to?

Solution:

$x = 2 + 2^{1/3} + 2^{2/3} \rightarrow x - 2 = 2^{1/3} + 2^{2/3}$ .... (1)

$\rightarrow (x - 2)^3 = (2^{1/3} + 2^{2/3})^3 \rightarrow x^3 - 2^3 - 3.x.2(x - 2) = 2 + 2^2 + 3.2^{1/3}.2^{2/3}(2^{1/3} + 2^{2/3})$.

$\rightarrow x^3 - 6x^2 + 12x - 8 = 6 + 6(x - 2)$ $\because x - 2 = 2^{1/3} + 2^{2/3}$

$\rightarrow x^3 - 6x^2 + 6x - 2 = 0 \rightarrow x^3 - 6x^2 + 6x + 3 = 5$. So, 5 is the required answer.

5. Let $p(x) = x^4 + 5x^3 + 4x^2 - 3x + 9 = (x - k)^2(ax^2 + bx + c)$ where $k \in I$ and a, b, c are real constants then find the value of $2a - b + 3c$.

Solution:

$p(x) = x^4 + 5x^3 + 4x^2 - 3x + 9 = (x - k)^2(ax^2 + bx + c)$ where $k \in I$

k is a repeated root $\rightarrow p'(x) = 4x^3 + 15x^2 + 8x - 3 = 0$ will also have root 'k'.

If we observe: $4x^3 + 15x^2 + 8x - 3 = (x + 1)(4x^2 + 11x - 3) = (x + 1)(4x - 1)(x + 3)$

But $p(-1) \neq 0$ so k must be equal to -3. ie $k = -3$ as $k \in I$.

$\rightarrow p(x) = x^4 + 5x^3 + 4x^2 - 3x + 9 = (x + 3)^2(x^2 - x + 1)$

$\rightarrow a = 1, b = -1$ and $c = 1$. Hence $2a - b + 3c = 6$.

6. If $f\left(2x + \dfrac{1}{x}\right) = x^2 + \dfrac{1}{4x^2} + 1 (x \neq 0)$, the value of $f(x)$ is

(A) $4x^2$ (B) $\dfrac{1}{4}\left(2x + \dfrac{1}{x}\right)^2$ (C) $x^2/4$ (D) $4\left(2x + \dfrac{1}{x}\right)^2$

Solution:

$f\left(2x + \dfrac{1}{x}\right) = x^2 + \dfrac{1}{4x^2} + 1 (x \neq 0)$

$\rightarrow f\left(2x + \dfrac{1}{x}\right) = \dfrac{1}{4}\left(4x^2 + \dfrac{1}{x^2}\right) + 1 = \dfrac{1}{4}\left[\left(2x + \dfrac{1}{x}\right)^2 - 4\right] + 1 = \dfrac{1}{4}\left(2x + \dfrac{1}{x}\right)^2$, so $f(x) = x^2/4$.

Option (C) is correct.

7. The least value of $\alpha \in R$ for which $4\alpha x^2 + \dfrac{1}{x} \geq 3, \forall x > 0$, is

(A) 1/9 (B) 3/32 (C) 1/27 (D) 1

Solution:

Case (I) $\alpha \in R^-$

$4\alpha x^2 + \dfrac{1}{x} \rightarrow -\infty$ as $x \rightarrow \infty$, hence $4\alpha x^2 + \dfrac{1}{x} \not\geq 3, \forall x > 0$ so $\alpha \notin R^-$

Case (II) $\alpha \in R^+$

Apply $AM \geq GM$ on $4\alpha x^2, \dfrac{1}{2x}$ and $\dfrac{1}{2x}$, we get

$$\dfrac{4\alpha x^2 + \dfrac{1}{2x} + \dfrac{1}{2x}}{3} \ge \sqrt[3]{4\alpha x^2 \cdot \dfrac{1}{2x} \cdot \dfrac{1}{2x}} = \sqrt[3]{\alpha} \;\rightarrow\; 4\alpha x^2 + \dfrac{1}{2x} + \dfrac{1}{2x} \ge 3\sqrt[3]{\alpha} \ge 3 \;\forall\, x > 0 \rightarrow \alpha \ge 1$$

So, $\alpha\,|_{MIN} = 1$

Alternate Method

Let $f(x) = 4\alpha x^2 + \dfrac{1}{x}, \forall x > 0 \;\rightarrow\; f'(x) = 8\alpha x - \dfrac{1}{x^2} = 0 \rightarrow x = \left(\dfrac{1}{8\alpha}\right)^{1/3}$

$f(x)$ attain minimum at $x = \left(\dfrac{1}{8\alpha}\right)^{1/3}$.

$f\left(\left(\dfrac{1}{8\alpha}\right)^{1/3}\right) = 3 \rightarrow 4\alpha\left(\dfrac{1}{8\alpha}\right)^{2/3} + (8\alpha)^{1/3} = 3 \rightarrow \alpha^{1/3} + 2\alpha^{1/3} = 3 \rightarrow \alpha = 1.$

Hence, Option (D) is correct.

8. If roots of the equation $z^2 + \alpha z + \beta = 0$ lie on $|z| = 1$ then

(A) $2\,|\,\text{Im}\alpha\,| = 1 - |\beta|^2$ 　　　　　　　　(B) $2\,|\,\text{Im}\alpha\,| = |\beta|^2 - 1$

(C) $\text{Im}\,\alpha = 0$ 　　　　　　　　　　　　　　(D) $|\alpha|_{Max} = 2\,|\,\beta\,|$

Solution:

Let z_1 & z_2 be the roots of above equation

So $z_1 = e^{i\theta_1}$, $\quad z_2 = e^{i\theta_2} \;\rightarrow\; e^{i(\theta_1 + \theta_2)} = \beta$ & $e^{i\theta_1} + e^{i\theta_2} = -\alpha$

$|\beta| = 1$ & $\text{Im}(\alpha) = (\sin\theta_1 + \sin\theta_2)$

$|\alpha| = 2\cos\left(\dfrac{\theta_1 - \theta_2}{2}\right)$

$|\alpha|_{max} = 2$. So option (D) is correct.

9. If $\alpha, \beta, \gamma, \delta, \dots.$ be the roots of the equation $f(x) = 0$ where $S_m =$ sum of mth power of roots, then

(A) $\dfrac{f'(x)}{f(x)} = \displaystyle\sum_{m=1}^{\infty} S_{m-1} x^{-m} \;\forall x > \max\{\alpha, \beta, \gamma, \dots\}$ 　　(B) $\dfrac{f'(x)}{f(x)} = \displaystyle\sum_{m=1}^{\infty} S_m x^m \;\forall x > \max\{\alpha, \beta, \gamma, \dots\}$

(C) $\dfrac{f'(x)}{f(x)} = \displaystyle\sum_{m=1}^{\infty} S_m x^{m+1} \;\forall x > \max\{\alpha, \beta, \gamma, \dots\}$ 　　(D) none of these

Solution:

$f(x) = (x - \alpha)(x - \beta)(x - \gamma)\dots$ n factors.

Taking log on both sides, $\forall\, x > \max\{\alpha, \beta, \gamma, \dots.\}$

$\log f(x) = \log(x - \alpha) + \log(x - \beta) + \log(x - \gamma) + \dots\dots$

Differentiating w.r.t. x, we get

$\dfrac{1}{f(x)} f'(x) = \dfrac{1}{x - \alpha} + \dfrac{1}{x - \beta} + \dfrac{1}{x - \gamma} + \dots(1)$

Now, $\dfrac{1}{x-\alpha} = \dfrac{1}{x}\left(1 - \dfrac{\alpha}{x}\right)^{-1}$

$= = \dfrac{1}{x} + \dfrac{\alpha}{x^2} + \dfrac{\alpha^2}{x^3} + \dfrac{\alpha^3}{x^4} + ...$

$\therefore \quad \dfrac{f'(x)}{f(x)} = \dfrac{1}{x} + \dfrac{\alpha}{x^2} + \dfrac{\alpha^2}{x^3} + \dfrac{\alpha^3}{x^4} + ... + \dfrac{1}{x} + \dfrac{\beta}{x^2} + \dfrac{\beta^2}{x^3} + \dfrac{\beta^3}{x^4} + ... +$ for n roots

$= n\dfrac{1}{x} + \dfrac{S_1}{x^2} + \dfrac{S_2}{x^3} + \dfrac{S_3}{x^4} + ...$

Now, $S_0 = 1 + 1 + 1 + ... = n$

$\dfrac{f'(x)}{f(x)} = s_0 x^{-1} + s_1 x^{-2} + s_2 x^{-3} + s_3 x^{-4} + ... = \displaystyle\sum_{m=1}^{\infty} S_{m-1} x^{-m}$

Hence, Option (A) is correct.

10. **Statement-1:** If $f^2(x)$ & $g^2(x)$ are integrable in [a, b] then $\left|\displaystyle\int_a^b f(x)g(x)dx\right| \le \sqrt{\left(\displaystyle\int_a^b f^2(x)dx\right)\left(\displaystyle\int_a^b g^2(x)dx\right)}$.

Statement-2: $a^2 x^2 + bx + c \ge 0$ if $b^2 - 4a^2c \le 0$

(A) Statement-1 is True, Statement-2 is a correct explanation for Statement-1.

(B) Statement-1 isTrue, Statement-2 isTrue, Statement-2 is not a correct explanation for Statement-1

(C) Statement-1 is True, Statement-2 is False.

(D) Statement-1 is False, Statement-2 is True.

Solution:

Let $F(x) = \{f(x) - \lambda g(x)\}^2 \ge 0 \ \forall \ \lambda \in R$ So $\displaystyle\int_a^b F(x) = \int_a^b \{f(x) - \lambda g(x)\}^2 dx \ge 0$

$\rightarrow \lambda^2 \displaystyle\int_a^b g^2(x)dx - 2\lambda \int_a^b f(x)g(x) + \int_a^b f^2(x)dx \ge 0 \ \forall \lambda \in R$.

Now $\lambda \in R$, hence $B^2 - 4AC \le 0$ to hold above inequality

$\rightarrow 4\left|\displaystyle\int_a^b f(x)g(x)\right|^2 - 4\displaystyle\int_a^b g^2(x)dx.\int_a^b f^2(x)dx \le 0$

$\rightarrow \left|\displaystyle\int_a^b f(x)g(x)dx\right| \le \sqrt{\displaystyle\int_a^b f^2(x)dx.\int_a^b g^2(x)dx}$. So, Option (A) is correct.

11. Roots of the quadratic equation $(2x^2 - 5x + 2) + \lambda(x^2 - 5x + 4) = 0$, $\lambda \in R$ will be

(A) Always real
(B) real only when λ is positive

(C) Real only when λ is negative
(D) always imaginary

Solution:

$(2x^2 - 5x + 2) + \lambda(x^2 - 5x + 4) = 0 \rightarrow (2 + \lambda)x^2 - 5(1 + \lambda)x + 2 + 4\lambda = 0$

$D = 9\lambda^2 + 10\lambda + 9 \rightarrow D > 0 \ \forall \lambda \in R - \{-2\}$

For $\lambda = -2$

Equation became $5x - 6 = 0 \rightarrow x = \dfrac{6}{5}$

So $\forall\ \lambda \in R$ quadratic equation has real root. Option (A) is correct.

12. If $\left|\displaystyle\int_a^b f(x)dx\right| < \displaystyle\int_a^b |f(x)|\,dx\ \forall\ a < b\ \&\ f'(x) > 0\ \forall\ x \in (a, b)$ then $f(x)$ has

   (A) At least 1 real root   (B) at most 1 real root

   (C) Exactly 1 real root   (D) can't say anything

Solution:

As $f'(x) > 0\ \forall\ x \in (a, b) \rightarrow f(x)$ is increasing function $\forall\ x \in (a, b)$

$\left|\displaystyle\int_a^b f(x)dx\right| < \displaystyle\int_a^b |f(x)|\,dx$ so exactly one real root. Hence, option (C) is correct.

13. Consider the biquadratic equation $x^4 + px^3 + qx^2 + rx + s = 0$

   Statement-1: The product of two roots of equation is equal to product of the other two then $r^2 = p^2 s$.

   Statement-2: If $\alpha\beta = \nu\delta$, then equation can be written at

   $x^4 + px^3 + qx^2 + rx + s \equiv (x^2 + \ell x + \sqrt{s})(x^2 + mx + \sqrt{s})$

   (A) Statement-1 is True, Statement-2 is a correct explanation for Statement-1.

   (B) Statement-1 isTrue, Statement-2 isTrue, Statement-2 is not a correct explanation for Statement-1

   (C) Statement-1 is True, Statement-2 is False.

   (D) Statement-1 is False, Statement-2 is True.

Solution:

Since $\alpha + \beta + \gamma + \delta = -p,\ \alpha\beta\gamma\delta = s \rightarrow \alpha\beta = \sqrt{s}$

Now $\sum \beta\gamma\delta = -r \rightarrow \alpha\beta(\alpha + \beta + \gamma + \delta) = -r$

Since $(\alpha\beta = r\delta) \rightarrow \sqrt{s}(p) = r \rightarrow r^2 = p^2 s$

Hence option (A) is correct.

14. $f(x)$ and $g(x)$ are quadratic polynomials such that $|f(x)| \geq |g(x)|\ \forall\ x \in R$. If $f(x) = 0$ have real roots then number of distinct roots of equation $h(x)\,h''(x) + (h'(x))^2 = 0$ are (where $h(x) = f(x)\,g(x)$).

   (A) 0   (B) 2   (C) 3   (D) 4

Solution:

$f(x),\ g(x)$ are quadratic and $|f(x)| \geq |g(x)|\ \forall\ x \in R\ \rightarrow f(x)\ \&\ g(x)$ have both root common

So, let $f(x) = a(x - \alpha)\,(x - \beta)$ then $g(x) = b(x - \alpha)\,(x - \beta)$

$\rightarrow h(x) = k(x - \alpha)^2 (x - \beta)^2 \rightarrow h(x)\,h'(x) = 2k(x - \alpha)^3 (x - \beta)^3 (2x - \alpha - \beta)$

So distinct roots of $\frac{d}{dx}(h(x)h'(x)) = 0$ are 4. So, option (D) is correct.

15. If $x \in [0, 10\pi]$ and $\lambda \in R$ such that $3 \sin x = \lambda^4 - 2\lambda^2 + 4$, then the maximum number of pairs (λ, x) is

(A) 4   (B) 6   (C) 8   (D) 10

Solution:

If we observe then we get RHS = $(\lambda^2 - 1)^2 + 3 \geq 3$ and LHS ≤ 3

$\rightarrow 3\sin x = (\lambda^2 - 1)^2 + 3 = 3$

$\rightarrow \sin x = 1 \rightarrow x = (4n + 1)\pi/2$

$x = \dfrac{\pi}{2}, \dfrac{5\pi}{2}, \dfrac{9\pi}{2}, \dfrac{13\pi}{2}, \dfrac{17\pi}{2}$, in $x \in [0, 10\pi]$ and $\lambda = \pm 1$

$\therefore$ maximum number of pairs = $5 \times 2 = 10$. Hence Option (D) is correct.

16. If equation $ax^2 + bx + c = 0$, $a, b, c \in R$ and $a \neq 0$ has imaginary roots then

(A) $(a + b + c)(a - b + c) > 0$ (B) $(a + b + c)(a - 2b + 4c) > 0$

(C) $(a - b + c)(4a - 2b + c) > 0$ (D) None of these

Solution:

Let $f(x) = ax^2 + bx + c$.

$ax^2 + bx + c = 0$ has imaginary roots $\rightarrow D < 0$.

So, two possibilities are there

(i) $f(x) = ax^2 + bx + c > 0 \ \forall \ x \in R$ Or

(ii) $f(x) = ax^2 + bx + c < 0 \ \forall \ x \in R$

Now observe $f(1).f(-1) > 0$, $f(1).f(-1/2) > 0$ and $f(-1).f(-2) > 0$

Hence, Options A, B and C are correct.

17. For the equation $\dfrac{40}{x-1} - \dfrac{160}{x-4} - \dfrac{200}{x-5} + \dfrac{320}{x-8} = 6x^2 - 27x$

(A) Number of real solutions of above equation is 3

(B) If E denotes the product of non - zero real or complex roots of the equation, then sum of divisors of E is 2904

(C) If S denotes the set of all real roots of the equation then, sum of elements of S taken two at a time is 81

(D) If $\alpha_1, \alpha_2 \in R$ be two roots of the equation such that $\log_{\alpha_2}(2\alpha_1)$ is defined then it must be 1.

Solution:

$\dfrac{40}{x-1} - \dfrac{160}{x-4} - \dfrac{200}{x-5} + \dfrac{320}{x-8} = 6x^2 - 27x$

Now, $\left(\dfrac{1}{x-1}+1\right)-\left(\dfrac{4}{x-4}+1\right)-\left(\dfrac{5}{x-5}+1\right)+\left(\dfrac{8}{x-8}+1\right)=\dfrac{6x^2-27x}{40}$

On clubbing 1st & last & 2nd & 3rd, we get

$\dfrac{2x-9}{(x-1)(x-8)}-\dfrac{2x-9}{(x-4)(x-5)}=\dfrac{3}{40}(2x-9) \rightarrow x=9/2$ is a solution

$\rightarrow \dfrac{1}{(x-1)(x-8)}-\dfrac{1}{(x-4)(x-5)}=\dfrac{3}{40}$

On solving this equation we get $x = 9$. Now we can verify the options.

So, Options (A) and (B) are correct.

18. If the sum of roots of the quadratic equation $(1 - a) x^2 + (a^2 - a + 4)x - 2a + 3 = 0$ is minimum, then the value of a (where a > 1) is _____

Solution:

Sum of roots $= \dfrac{a^2-a+4}{a-1}=a+\dfrac{4}{a-1}=(a-1)+\dfrac{4}{(a-1)}+1\geq 5$

Equality holds when $a-1=\dfrac{4}{a-1}\rightarrow a=3$.

19. Consider an equation in x, $8x^4 - 16x^3 + 16x^2 - 8x + a = 0$, then the sum of all the non-real roots of the equation can be (a $\in$ R)

(A) 1   (B) 2   (C) 1/2   (D) None of these

Solution:

Put $x = y + (1/2) \rightarrow 8y^2 + 4y^2 + a - 3/2 = 0$. Now put $y^2 = z$

$\therefore 8z^2 + 4z + a - (3/2) = 0$.

Case – I If $a > 3/2 \rightarrow$ All roots are non real $\therefore$ sum of roots = 2

Case – II If $a = 3/2 \rightarrow z = 0$ or $- 1/2 \therefore y = 0, 0$ or $x = 1/2, 1/2$

     If $z = -1/2 \rightarrow$ two non real roots $\therefore z = 1/2 + 1/2 + \alpha + \beta \rightarrow \alpha + \beta = 1$.

Case – III: If $a < 3/2 \rightarrow z \rightarrow x_1, - x_2$

$y=\pm\sqrt{x_1} \therefore x=\dfrac{1}{2}\pm\sqrt{x_1}$. Again $\alpha + \beta = 1$. So, options (A) and (B) are correct.

20. If $f(x)$ is a polynomial of degree 4 with rational coefficients and touches x-axis at $(\sqrt{2}, 0)$, then for the equation $f(x) = 0$ -

(A) Sum of roots is 0   (B) sum of roots is $4\sqrt{2}$

(C) Product of roots is 4   (D) product of roots is – 4

Solution:

Since f(x) touches x – axis at $(\sqrt{2}, 0) \rightarrow$ f(x) has repeated roots $x = \sqrt{2}$

But f(x) = 0 has rational coefficients $\rightarrow$ f(x) much touch the x-axis at $x = - \sqrt{2}$

→ $f(x) = k(x - \sqrt{2})^2 (x + \sqrt{2})^2$ → roots are $\sqrt{2}, \sqrt{2}, -\sqrt{2}$ and $-\sqrt{2}$.

Hence, Option (A) and (C) are correct.

21. Let a, b, c are the roots of the equation $x^3 - 2x^2 + 3x - 4 = 0$, then the value of $\left(a + \dfrac{1}{a}\right)\left(b + \dfrac{1}{b}\right)\left(c + \dfrac{1}{c}\right)$

is _____

Solution:

$\left(a + \dfrac{1}{a}\right)\left(b + \dfrac{1}{b}\right)\left(c + \dfrac{1}{c}\right) \rightarrow \left(\dfrac{a^2 + 1}{a}\right)\left(\dfrac{b^2 + 1}{b}\right)\left(\dfrac{c^2 + 1}{c}\right)$ and abc = 4

Now, $x^3 - 2x^2 + 3x - 4 = (x - a)(x - b)(x - c)$

Now Put x = i and x = - i and multiply together

→ $(- i + 2 + 3i - 4)(i + 2 - 3i - 4) = (i - a)(i + a)(i - b)(i + b)(i - c)(i + c)$

→ $(2i - 2)(-2 - 2i) = (1 + a^2)(1 + b^2)(1 + c^2)$

→ $8 = (1 + a^2)(1 + b^2)(1 + c^2) \rightarrow \left(\dfrac{a^2 + 1}{a}\right)\left(\dfrac{b^2 + 1}{b}\right)\left(\dfrac{c^2 + 1}{c}\right) = \dfrac{8}{4} = 2$

22. Let a, b, c be the positive integers such that a < b < c. If the two curves $y = |x - a| + |x - b| + |x - c|$ and $2x + y = 2003$ have exactly one point in common, then -

(A) Least possible value of c is 1002
(B) Greatest possible value of b is 1001
(C) Least possible value of b is 1002
(D) Greatest possible value of a is 1000

Solution:

By compairing the slopes, condition in problem is satisfied for (a, b + c - 2a) to lie on the line.

Putting in line we get

b + c = 2003,

$b_{max} = 1004$ and $c_{min} = 1002$

Hence, Options (A), (B) and (D) are correct

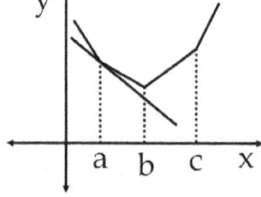

23. Let x_1 and x_2 are the roots of the equation $x^2 + x + 1 = 0$, then

(A) $x_1^{2016} + x_2^{2016} = 2$ (B) $x_1^{2016} + x_2^{2016} = -1$ (C) $x_1^{2017} + x_2^{2017} = -1$ (D) $x_1^{2018} + x_2^{2018} = -2$

Solution:

Let $f(x) = x^2 + x + 1 = 0$ have roots x_1 and x_2

→ $(x - 1)(x^2 + x + 1) = 0 \rightarrow x^3 - 1 = 0 \rightarrow x_1^3 = 1$ and $x_2^3 = 1$

→ $x_1^{2016} + x_2^{2016} = (x_1^3)^{672} + (x_2^3)^{672} = 1 + 1 = 2$

So, option (A) is correct and (B) is wrong

Similarly, $x_1^{2017} + x_2^{2017} = x_1(x_1^{2016}) + x_2(x_2^{2016}) = x_1 + x_2 =$ Sum of roots $= -1 \rightarrow$ Option (C) is correct.

$x_1^{2018} + x_2^{2018} = x_1^2(x_1^{2016}) + x_2^2(x_2^{2016}) = x_1^2 + x_2^2 = (x_1 + x_2)^2 - 2x_1x_2 = (-1)^2 - 2(1) = -1$

Hence, option (D) is wrong. So, Options (A) and (C) are correct.

You can also solve the problem by complex no. As $x^2 + x + 1 = (x - \omega)(x - \omega^2)$ *where* $\omega = e^{i2\pi/3}$.

24. Let $a, b \in R$ and $ab \neq 1$. If $6a^2 + 20a + 15 = 0$ and $15b^2 + 20b + 6 = 0$ and $\dfrac{b^3}{ab^2 - 9(ab+1)^3} = \dfrac{p}{q}$ (where

p, q are co-prime naturals) then

(A) The digit at unit's place of 'p' is 6   (B) Digit at unit's place of q is 6

(C) The digit at unit's place of p is 5   (D) Digit at unit's place of q is 5.

Solution:

If we observe $6a^2 + 20a + 15 = 0$ and $15b^2 + 20b + 6 = 0$

$\rightarrow$ a and 1/b are roots of equation $6x^2 + 20x + 15 = 0$

$\rightarrow a + \dfrac{1}{b} = -\dfrac{10}{3}$ and $\dfrac{a}{b} = \dfrac{5}{2}$. Now, $\dfrac{b^3}{ab^2 - 9(ab+1)^3} = \dfrac{1}{a \cdot \dfrac{1}{b} - 9\left(a + \dfrac{1}{b}\right)^3} = \dfrac{1}{\dfrac{5}{2} - 9\left(-\dfrac{10}{3}\right)^3} = \dfrac{6}{2015} = \dfrac{p}{q}$

Hence, Option (A) and (D) are correct.

MATHSARC EDUCATION

A learning place to fulfill your dream of success!

MATHEMATICS   **PRACTICE PROBLEMS**

SUBJECTIVE PROBLEMS

1. If α, β are real roots of $\sqrt{x+3}-1=\sqrt{x-\sqrt{x-2}}$ then find the value of $9(\alpha + \beta)$

2. Let r, s & t are roots of equation $8x^3 + 1001x + 2008 = 0$. Then value of $(r + s)^3 + (s + t)^3 + (t + r)^3$ is

3. Let k be an integer and p is a prime number such that the quadratic equation $x^2 + kx + p = 0$ has two distinct positive integer solutions. Then the value of $-(p + k)$ is

4. α, β are roots of $2x^2 + 24x - k = 0$, such that $\alpha > \beta$. If $\alpha - \beta = 14$ then find the value of k

5. If both roots of equation $4x^2 - 20px + 25p^2 + 15p - 66 = 0$ are greater than 2, then sum of all possible integral values of p is _____

6. Solve: $x^2 - 18x + 77 = |3x - 13|$ where $x \in R$.

7. Let t_1 & t_2 are values of 't' for which the roots of equation $x^2 + tx + t + 2 = 0$ are proportion to $2 : 1$. Then find the value of $2(t_1 + t_2)$ _____

8. If $\sqrt{x} + y = 7$ & $x + \sqrt{y} = 11$. Then find the value of x and $y \in I$.

9. Let α satisfy the equation $x^3 + 3x^2 + 4x + 5 = 0$ and β satisfy the equation $x^3 - 3x^2 + 4x - 5 = 0$, α, $\beta \in R$ the value of $\alpha + \beta$ is _____

10. If $\dfrac{a}{b} \in Q$ in lowest form, is a root of the equation $\sqrt[3]{20x + \sqrt[3]{20x + 13}} = 13$, then the value of b is

11. If $x^2 + tx + 1$ is a factor of $ax^3 + bx + c$ then prove that $a^2 - c^2 = ab$

12. If α, β, γ are roots of equation $x^3 + 4x + 1 = 0$ then find the value of $(\alpha + \beta)^{-1} + (\beta + \gamma)^{-1} + (\gamma + \alpha)^{-1}$.

13. If a, a^2 are the roots of $x^2 + x + 1 = 0$. Find the quadratic equation having roots a^{19} and a^{20}.

14. If α_1, α_2, α_3, α_4 and α_5 are roots of the equation $x^5 - 4x^2 + x - 13 = 0$. Find the value of $\displaystyle\sum_{i=1}^{5}\left(\dfrac{1}{2-\alpha_i}\right)$.

15. The set of values of 'a' for which the inequality $x^2 - (a + 2)x - (a + 3) < 0$ is satisfied for at least one positive real x is _____.

OBJECTIVE PROBLEMS

1. Let a, b be non - zero real numbers. Which of the following statements about the quadratic equation $ax^2 + (a + b)x + b = 0$ is necessarily true ?

    (I) It has at least one negative root

    (II) It has at least one positive root

    (III) Both its roots are real

    (A) (I) and (II) only   (B) (I) and (III) only   (C) (II) and (III) only   (D) All of them

2. If the product of the roots of the equation $x^2 - 3kx + 2e^{2 \ln(k)} - 1 = 0$ is 7, then the roots are real for k equal to

    (A) 2   (B) 4   (C) – 2   (D) none of these

3. If both roots of the equation $x^2 - 2ax + a^2 - 1 = 0$ lies between – 3 and 4, then [a] is ____, where [.] denotes greatest integer function.

    (A) 0, 1, 2   (B) –1, 0, 1, 2   (C) 0, 1, 2, 3   (D) –3, –2, –1, 0

4. If α, β and γ are roots of the equation, $x^3 + P_0x^2 + P_1x + P_2 = 0$, then $(1 - \alpha^2)(1 - \beta^2)(1 - \gamma^2)$ is equal to

    (A) $(1 + P_1)^2 - (P_0 + P_2)^2$ (B) $(1 + P_1)^2 + (P_0 + P_2)^2$ (C) $(1 - P_1)^2 - (P_0 - P_2)^2$ (D) None of these

5. Suppose a, b, c are real numbers, and each of the equations $x^2 + 2ax + b^2 = 0$ and $x^2 + 2bx + c^2 = 0$ has two distinct real roots. Then the equation $x^2 + 2cx + a^2 = 0$ has:

    (A) Two distinct positive real roots   (B) two equal roots

    (C) One positive and one negative root   (D) no real roots

6. The set of all the possible values of a, so that 6 lies between the roots of the equation $x^2 + 2(a - 3) x + 9 = 0$ is

    (A) $(- \infty, 0) \cup (6, 0)$ (B) $(- \infty, - 3/4)$ (C) $(0, \infty)$ (D) none of these

7. The roots of the equation $\dfrac{1+x^4}{(1+x)^4} = 7$ is/are

    (A) $\dfrac{3+\sqrt{5}}{2}$ (B) $\dfrac{-3-\sqrt{5}}{2}$ (C) $\dfrac{-3+\sqrt{5}}{2}$ (D) $\dfrac{3-\sqrt{5}}{2}$

8. Suppose that $|x+y|+|x-y| = 2$. What is the maximum possible value of $x^2 - 6x + y^2$?

    (A) 5   (B) 6   (C) 7   (D) 8

9. If $ax^2 + bx + 6 = 0$ $(a \in R - \{0\}$ & $b \in R)$ do not have two distinct real roots, then least value of 3a + b is

    (A) 4   (B) – 1   (C) 1   (D) – 2

10. If α, β are roots of $4x^2 - 16x + \lambda = 0$, $\lambda \in R$, such that $1 < \alpha < 2$ and $2 < \beta < 3$ then number of integral solutions of λ is

    (A) 5   (B) 6   (C) 2   (D) 3

11. If $(1 + m) x^2 - 2(1 + 3m) x + (1 + 8m) = 0$ has equal roots, then m is equal to

    (A) 0, 1   (B) 1, 2   (C) 0, 3   (D) none of these

12. If $3x^2 - 2mx - 4 = 0$ and $x^2 - 4m + 2 = 0$ have a common root, then m is

(A) $\pm 1/2$ (B) $\pm 1/\sqrt{3}$ (C) $\pm 1/3$ (D) $\pm 1/\sqrt{2}$

13 If $ax^2 + bx + c = 0$ & $bx^2 + cx + a = 0$ have one common root & a, b, c are non-zero real numbers, then which of the following can be true

(A) $\dfrac{a^3 + b^3 + c^3}{3abc} = 1$ (B) $a = b = c$

(C) $a + b + c = 0$ (D) $a^2 + b^2 + c^2 - ab - bc - ca = 0$

14. If α is a root of $x^2 + x + 2 = 0$ then the other root is/are

(A) $2/\alpha$ (B) $\alpha^2 + 1$ (C) $-\alpha^3 - 2\alpha + 1$ (D) $\bar{\alpha}$

15. The roots of equation $z^4 + az^3 + (12 + 9i) z^2 + bz = 0$ are vertices of a square then

(A) a may be $6 + 2i$ (B) a may be $-6 - 2i$ (C) b may be $9 + 13 i$ (D) b may be $-9 - 13 i$

ARCHIVE – JEE MAINS

1. The sum of all real values of x satisfying the equation $\left(x^2 - 5x + 5\right)^{x^2 + 4x - 60} = 1$ is:-   [2016]

   (1) 5   (2) 3   (3) – 4   (4) 6

2. If the equations $x^2 + bx - 1 = 0$ and $x^2 + x + b = 0$ have a common root different from -1, then $|b|$ is equal to: [2016]

   (1) $\sqrt{2}$ (2) 2   (3) 3   (4) $\sqrt{3}$

3. If x is a solution of the equation, $\sqrt{2x+1} - \sqrt{2x-1} = 1$, $\left(x \geq \dfrac{1}{2}\right)$ then $\sqrt{4x^2 - 1}$ is equal to: [2016]

   (1) 3/4   (2) 1/2   (3) 2   (4) $2\sqrt{2}$

4. Let α and β be the roots of equation $x^2 - 6x - 2 = 0$. If $a_n = \alpha^n - \beta^n$, for $n \geq 1$, then the value of $\dfrac{a_{10} - 2a_8}{2a_9}$ is equal to: [2015]

   (1) – 6   (2) 3   (3) - 3   (4) 6

5. If 2 + 3i is one of the roots of the equation $2x^3 - 9x^2 + kx - 13 = 0$, $k \in R$, then the real root of this equation: [2015]

   (1) does not exist.

   (2) exists and is equal to 1/2

   (3) exists and is equal to – 1/2

   (4) exists and is equal to 1.

6. If the two roots of the equation, $(a - 1)(x^4 + x^2 + 1) + (a + 1)(x^2 + x + 1)^2 = 0$ are real and distinct, then the set of all values of 'a' is: [2015]

   (1) $\left(-\dfrac{1}{2}, 0\right)$

   (2) $(-\infty, -2) \cup (2, \infty)$

   (3) $\left(-\dfrac{1}{2}, 0\right) \cup \left(0, \dfrac{1}{2}\right)$

   (4) $\left(0, \dfrac{1}{2}\right)$

7. If $a \in R$ and the equation $- 3(x - [x])^2 + 2(x - [x]) + a^2 = 0$ (where [x] denotes the greatest integer $\leq x$) has no integral solution, then all possible values of a lie in the interval: [2014]

   (1) (- 2, - 1)   (2) $(-\infty, -2) \cup (2, \infty)$ (3) (- 1, 0) $\cup$ (0, 1)   (4) (1, 2)

8. Let α and β be the roots of equation $px^2 + qx + r = 0$, $p \neq 0$. If p, q, r are in A. P. and $\dfrac{1}{\alpha} + \dfrac{1}{\beta} = 4$, then the value of $|\alpha - \beta|$ is: [2014]

   (1) $\dfrac{\sqrt{34}}{9}$ (2) $\dfrac{2\sqrt{13}}{9}$ (3) $\dfrac{\sqrt{61}}{9}$ (4) $\dfrac{2\sqrt{17}}{9}$

9. If equation $ax^2 + bx + c = 0$, $(a, b, c \in R, a \neq 0)$ and $2x^2 + 3x + 4 = 0$ have a common root, then $a : b : c$ equals: [2014]
   (1) $1 : 2 : 3$ (2) $2 : 3 : 4$ (3) $4 : 3 : 2$ (4) $3 : 2 : 1$

10. If $\dfrac{1}{\sqrt{\alpha}}$ and $\dfrac{1}{\sqrt{\beta}}$ are the roots of the equation $ax^2 + bx + 1 = 0$ ($a \neq 0$, $a, b \in R$), then the equation,

    $x(x + b^3) + (a^3 - 3abx) = 0$ has roots: [2014]
    (1) $\alpha^{3/2}$ and $\beta^{3/2}$ (2) $\alpha\beta^{1/2}$ and $\alpha^{1/2}\beta$ (3) $\sqrt{\alpha\beta}$ and $\alpha\beta$ (4) $\alpha^{-3/2}$ and $\beta^{-3/2}$

11. The sum of the roots of the equation $x^2 + |2x - 3| - 4 = 0$, is [2014]
    (1) 2 (2) -2 (3) $\sqrt{2}$ (4) $-\sqrt{2}$

12. The equation $\sqrt{3x^2 + x + 5} = x - 3$, where x is real, has: [2014]
    (1) No solution   (2) Exactly one solution
    (3) exactly two solution   (4) exactly four solution

13. If non - zero real number b and c are such that min $f(x)$ > max $g(x)$, where $f(x) = x^2 + 2bx + 2c^2$ and

    $g(x) = -x^2 - 2cx + b^2$ ($x \in R$); then $\left|\dfrac{c}{b}\right|$ lies in the interval [2014]

    (1) $\left(0, \dfrac{1}{2}\right)$ (2) $\left[\dfrac{1}{2}, \dfrac{1}{\sqrt{2}}\right)$ (3) $\left[\dfrac{1}{\sqrt{2}}, \sqrt{2}\right]$ (4) $(\sqrt{2}, \infty)$

14. The values of 'a' for which one root of the equation $x^2 - (a + 1)x + a^2 + a - 8 = 0$ exceeds 2 and the other is lesser than 2, are given by : [2013]
    (1) $3 < a < 10$ (2) $a \geq 10$ (3) $-2 < a < 3$ (4) $a \leq -2$

15. If α and β are roots of the equation $x^2 + px + \dfrac{3p}{4} = 0$, such that $|\alpha - \beta| = \sqrt{10}$, then p belongs to the set: [2013]
    (1) $\{2, -5\}$ (2) $\{-3, 2\}$ (3) $\{-2, 5\}$ (4) $\{3, -5\}$

16. If p and q are non – zero real numbers and $\alpha^3 + \beta^3 = -p$, $\alpha\beta = q$, then a quadratic equation whose roots are $\dfrac{\alpha^2}{\beta}, \dfrac{\beta^2}{\alpha}$ is: [2013]
    (1) $px^2 - qx + p^2 = 0$ (2) $qx^2 + px + q^2 = 0$
    (3) $px^2 + qx + p^2 = 0$ (4) $qx^2 - px + q^2 = 0$

17. The number of values of k, for which the system of equations:

$$(k + 1) x + 8y = 4k$$
$$kx + (k + 3)y = 3k - 1$$

has no solution is: [2013]

(1) infinite   (2) 1   (3) 2   (4) 3

18. The equation $e^{\sin x} - e^{-\sin x} - 4 = 0$ has [2012]

(1) infinite number of real roots   (2) no real roots

(3) exactly one real root   (4) exactly four real roots

19. Let α, β be real and z be a complex number. If $z^2 + \alpha z + \beta = 0$ has two distinct roots on the line Re(z) = 1, then it is necessary that [2011]

(1) $\beta \in (-1, 0)$ (2) $|\beta| = 1$ (3) $\beta \in (1, \infty)$ (4) $\beta \in (0, 1)$

20. If α and β are the roots of the equation $x^2 - x + 1 = 0$, then $\alpha^{2009} + \beta^{2009} =$ [2010]

(1) - 1   (2) 1   (3) 2   (4) – 2

21. If the roots of the equation $bx^2 + cx + a = 0$ be imaginary, then for all real values of x, the expression $3b^2x^2 + 6bcx + 2c^2$ is [2009]

(1) greater than 4ab   (2) less than 4ab   (3) greater than – 4ab   (4) less than – 4ab

22. How many real solutions does the equation $x^7 + 14x^5 + 16x^3 + 30x - 560 = 0$ have? [2008]

(1) 7   (2) 1   (3) 3   (4) 5

23. The quadratic equations $x^2 - 6x + a = 0$ and $x^2 - cx + 6 = 0$ have one root in common. The other roots of the first and second equations are integers in the ratio 4 : 3. Then the common root is [2008]

(1) 1   (2) 4   (3) 3   (4) 2

24. If the difference between the roots of the equation $x^2 + ax + 1 = 0$ is less than $\sqrt{5}$, then the set of possible values of a is [2007]

(1) $(-3, 3)$ (2) $(-3, \infty)$ (3) $(3, \infty)$ (4) $(-\infty, -3)$

25. If the roots of the quadratic equation $x^2 + px + q = 0$ are tan30° and tan15°, respectively then the value of 2 + q - p is [2006]

(1) 2   (2) 3   (3) 0   (4) 1

26. All the values of m for which both roots of the equations $x^2 - 2mx + m^2 - 1 = 0$ are greater than - 2 but less than 4, lie in the interval [2006]

(1) $- 2 < m < 0$ (2) $m > 3$ (3) $- 1 < m < 3$ (4) $1 < m < 4$

27. If x is real, the maximum value of $\dfrac{3x^2 + 9x + 17}{3x^2 + 9x + 7}$ is: [2006]

    (1) 1/4   (2) 41   (3) 1   (4) 17/7

28. In a triangle PQR, $\angle R = \dfrac{\pi}{2}$, If tan(P/2) & tan(Q/2) are roots of $ax^2 + bx + c = 0$, $a \neq 0$ then [2005]

    (1) a = b + c   (2) c = a + b   (3) b = c   (4) b = a + c

29. The value of α for which the sum of the squares of the roots of the equation $x^2 - (a - 2)x - a - 1 = 0$ assume the least value is [2005]

    (1) 1   (2) 0   (3) 3   (4) 2

30. If roots of the equation $x^2 - bx + c = 0$ be two consecutive integers, then $b^2 - 4c$ equals [2005]

    (1) - 2   (2) 3   (3) 2   (4) 1

31. Let α and β be the distinct roots of $ax^2 + bx + c = 0$, then $\lim\limits_{x \to \alpha} \dfrac{1 - \cos\left(ax^2 + bx + c\right)}{(x - \alpha)^2}$ is equal to [2005]

    (1) $\dfrac{a^2}{2}(\alpha - \beta)^2$ (2) 0   (3) $-\dfrac{a^2}{2}(\alpha - \beta)^2$ (4) $\dfrac{1}{2}(\alpha - \beta)^2$

32. If both the roots of the quadratic equation $x^2 - 2kx + k^2 + k - 5 = 0$ are less than 5, then k lies in the interval [2005]

    (1) (5, 6]   (2) (6, ∞)   (3) (- ∞, 4)   (4) [4, 5]

33. If the equation $a_n x^n + a_{n-1} x^{n-1} + \ldots\ldots + a_1 x = 0$, $a_1 \neq 0$, $n \geq 0$, has a positive root $x = \alpha$, then the equation $na_n x^{n-1} + (n-1)a_{n-1} x^{n-2} + \ldots\ldots + a_1 = 0$ has a positive root, which is [2005]

    (1) greater than α   (2) smaller than α

    (3) greater than or equal to α   (4) equal to α

34. Let two numbers have arithmetic mean 9 and geometric mean 4. Then these numbers are the roots of the quadratic equation [2004]

    (1) $x^2 + 18x + 16 = 0$ (2) $x^2 - 18x - 16 = 0$ (3) $x^2 + 18x - 16 = 0$ (4) $x^2 - 18x + 16 = 0$

35. If one root of the equation $x^2 + px + 12 = 0$ is 4, while the equation $x^2 + px + q = 0$ has equal roots, then the value of 'q' is [2004]

    (1) 49/4   (2) 4   (3) 3   (4) 12

36. If (1 - p) is a root of quadratic equation $x^2 + px + (1 - p) = 0$ then its roots are: [2003]

    (1) - 1, 2   (2) - 1, 1   (3) 0, - 1   (4) 0, 1

37. If the sum of the roots of the quadratic equation $ax^2 + bx + c = 0$ is equal to the sum of the squares of their reciprocals, then a/c, b/a and c/b are in   [2003]

   (1) Arithmetic progression   (2) geometric progression

   (3) Harmonic progression   (4) arithmetic – geometric – progression

38. The number of real solutions of the equation $x^2 - 3|x| + 2 = 0$ is   [2003]

   (1) 2   (2) 4   (3) 1   (4) 3

39. The value of 'a' for which one root of the quadratic equation $(a^2 - 5a + 3)x^2 + (3a - 1)x + 2 = 0$ is twice as large as the other, is   [2003]

   (1) 2/3   (2) -2/3   (3) 1/3   (4) -1/3

40. If $\alpha \neq \beta$ but $\alpha^2 = 5\alpha - 3$ and $\beta^2 = 5\beta - 3$ then the equation having α/β and β/α as its roots is   [2002]

   (1) $3x^2 - 19x + 3 = 0$ (2) $3x^2 + 19x - 3 = 0$ (3) $3x^2 - 19x - 3 = 0$ (4) $x^2 - 5x + 3 = 0$

41. Product of real roots of the equation $t^2x^2 + |x| + 9 = 0$ [2002]

   (1) Is always positive   (2) is always negative   (3) does not exist   (4) none of these

42. If p and q are the roots of the equation $x^2 + px + q = 0$, then   [2002]

   (1) $p = 1, q = -2$ (2) $p = 0, q = 1$ (3) $p = -2, q = 0$ (4) $p = -2, q = 1$

43. If a, b, c are distinct + ve real numbers and $a^2 + b^2 + c^2 = 1$ then $ab + bc + ca$ is   [2002]

   (1) less than 1   (2) equal to 1   (3) greater than 1   (4) any real number

44. Difference between the corresponding roots of $x^2 + ax + b = 0$ and $x^2 + bx + a = 0$ is same and $a \neq b$, then   [2002]

   (1) $a + b + 4 = 0$ (2) $a + b - 4 = 0$ (3) $a - b - 4 = 0$ (4) $a - b + 4 = 0$

ARCHIVE – IIT JEE ADVANCED

SINGLE OPTION CORRECT

1. The least value of $\alpha \in R$ for which $4\alpha x^2 + (1/x) \geq 1$, for all $x > 0$, is –   [2016]

     (A) $1/64$ (B) $1/32$ (C) $1/27$ (D) $1/25$

2. Let $-\dfrac{\pi}{6} < \theta < -\dfrac{\pi}{12}$. Suppose α_1 and β_1 are the roots of the equation $x^2 - 2x\sec\theta + 1 = 0$ and α_2 and β_2 are the roots of the equation $x^2 + 2x\tan\theta - 1 = 0$. If $\alpha_1 > \beta_1$ and $\alpha_2 > \beta_2$, then $\alpha_1 + \beta_2$ equals   [2016]

     (A) $2(\sec\theta - \tan\theta)$ (B) $2\sec\theta$ (C) $-2\tan\theta$ (D) 0

3. Let z be a complex number such that the imaginary part of z is nonzero and $a = z^2 + z + 1$ is real. Then a cannot take the value   [2012]

     (A) -1 (B) $1/3$ (C) $1/2$ (D) $3/4$

4. Let $\alpha(a)$ and $\beta(b)$ be the roots of the equation $\left(\sqrt[3]{1+a} - 1\right)x^2 + \left(\sqrt{1+a} - 1\right)x + \left(\sqrt[6]{1+a} - 1\right) = 0$ where $a > -1$. Then $\lim\limits_{a \to 0^+} \alpha(a)$ and $\lim\limits_{a \to 0^+} \beta(a)$ are   [2012]

     (A) $-5/2$ and 1 (B) $-1/2$ and -1 (C) $-7/2$ and 2 (D) $-9/2$ and 3

5. Let (x_0, y_0) be solution of the following equations

$$(2x)^{\ln 2} = (3y)^{\ln 3}$$
$$3^{\ln x} = 2^{\ln y}$$

     Then x_0 is   [2011]

     (A) $1/6$ (B) $1/3$ (C) $1/2$ (D) 6

6. Let α and β be the roots of $x^2 - 6x - 2 = 0$, with $\alpha > \beta$. If $a_n = \alpha^n - \beta^n$ for $n \geq 1$, then the value of $\dfrac{a_{10} - 2a_8}{2a_9}$ is   [2011]

     (A) 1 (B) 2 (C) 3 (D) 4

7. Let a, b and c be three real numbers satisfying $\begin{bmatrix} a & b & c \end{bmatrix} \begin{bmatrix} 1 & 9 & 7 \\ 8 & 2 & 7 \\ 7 & 3 & 7 \end{bmatrix} = \begin{bmatrix} 0 & 0 & 0 \end{bmatrix}$ ....(E)

     Let $b = 6$, with a & c satisfying (E). If α and β are roots of the quadratic equation $ax^2 + bx + c = 0$, then $\sum\limits_{n=0}^{\infty} \left(\dfrac{1}{\alpha} + \dfrac{1}{\beta}\right)^n$ is   [2011]

     (A) 6 (B) 7 (C) $6/7$ (D) ∞

8. A value of b for which the equations $x^2 + bx - 1 = 0$ & $x^2 + x + b = 0$, have one root in common is

(A) $-\sqrt{2}$ (B) $-i\sqrt{3}$ (C) $i\sqrt{5}$ (D) $\sqrt{2}$ [2011]

9. Let p and q be real numbers such that $p \neq 0$, $p^3 \neq q$ and $p^3 \neq -q$. If α and β are nonzero complex numbers satisfying $\alpha + \beta = -p$ and $\alpha^3 + \beta^3 = q$, then a quadratic equation having α/β and β/α as its roots is   [2010]

(A) $(p^3 + q)x^2 - (p^3 + 2q)x + (p^3 + q) = 0$ (B) $(p^3 + q)x^2 - (p^3 - 2q)x + (p^3 + q) = 0$

(C) $(p^3 - q)x^2 - (5p^3 - 2q)x + (p^3 - q) = 0$ (D) $(p^3 - q)x^2 - (5p^3 + 2q)x + (p^3 - q) = 0$

10. Let a, b, c, p, q be real numbers. Suppose α, β are the roots of the equation $x^2 + 2px + q = 0$ and α, $1/\beta$ are the roots of the equation $ax^2 + 2bx + c = 0$, where $\beta^2 \notin \{-1, 0, 1\}$.   [2008]

STATEMENT-1: $(p^2 - q)(b^2 - ac) \geq 0$

STATEMENT-2: $b \neq pa$ or $c \neq qa$

(A) Statement-1 is True, Statement-2 is True, Statement-2 is a correct explanation for Statement-1.

(B) Statement-1 isTrue, Statement-2 isTrue, Statement-2 is not a correct explanation for Statement-1

(C) Statement-1 is True, Statement-2 is False.

(D) Statement-1 is False, Statement-2 is True.

11. Let α, β be the roots of the equation $x^2 - px + r = 0$ and $\alpha/2$, 2β be the roots of the equation $x^2 - qx + r = 0$. Then the value of r is   [2007]

(A) $\dfrac{2}{9}(p-q)(2q-p)$ (B) $\dfrac{2}{9}(q-p)(2p-q)$ (C) $\dfrac{2}{9}(q-2p)(2q-p)$ (D) $\dfrac{2}{9}(2p-q)(2q-p)$

12. Let a, b, c be the sides of a triangle. No two of them are equal and $\lambda \in R$. If the roots of the equation $x^2 + 2(a + b + c)x + 3\lambda(ab + bc + ca) = 0$ are real, then   [2006]

(A) $\lambda < 4/3$ (B) $\lambda > 5/3$ (C) $\lambda \in (1/3, 5/3)$ (D) $\lambda \in (4/3, 5/3)$

13. In the quadratic equation $ax^2 + bx + c = 0$, if $\Delta = b^2 - 4ac$ and $\alpha + \beta$, $\alpha^2 + \beta^2$, $\alpha^3 + \beta^3$ are in G.P. where α, β are the roots of $ax^2 + bx + c = 0$, then   [2005]

(A) $\Delta \neq 0$ (B) $b\Delta = 0$ (C) $c\Delta = 0$ (D) $\Delta = 0$

14. For all 'x', $x^2 + 2ax + 10 - 3a > 0$, then the interval in which 'a' lies is   [2004S]

(A) $a < -5$ (B) $-5 < a < 2$ (C) $a > 5$ (D) $2 < a < 5$

15. If $\alpha \in (0, \pi/2)$ then $\sqrt{x^2 + x} + \dfrac{\tan^2 \alpha}{\sqrt{x^2 + x}}$ is always greater than or equal to   [2003S]

(A) $2\tan\alpha$ (B) 1 (C) 2 (D) $\sec^2\alpha$

16. The set of all real numbers x for which $x^2 - |x + 2| + x > 0$, is [2002S]

   (A) $(-\infty, -2) \cup (2, \infty)$ (B) $(-\infty, -\sqrt{2}) \cup (\sqrt{2}, \infty)$ (C) $(-\infty, -1) \cup (1, \infty)$ (D) $(\sqrt{2}, \infty)$

17. For the equation $3x^2 + px + 3 = 0$, $p > 0$, if one of the root is square of the other, then p is equal to

   (A) $1/3$ (B) 1 (C) 3 (D) $2/3$ [2005]

18. If $b > a$, then the equation $(x - a)(x - b) - 1 = 0$ has   [2000S]

   (A) both roots in (a, b)

   (B) both root in $(-\infty, a)$

   (C) both roots in (b, ∞)

   (D) one root in $(-\infty, a)$ and the other in (b, ∞)

19. If α and β $(\alpha < \beta)$ are the roots of the equation $x^2 + bx + c = 0$, where $c < 0 < b$, then [2000S]

   (A) $0 < \alpha < \beta$ (B) $\alpha < 0 < \beta < |\alpha|$ (C) $\alpha < \beta < 0$ (D) $\alpha < 0 < |\alpha| < \beta$

MULTI OPTION CORRECT

1. Let S be the set of all non-zero numbers α such that the quadratic equation $ax^2 - x + a = 0$ has two distinct real roots x_1 and x_2 satisfying the inequality $|x_1 - x_2| < 1$. Which of the following intervals is(are) a subset(s) of S? [2015]

   (A) $\left(-\dfrac{1}{2}, -\dfrac{1}{\sqrt{5}}\right)$ (B) $\left(-\dfrac{1}{\sqrt{5}}, 0\right)$ (C) $\left(0, \dfrac{1}{\sqrt{5}}\right)$ (D) $\left(\dfrac{1}{\sqrt{5}}, \dfrac{1}{2}\right)$

2. Let $a \in R$ and let $f : R \to R$ be given by $f(x) = x^5 - 5x + a$, then [2014]

   (A) $f(x)$ has three real roots if $a > 4$ (B) $f(x)$ has only one real roots if $a > 4$

   (C) $f(x)$ has three real roots if $a < -4$ (D) $f(x)$ has three real roots if $-4 < a < 4$

3. If $3^x = 4^{x-1}$, then $x =$ [2013]

   (A) $\dfrac{2\log_3 2}{2\log_3 2 - 1}$ (B) $\dfrac{2}{2 - \log_2 3}$ (C) $\dfrac{1}{1 - \log_4 3}$ (D) $\dfrac{2\log_2 3}{2\log_2 3 - 1}$

SUBJECTIVE / INTEGER TYPE

1. The number of distinct real roots of $x^4 - 4x^3 + 12x^2 + x - 1 = 0$ is [2011]

2. The minimum value of the sum of real number a^{-5}, a^{-4}, $3a^{-3}$, 1, a^8 and a^{10} where $a > 0$ is [2011]

3. The smallest value of k, for which both the roots of the equation $x^2 - 8kx + 16(k^2 - k + 1) = 0$ are real, distinct and have values at least 4, is [2009]

4. If roots of the equation $x^2 - 10cx - 11d = 0$ are a, b and those of $x^2 - 10ax - 11b = 0$ are c, d, then the value of $a + b + c + d$ is (a, b, c and d are distinct numbers) [2006]

5. If $x^2 + (a - b)x + (1 - a - b) = 0$ where a, b $\in$ R then find the values of a for which equation has unequal real roots for all values of b [2003]

6. Let a, b, c be real numbers with a $\neq$ 0 and let α, β be the roots of the equation $ax^2 + bx + c = 0$. Express the roots of $a^3 x^2 + abc \, x + c^3 = 0$ in terms of α, β. [2001]

7. If α, β are the roots of $ax^2 + bx + c = 0$, (a $\neq$ 0) and $\alpha + \delta$, $\beta + \delta$ are the roots of $Ax^2 + Bx + C = 0$, (A $\neq$ 0) for some constant δ, then prove that $\dfrac{b^2 - 4ac}{a^2} = \dfrac{B^2 - 4AC}{A^2}$. [2000]

ULTIMATE FINISH !

The section contains H.O.T. (Higher Order Thinking) questions i.e. IIT ++. If you are confident enough then solve this section to get thinking of top 500 IIT JEE aspiring students. The section may be a part of disappointment but learning is at top priority. The section will bring your best.

UF1. Evaluate $S_{100} = \left\lfloor \dfrac{1}{\sqrt{2}} + \dfrac{1}{\sqrt{3}} + \dfrac{1}{\sqrt{4}} + \cdots\cdots + \dfrac{1}{\sqrt{100}} \right\rfloor$ where [.] = G.I.F.

UF2. The maximum value M of $3^x + 5^x - 9^x + 15^x - 25^x$, as x varies over real, satisfies:

(A) $3 < M < 5$ (B) $0 < M < 2$ (C) $9 < M < 25$ (D) $5 < M < 9$

UF3. Let $P(x) = (x - 1)(x - 2)(x - 3)\ldots\ldots(x - 50)$ and $Q(x) = (x + 1)(x + 2)(x + 3)\ldots\ldots(x + 50)$.

If $P(x). Q(x) = a_0 + a_1 x + a_2 x^2 + \ldots\ldots + a_{100} x^{100}$, then the value of $a_{100} - a_{99} + a_{98} - a_{97}$

(A) $- 42925$ (B) 0 (C) $- 42924$ (D) $- 42000$

UF4. If $f(x) = \displaystyle\prod_{r=1}^{999}\left(x^2 - 47x + r\right)$ then product of all real roots of f(x) = 0.

(A) $550!$ (B) $551!$ (C) $552!$ (D) $999!$

UF5. The number of points P(x, y) satisfying $y^2\left(\log_2\left(x^2 + 1\right) + \log_{x^2+1} 16\right) = 6y^2 - y^4 - 1$, are _____

UF6. If $4^x - 2^{x+2} + 5 + \left|\,|b - 1| - 3\,\right| = |\sin y|$, x, y, b $\in$ R, then the number of possible value of b is _____

UF7. The number of positive real roots of equation
$\log\left(x^{2004} + 1\right) + \log\left(1 + x^2 + x^4 + .. + x^{2012}\right) = \log 2014 + 2013 \cdot \log(x)$ is/are

(A) 1 (B) 2 (C) 3 (D) infinite

UF8. If f is a continuous and differentiable function in x $\in$ (0, 1) such that $\displaystyle\sum_{r=0}^{10}\left(f(x+r) - \left|e^x - r - 1\right|\right) = 0$

where $|.|$ is absolute value function then $\displaystyle\int_0^{11} f(x)dx$ is

(A) $65 + 4 \ln2 - 7e$ (B) $63 + 4 \ln2 - 9e$ (C) $69 - 9e$ (D) $29 - 23 e$

UF9. If the equation $|x^2 - 5x + 6| - mx + 7m = 0$ has p distinct solutions, then

(A) if p = 3, then m $\in \{- 9 - 4\sqrt5\}$ (B) if p = 4, then m $\in (- 9 + 4\sqrt5, 0)$

(C) if p = 2, then m $\in (- 9 + 4\sqrt5, 0)$ (D) if p = 3, then m $\in \{- 9 + 4\sqrt5\}$

UF10. Consider $f(x) = x^{51} + \log_7\left(x + \sqrt{x^2 + 1}\right)$, then $\forall$ a, b $\in$ R such that a + b > 0

(A) f(a) + f (b) < 0   (B) f(a) + f (b) > 0   (C) f(a + 7) + f (b) < 0   (D) f(a) + f (b + 1) = 0

UF11. In the identity $\dfrac{25!}{x(x+1)(x+2)(x+3)\ldots\ldots(x+25)} = \sum\limits_{i=0}^{25}\left(\dfrac{A_i}{x+i}\right)$, then sum of digits of numbers A_{24} is _

UF12. Match the following for the equation x^2 + a $|x|$ + 1 = 0 where a is real parameter

	COLUMN - I		COLUMN - II
A	No real roots	P	a < - 2
B	Exactly two real and distinct roots	Q	ϕ
C	Exactly three real and distinct roots	R	a = - 2
D	Four distinct real roots	S	a > - 2
		T	None of these

UF13. The value of parameter a for which $1 + \log_5\left(x^2 + 1\right) \le \log_5\left(ax^2 + 4x + a\right)$ is true $\forall$ x $\in$ R is

(A) [2, 9]   (B) [3, 7]   (C) [7, ∞)   (D) [5, ∞)

UF14. Let f(x) = x^3 + x + 1 and P(x) be a cubic polynomial such that P(0) = - 1 and the roots of P(x) = 0 are the squares of the roots of f(x) = 0, then

(A) P (4) = 98   (B) P (4) = 100   (C) P (9) = 899   (D) P (9) = 900

UF15. If one root of kx^2 + x + 1 = 0 is real and other is imaginary (k $\neq$ 0) then k^2 cannot be equal to

(A) - 4   (B) -1   (C) 1   (D) 4

UF16. If $3^{\sqrt{\log_3 7}}$ and $7^{\sqrt{\log_7 3}}$ are roots of equation ax^2 + bx + c = 0. Then its Discriminant is _____

UF17. The number of real roots of the equation $x^2 - 2x - \log_2|1-x| = 3$ is _____

UF18. The number of integral values of m satisfying $17^2 + n^4 = m^2$, is _____

UF19. The value of a for which the equation $(x - 1)^2$ = $|x - a|$ has exactly three real solutions is/are

(A) 1/4   (B) 3/4   (C) 1   (D) 5/4

UF20. It is given that f(x) is a function defined on R, satisfying f (1) =1, and for any x$\in$ R, f (x + 5) $\ge$ f(x) + 5 and f(x + 1) $\le$ f(x) + 1. If g(x) = f(x) + 1- x, then g (2016) = _____

UF21. The number of natural numbers n in the interval [1005, 2010] for which the polynomial $1 + x + x^2 + x^3 + ... + x^{n-1}$ divides the polynomial $1 + x^2 + x^4 + x^6 + ... + x^{2010}$ is

(A) 0   (B) 100   (C) 503   (D) 1006

UF22. Let r be a real number and $n \in N$ be such that the polynomial $2x^2 + 2x + 1$ divides the polynomial $(x + 1)^n - r$. Then (n , r) can be

(A) $(4000, 4^{1000})$ (B) $\left(4000, \dfrac{1}{4^{1000}}\right)$ (C) $\left(4^{1000}, \dfrac{1}{4^{1000}}\right)$ (D) $\left(4000, \dfrac{1}{4000}\right)$

UF23. Solve the system: $x + y + z = 1$ and $xyz = 1$, knowing that $x, y, z \in C$ of absolute value 1.

UF24. If $55^{f(x)} + 5^x - 2012 = 0$ and f(x) is defined. Then possible integral value(s) of x is/are

(A) –1   (B) 2   (C) 3   (D) 5

UF25. Let $f(x) = x^4 + ax^3 + bx^2 + cx + d$ and x_1, x_2, x_3 are three consecutive positive integer such that $f(x_i) = 10x_i$, $\forall i = 1, 2, 3$, If $f(12) + f(-8)$ is independent of a, b, c, d then '$x_1 + x_2 + x_3$' is equal to

(A) 6   (B) 9   (C) 12   (D) 15

UF26. If $f\left(x + \dfrac{y}{8}, x - \dfrac{y}{8}\right) = xy$, then $f(m, n) + f(n, m) = 0$

(A) only when m = n   (B) only when m ≠ n   (C) only when m = - n   (D) $\forall$ m and n

UF27. Let f(x) be a function satisfying $f(x)f(y) = f(xy)$ $\forall$ real x, y. If f(2) = 4, then what is the value of f(1/2)?

(A) 0   (B) 1/4   (C) 1/2   (D) can't be determined

UF28. A function f(x) satisfies f (1) = 3600, and $f(1) + f(2) + f(3) + ... + f(n) = n^2 f(n)$ $\forall$ positive integer n>1. What is the value of f (9)?

(A) 80   (B) 240   (C) 200   (D) 100

EXERCISE - HINT AND SOLUTIONS

GEO PROBLEM # 1: 3

Sol. Diagonal bisect the rectangle in two equal areas. So from geometry both areas must be same.

MIND REFRESHMENT # 1: 3 + 4π

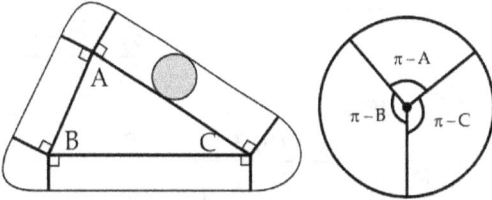

Sol. Here you can see:

π + A + π + B + π + c = 2π

→ the circular part will form a complete circle.

So, outer perimeter = 3 + 2π (R)

Where R = diameter of coin. Why? Think!

So, required perimeter = 3 + 4π

MIND REFRESHMENT # 2: $x = \log_{\left(\frac{3}{2}\right)}\left(\frac{1+\sqrt{5}}{2}\right)$

Sol. $9^x = 6^x + 4^x$

On dividing both side by 4^x we get

$\left(\frac{3}{2}\right)^{2x} = \left(\frac{3}{2}\right)^x + 1$. Let $\left(\frac{3}{2}\right)^x = t$ then equation becomes $t^2 - t - 1 = 0 \rightarrow t = \frac{1 \pm \sqrt{5}}{2}$

but $\left(\frac{3}{2}\right)^x = t > 0$ so $t = \frac{1+\sqrt{5}}{2} = \left(\frac{3}{2}\right)^x \rightarrow x = \log_{\left(\frac{3}{2}\right)}\left(\frac{1+\sqrt{5}}{2}\right)$

MIND REFRESHMENT # 3: 042

ANSWER KEY – EXERCISE 1

Q. NO.	ANSWER	Q. NO.	ANSWER	Q. NO.	ANSWER
1	x	4(iii)	(x + 1)(3x - 2)(2x - 1)	8	A
2	a = 2, b = 1	4(iv)	$(x - y)(x + y)(x^2 + y^2)$	9	5
3	A, C, D	5	B	10	B, C, D
4(i)	$(x^2 + x + 1)(x^2 - x + 1)$	6	D	11	$\lambda \in (-4\sqrt{2}, 4\sqrt{2})$
4(ii)	(x - 1)(2x + 1)(x - 3)	7	B		

ANSWER KEY – EXERCISE 2

Q. NO.	ANSWER	Q. NO.	ANSWER	Q. NO.	ANSWER
1	m = 8	8	(i) $a^2 x^2 - (b^2 - 2ac) x + c^2 = 0$ (ii) $ac x^2 - (b^2 - 2ac) x + ac = 0$	14	0, - 1, 2, - 2
2	a = 1/3	10	k = 2/3	15	- 15
3	2	11	32	16	5x
5	x = - 1, 6	12	7	17	a = 2, b = 9, c = - 3
7	$k = \frac{a-b}{a+b}$	13	2	18	$k \in R - \{1\}$

ANSWER KEY – EXERCISE 3

Q. NO.	ANSWER	Q. NO.	ANSWER	Q. NO.	ANSWER
1	0	4(iv)	1, 2, 1/2	6	1
2	2	4(v)	2, 3, - 1/2, - 1/3	7(i)	(1, 1), (2, 3)
3(i)	8	4(vi)	0, -2	7(ii)	$(x, y, z) = (0, 0, 0)$
3(ii)	2, 6, -10/3	4(vii)	4, 4/3	7(iii)	4, 5
3(iii)	3	4(viii)	- 3, - 1/3, $5 \pm 2\sqrt{6}$	7(iv)	- 2, 4, - 7, 8
3(iv)	3, - 4	4(ix)	27, 25/147	7(v)	± 13
3(v)	$x = 2 \pm \dfrac{2}{\sqrt{3}}$	4(x)	$0, - 5, \dfrac{-5 \pm i\sqrt{15}}{2}$	7(vi)	$\log_3 4$
3(vi)	8	4(xi)	-1, 6	8	$x = 2, y = - 3,$ $u = 5, v = -7$
4(i)	0, - 9	4(xii)	2, 1/2, 2/3, 3/2		
4(ii)	4, 1/4	5(i)	1/3	9	$(x, y, z) = (4, 1, 2)$ or $(-4, -1, -2)$
4(iii)	5/7, - 12/7	5(ii)	x = 1, 36		

ANSWER KEY – EXERCISE 4

Q. NO.	ANSWER	Q. NO.	ANSWER
1	$2 + i\sqrt{3}, 2 - i\sqrt{3}, - 2 + i$ and $- 2 - i$	12	3/7
2	$1 + \sqrt{5}, 1 - \sqrt{5}, 1, 2$	13	3
3	$x^4 - 2x^3 + 2x^2 + 6x - 15 = 0$	14	A
4	$x = 1/2, 1/2, - 1, \pm i$	15	$1, - 3, 1 - \sqrt{5}, 1 + \sqrt{5}$
6	$x = 1/3, 1, 1/5$	16	$x^3 + x^2 - 2x - 1 = 0$
7	4, 2, 5	17	B
8	1/2, 1/2, - 6	18	k = 7/ 4 or – 11/4
9	3	19	$\alpha = 0, 1$
11	- 7	20	a = 1/3 or a < 0

ANSWER KEY – EXERCISE 5

Q. NO.	ANSWER	Q. NO.	ANSWER
2	(i) $(- \infty, 19]$ (ii) $[- 13, 19]$ (iii) $[- 31, 17)$	7	For $b \in (- \infty, 2]$, $f(x) \mid_{\text{LEAST}} = f(4) = 8b - 3b^2 - 16$ For $b \in [2, \infty)$, $f(x) \mid_{\text{LEAST}} = f(0) = - 3b^2$
3	A	8	$p = 1 - \sqrt{2}$ or $5 + \sqrt{10}$
4	A, C	10	Max = 4 & Min = - 5
5	3	11	$a \in [-12, 2]$
6	$p = \pm 3, \pm 4$	12	$D_f = R - \{1\}, D_g = R \ \& \ R_f = \left(1, \dfrac{7}{3}\right], R_g = \left(1, \dfrac{7}{3}\right]$

ANSWER KEY – EXERCISE 6

Q. NO.	ANSWER	Q. NO.	ANSWER
1	$p \in (-\infty, 0] \cup (5, \infty)$	9	A
2	$a \in \left(-\infty, -\dfrac{1}{4}\right) \cup (5, 6]$	10	D
3	$k \in (-1, 0)$	11	C
4	$x = -2/3 \ \& \ 0$	12	A
6	$a \in (-\infty, 4) \cup (12, \infty)$	13	5011
7	C	14	D
8	B, C		

HINT / SOLUTIONS – EXERCISE · 1

1. As per remainder theorem, the type of remainder must be of ax + b (linear) nature.

    Hence, $x^{100} = Q(x) \times (x - 1) + a\,x + b \ \forall \ x \in R$, where $Q(x)$ is quotient polynomial.

    If we put x = 0, then we get $0 = 0 + b \rightarrow b = 0$.

    Now, if we put x = 1, then we get $1 = 0 + a + b \rightarrow a = 1$ as b = 0. Hence remainder is 'x'.

2. a = 2, b = 1.

3. A, C, D

4. (i) $x^4 + x^2 + 1 = x^4 + 2x^2 + 1 - x^2$

    $= (x^2 + 1)^2 - x^2 = (x^2 + x + 1)(x^2 - x + 1)$

    (ii) $2x^3 - 7x^2 + 2x + 3 = (x - 1)(2x^2 - 5x - 3)$

    $= (x - 1)(2x + 1)(x - 3)$

    (iii) $6x^3 - x^2 - 5x + 2 = (x + 1)(6x^2 - 7x + 2)$

    $= (x + 1)(3x - 2)(2x - 1)$

    (iv) $x^8 - y^8 = (x - y)(x + y)(x^2 + y^2)$

    (v) $x^{10} + x^5 + 1 = (x^3 - 1)(x^2 + x + 1)$

    (vi) $x^8 + x^4 + 1 = (x^4 - x^2 + 1)(x^2 + x + 1)(x^2 - x + 1)$

5. B

    As per question, $f(0)\, f(1) < 0 \rightarrow p\,(p - 3) < 0 \rightarrow p \in (0, 3)$,

6. D

    $f(x^5) = (x^{20} - 1)(x^{15} - 1)(x^{10} - 1)(x^5 - 1) + 5$. Hence remainder is 5

7. B

    Let $P(x) = Q(x)(x - 19)(x - 94) + 1994$ ....(As per remainder theorem)

    $P(0) = Q(0) \times (1786) + 1994$. Here $P(0)$ is the constant term

    $P(x)$ has integral coefficients that's why $Q(0) = -1$

    $P(0) = -1 \times 1786 + 1994 = 208$

8. A

    $x^3, \ y^3$ are of the form 9n, 9n+1, 9n+8

$\therefore$ required ways $= 33 \times 50 = 1650$

9. 5

$5x^3 + Mx + N = (5x - 5)(x^2 + x + 1) + Mx + N + 5$

Remainder $= 0 \rightarrow |M + N| = 5$

10. B, C, D

As $f([i]) = \dfrac{1}{1+[i]} \rightarrow f(0) = 1, f(1) = \dfrac{1}{2}, f(2) = \dfrac{1}{3}, f(3) = \dfrac{1}{4}, f(4) = \dfrac{1}{5}$

$\rightarrow g(x) = (1+x) f(x) - 1$, is a polynomial of degree 5 with roots 0, 1, 2, 3, 4

$\rightarrow g(x) = Ax(x-1)(x-2)(x-3)(x-4) \rightarrow A = \dfrac{1}{5!}$

$\rightarrow (1+x)f(x) - 1 = \dfrac{1}{5!}x(x-1)(x-2)(x-3)(x-4)$

11. $\lambda \in (-4\sqrt{2}, 4\sqrt{2})$

HINT / SOLUTIONS – EXERCISE - 2

1. m = 8

$\alpha + \beta = 6$ and $2\alpha + \beta = 8 \rightarrow \alpha = 2$ which is a root

$\rightarrow 2^2 - 6 \times 2 + m = 0 \rightarrow m = 8$.

2. a = 1/3

If $a = 0$ then $ax^2 - 4x + 9 = 0 \rightarrow x = 9/4 \notin I$

If $a \neq 0$ then $ax^2 - 4x + 9 = 0$ can be rewritten as $x^2 - \dfrac{4}{a}x + \dfrac{9}{a} = 0$ where $\dfrac{4}{a}, \dfrac{9}{a} \in I$

(for $a = 1, D < 0$) $\rightarrow a \notin I, a \notin \dfrac{p}{q}$ where $p \neq 1, \rightarrow a = \dfrac{1}{b}$ where $b \in I$

$\rightarrow x^2 - \dfrac{4}{a}x + \dfrac{9}{a} = 0 \rightarrow x^2 - 4bx + 9b = 0$

As per condition of integral roots D must be a perfect square

$\rightarrow D = 16 b^2 - 36 b = 4(4b^2 - 9b) \rightarrow 4b^2 - 9b = k^2$ where $k \in I$

$\rightarrow (2b)^2 - 2 \times (9/4) \times 2b + (81/16) = k^2 + (81/16)$

$\rightarrow (2b - (9/4))^2 - k^2 = (81/16) \rightarrow (8b - 9)^2 - 4k^2 = 81 \rightarrow (8b - 9 - 2k)(8b - 9 + 2k) = 81$

$\rightarrow$ as both factors are odd (observe)

$\rightarrow (8b - 9 - 2k)(8b - 9 + 2k) = 81 = 1 \times 81 = 3 \times 27 = -1 \times (-81) = (-3) \times (-27)$

We are getting b an integer only in the case $(8b - 9 - 2k)(8b - 9 + 2k) = 3 \times 27$

$\rightarrow 8b - 9 - 2k = 3$ and $8b - 9 + 2k = 27 \rightarrow b = 3, k = 6$

So, $a = 1/b = 1/3$.

3. 2

5. x = -1, 6

The correct quadratic is $x^2 - 5x - 6 = 0$.

7. $k = \dfrac{a-b}{a+b}$

The equation can be written as $(k+1)(x^2 - bx) = (k-1)(ax - c)$

$\rightarrow (k+1)x^2 - (b(k+1) + a(k-1))x + c(k-1) = 0$

As per question sum of roots $= 0 \rightarrow b(k+1) + a(k-1) = 0$

$\rightarrow k(a+b) = a - b \rightarrow k = \dfrac{a-b}{a+b}$.

8. (i) $a^2 x^2 - (b^2 - 2ac) x + c^2 = 0$

(ii) $acx^2 - (b^2 - 2ac) x + ac = 0$

10. $k = 2/3$

$(k^2 - 5k + 3)x^2 + (3k - 1)x + 2 = 0$

Let the roots are α and $2\alpha \rightarrow \alpha + 2\alpha = 3\alpha = \dfrac{3k-1}{k^2 - 5k + 3}$ ....(1)

Product of roots $2\alpha^2 = \dfrac{2}{k^2 - 5k + 3}$ ....(2)

From (1) and (2) we get

$2\left(\dfrac{3k-1}{3\left(k^2 - 5k + 3\right)}\right)^2 = \dfrac{2}{k^2 - 5k + 3} \rightarrow (3k-1)^2 = 9(k^2 - 5k + 3) \rightarrow 9k^2 - 6k + 1 = 9k^2 - 45k + 27$

$\rightarrow 39k = 26 \rightarrow k = 2/3$.

11. 32

$\alpha + \beta = 2$ and $\alpha\beta = 4$

$\alpha^5 + \beta^5 = (\alpha + \beta)(\alpha^4 - \alpha^3\beta + \alpha^2\beta^2 - \alpha\beta^3 + \beta^4)$

$\alpha^5 + \beta^5 = (\alpha + \beta)(\alpha^4 + \beta^4 - \alpha^3\beta - \alpha\beta^3 + \alpha^2\beta^2)$

$\alpha^5 + \beta^5 = (\alpha + \beta)(\alpha^4 + \beta^4 - \alpha\beta(\alpha^2 + \beta^2) + \alpha^2\beta^2)$ ....(1)

$\alpha^2 + \beta^2 = (\alpha + \beta)^2 - 2\alpha\beta = 4 - 8 = -4$........(2)

$\alpha^4 + \beta^4 = (\alpha^2 + \beta^2)^2 - 2\alpha^2\beta^2 = 16 - 2 \times 16 = -16$ ....(3)

From (1), (2) and (3) we get $\alpha^5 + \beta^5 = 2(-16 - 4(-4) + 16) = 32$

12. 7

$px^2 + qx + 1 = 0, D = q^2 - 4p \geq 0$

$q = 1, p = $ No value

$q = 2,$ and $p = 1$

$q = 3, \quad p = 1, 2$

$q = 4 \quad p = 1, 2, 3, 4$

so that No of pair is equal to 7. Hence there are 7 such quadratic equations.

13. 2

$x^3 - 2x + 1 = (x^2 - x - 1)(x + 1) + 2$

14. $0, -1, 2, -2$

Apply $(0^0 = 1, 1^{(\text{anything})} = 1, (\text{anything})^0 = 1, (-1)^{(\text{even})} = 1)$

15. -15

$f(x) = ax^7 + bx^5 + cx^3 - 6$

$f(-9) = 3 \rightarrow -a(9)^7 - b(9)^5 - c(9)^3 - 6 = 3 \rightarrow a(9)^7 + b(9)^5 + c(9)^3 = -9$

$f(9) = a(9)^7 + b(9)^5 + c(9)^3 - 6 = -9 - 6 = -15$

16. $5x$

$x^{81} + x^{49} + x^{25} + x^9 + x$ is divided by $x^3 - x$.

as per remainder theorem: $x^{81} + x^{49} + x^{25} + x^9 + x = x(x-1)(x+1)Q(x) + ax^2 + bx + c, \forall x \in R$

so, if $x = 0 \rightarrow c = 0$

if $x = 1 \rightarrow 5 = a + b$ ......(1)

if $x = -1 \rightarrow -5 = a - b$ .....(2)

from (1) and (2) we get $a = 0$ and $b = 5$

so, remainder is $5x$.

17. $a = 2, b = 9, c = -3$

$p(x) = a(x-3)^2 + bx + 1$ and $q(x) = 2x^2 + c(x-2) + 13$ are same, so

$p(0) = q(0) \rightarrow 9a + 1 = -2c + 13 \rightarrow 2c + 9a = 12$ ....(1)

$p(3) = q(3) \rightarrow 3b + 1 = 18 + c + 13 \rightarrow 3b - c = 30$ ....(2)

$p(2) = q(2) \rightarrow a + 2b + 1 = 8 + 13 \rightarrow a + 2b = 20$ ....(3)

from (1) and (2) we get $3a + 2b = 24$ .....(4)

from (3) and (4) we get $a = 2, b = 9 \rightarrow c = -3$.

18. $k \in R - \{1\}$

Intersection at $x - $ axis $\rightarrow y = 0$

$\rightarrow x^2 + (2x + 3)k + 4(x + 2) + 3k - 5 = 0$, has distinct real roots

$\rightarrow x^2 + (2k + 4)x + 6k + 3 = 0$ has distinct real roots

$\rightarrow D > 0 \rightarrow (2k + 4)^2 - 4(6k + 3) > 0$

$\rightarrow (k + 2)^2 - (6k + 3) > 0 \rightarrow k^2 - 2k + 1 > 0 \rightarrow (k - 1)^2 > 0$

$\rightarrow k \in R - \{1\}$

HINT / SOLUTIONS – EXERCISE - 3

3. (ii) $x = 6, 2, -10/3$ Hint: square both side & get $(x-2)\left(x + 6 - 2\sqrt{x^2 + x - 6}\right) = 0$

5. (i) $x = \dfrac{1}{3}$ & $x \neq 0, \dfrac{56}{65}$ Hint: square both side: $\sqrt{4x^2 + 5x + 1} = 2\sqrt{x^2 - x + 1} + (9x - 3)$

And get the equation: $(3x - 1)\left(9x - 4 + 4\sqrt{x^2 - x + 1}\right) = 0$

(ii) $x = 1, 36$: Hint, Let $\sqrt{x} = t$ then given equation become $(t-1)(t-6)(t^2 + 19t + 6) = 0$

8. Adding all the equations and dividing by 3, we obtain

$x + y + u + v = -3$

$\rightarrow 4 + v = -3, -5 + x = -3, 0 + y = -3, -8 + u = -3$.

Thus: $x = 2, y = -3, u = 5, v = -7$.

9. Let $u = y + z, v = z + x, w = x + y$.

   Then the system becomes: $vw = 30, wu = 15, uv = 18$.

   On multiplying these equations we obtain

   $u^2 v^2 w^2 = 8100$, which we solve to obtain $uvw = \pm 90$.

   Next, we solve for u, v, and w to obtain

   $u = 3, v = 6, w = 5$, or $u = -3, v = -6, w = -5$. Then we have:

   $y + z = \pm 3, z + x = \pm 6$ and $x + y = \pm 5$.

   Thus we conclude that

   $x = 4, y = 1, z = 2$ or $x = -4, y = -1, z = -2$.

ANSWER KEY – PRACTICE PROBLEMS

SUBJECTIVE

1.	76; Roots $\alpha, \beta = 6, 22/9$.	2.	753	3.	1
4.	26	5.	7; where $p \in (2, 22/5]$	6.	$x = 6, 15$
7.	9	8.	$x = 9, y = 4$	9.	0	10.	5
12.	4	13.	$x^2 + x + 1 = 0.$	14.	13	15.	$(-2, \infty)$

OBJECTIVE

1.	B	2.	A	3.	B	4.	A
5.	D	6.	B	7.	B, C	8.	D
9.	C	10.	D	11.	C	12.	C
13.	A, B, C, D	14.	A, B, C, D	15.	A, B, C, D.		

ARCHIVE – JEE MAINS

1.	(2)	2.	(4)	3.	(1)	4.	(2)
5.	(2)	6.	(3)	7.	(3)	8.	(2)
9.	(2)	10.	(1)	11.	(3)	12.	(1)
13.	(4)	14.	(3)	15.	(3)	16.	(2)
17.	(2)	18.	(2)	19.	(3)	20.	(2)
21.	(3)	22.	(2)	23.	(4)	24.	(1)
25.	(2)	26.	(3)	27.	(2)	28.	(2)
29.	(1)	30.	(4)	31.	(1)	32.	(3)
33.	(2)	34.	(4)	35.	(1)	36.	(3)
37.	(3)	38.	(2)	39.	(1)	40.	(1)
41.	(1)	42.	(1)	43.	(1)	44.	(1)

ARCHIVE – IIT JEE ADVANCED

SINGLE OPTION CORRECT

1. C	2. C	3. D	4. B
5. C	6. C	7. B	8. B
9. B	10. B	11. D	12. A
13. C	14. B	15. A	16. B
17. C	18. D	19. B	

MULTI OPTION CORRECT

1. A, D	2. B, D	3. A, B, C

SUBJECTIVE / INTEGER TYPE

1. 2	2. 8	3. 2	4. 1210
5. $a > 1$	6. $\alpha^2\beta,\ \beta^2\alpha$		

SOLUTION - ULTIMATE FINISH

UF1   17

Concept:

$$\frac{1}{\sqrt{k}} = \frac{2}{\sqrt{k}+\sqrt{k}} \geq \frac{2}{\sqrt{k}+\sqrt{k+1}} = 2\left(\sqrt{k+1}-\sqrt{k}\right) \text{ and } \frac{1}{\sqrt{k}} = \frac{2}{\sqrt{k}+\sqrt{k}} \leq \frac{2}{\sqrt{k-1}+\sqrt{k}} = 2\left(\sqrt{k}-\sqrt{k-1}\right)$$

$$\rightarrow 2\left(\sqrt{k+1}-\sqrt{k}\right) \leq \frac{1}{\sqrt{k}} \leq 2\left(\sqrt{k}-\sqrt{k-1}\right) \ \forall\ k \in N$$

$$2\left(\sqrt{3}-\sqrt{2}\right) \leq \frac{1}{\sqrt{2}} \leq 2\left(\sqrt{2}-\sqrt{1}\right)$$

$$2\left(\sqrt{4}-\sqrt{3}\right) \leq \frac{1}{\sqrt{3}} \leq 2\left(\sqrt{3}-\sqrt{2}\right)$$

$$2\left(\sqrt{5}-\sqrt{4}\right) \leq \frac{1}{\sqrt{2}} \leq 2\left(\sqrt{4}-\sqrt{3}\right)$$

So, on addition we get ....

+..

$$+\ 2\left(\sqrt{101}-\sqrt{100}\right) \leq \frac{1}{\sqrt{100}} \leq 2\left(\sqrt{100}-\sqrt{99}\right)$$

$$\overline{2\left(\sqrt{101}-\sqrt{2}\right) \leq S_{100} \leq 2\left(\sqrt{100}-\sqrt{1}\right)}$$

$$\rightarrow 17 < 2\left(\sqrt{101}-\sqrt{2}\right) < S_{100} \rightarrow S_{100} = \left\lfloor \frac{1}{\sqrt{2}}+\frac{1}{\sqrt{3}}+\frac{1}{\sqrt{4}}+\cdots\cdots+\frac{1}{\sqrt{100}} \right\rfloor = 17.$$

UF2   B

Let $3^x = a$, $5^x = b \rightarrow M = a + b - a^2 + ab - b^2$. As $(a - b)^2 \geq 0 \rightarrow a^2 + b^2 \geq 2ab \rightarrow -(a^2 + b^2) \leq -2ab$

$\rightarrow M \leq a + b + ab - 2ab \rightarrow M \leq a + b - ab \rightarrow M < 1 - (1 - a)(1 - b)$. Hence, $0 < M < 2$

UF3:   C

$P(x). Q(x) = (x^2 - 1^2)(x^2 - 2^2)(x^2 - 3^2)\ldots\ldots(x^2 - 50^2)$

Now there is no odd powers of x so, $a_{97} = a_{99} = 0$ and $a_{100} = 1$.

$a_{98} = -(1^2 + 2^2 + 3^2 + \ldots + 50^2) = -\dfrac{50 \times 51 \times 101}{6} = -42925$

so, $a_{100} - a_{99} + a_{98} - a_{97} = 1 - 42925 = -42924$.

UF4 C

Let $x^2 - 47x + r = 0$. For real roots $D \geq 0 \rightarrow (47)^2 - 4r \geq 0 \rightarrow r \leq 552$

$\therefore r = 1, 2, 3, 4, \ldots\ldots, 552$

Product of real roots $= 1 \times 2 \times 3 \times 4 \times \ldots\ldots \times 551 \times 552 = 552!$

UF5 4

$y^2 \left(\log_2 \left(x^2 + 1 \right) + \log_{x^2+1} 16 \right) = 6y^2 - y^4 - 1$

$\rightarrow \log_2 \left(x^2 + 1 \right) + \dfrac{4}{\log_2 \left(x^2 + 1 \right)} = 6 - \left(y^2 + \dfrac{1}{y^2} \right)$

$\log_2 \left(x^2 + 1 \right) + \dfrac{4}{\log_2 \left(x^2 + 1 \right)} \geq 4$ and $6 - \left(y^2 + \dfrac{1}{y^2} \right) \leq 4$ so equality holds only if

$\log_2 \left(x^2 + 1 \right) = \dfrac{4}{\log_2 \left(x^2 + 1 \right)}$ and $y^2 = \dfrac{1}{y^2} \rightarrow \log_2 \left(x^2 + 1 \right) = \pm 2$ and $y = \pm 1$

$\rightarrow$ Points are $(\sqrt{3}, 1), (\sqrt{3}, -1), (-\sqrt{3}, 1)$ and $(-\sqrt{3}, -1)$

UF6 2

$4^x - 2^{x+2} + 5 + ||b-1| - 3| = |\sin y| \rightarrow (2^{2x} - 4 \cdot 2^x + 4) + 1 + ||b-1| - 3| = |\sin y|$

$\rightarrow (2^x - 2)^2 + 1 + ||b-1| - 3| = |\sin y|$

Possible only if $2^x - 2 = 0$, $|b-1| - 3 = 0$ & $\sin y = \pm 1$

So, $|b-1| - 3 = 0 \rightarrow b = 4$ & -2

UF7 A

The given equation can be rewritten as $\left(x^{2014} + 1 \right)\left(1 + x^2 + x^4 + \ldots\ldots + x^{2012} \right) = 2014 \cdot x^{2013}$

$\rightarrow \left(x + \dfrac{1}{x^{2013}} \right)\left(1 + x^2 + x^4 + \ldots\ldots + x^{2012} \right) = 2014$

$\rightarrow x + x^3 + x^5 + \ldots\ldots + x^{2013} + \dfrac{1}{x^{2013}} + \dfrac{1}{x^{2011}} + \ldots\ldots + \dfrac{1}{x} = 2014$

$\rightarrow \left(x + \dfrac{1}{x} \right) + \left(x^3 + \dfrac{1}{x^3} \right) + \ldots\ldots + \left(x^{2013} + \dfrac{1}{x^{2013}} \right) \geq 2 \times 1007 = 2014$ (addition after AM $\geq$ GM pair wise)

So, equality holds only if $x = \dfrac{1}{x}, x^3 = \dfrac{1}{x^3}, \ldots\ldots, x^{2013} = \dfrac{1}{x^{2013}} \rightarrow x = 1$

UF8 A

$f(x) + f(x+1) + f(x+2) + \ldots\ldots + f(x+10) = |e^x - 1| + |e^x - 2| + |e^x - 3| + \ldots + |e^x - 11|$

$$\rightarrow \int_{0}^{11} f(x)dx = 65 + 4\ln 2 - 7e$$

UF9 B, D

For, 3 distinct real solutions

y = m (x - 7) will be tangent to

y = - x² + 5x - 6 and $2 \le \dfrac{5-m}{2} \le 3 \rightarrow m \in (-1, 1)$

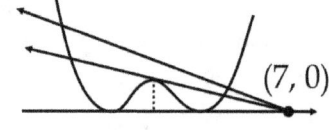

(7, 0)

UF10. B

f(x) is an odd function so f(a) + f(- a) = 0

f(x) is monotonically increasing function so

if a + b > 0 → a > - b → f(a) > f(- b) → f(a) + f (b) > 0.

UF11. 7

Multiply both side by x (x + 1)(x + 2)......(x + 25) and equating numerator and putting x = - 24 we get A_{24} = 25

UF12. A → S, B → R, C → Q, D → P

$a = -\left(|x| + \dfrac{1}{|x|}\right) \le -2$ for real x.

(A) So, a > - 2 for no real root

(B) a = - 2 at x = ± 1

(C) a ∈ φ

(D) a < - 2

UF13. C

ax² + 4x + a > 0 → a > 0 & D < 0 → a > 2 ....(1)

Also (a - 5)x² + 4x + (a - 5) ≥ 0

→ a > 5 & D ≤ 0 → a ≤ 3 or a ≥ 7 ....(2)

From (1) and (2) a ≥ 7

UF14. C

Let f(x) = x³ + x + 1 = (x - a)(x - b)(x - c) where a, b, c are roots of f(x) = 0

∴ P(x) = k(x – a²)(x – b²)(x – c²) for some k

abc = - 1, P(0) = - k a²b²c² = - 1 → k = 1

∴ P(x²) = (x² – a²)(x² – b²)(x² – c²) =(x - a)(x - b)(x - c)(x + a)(x + b)(x + c)

→ P(x²) = - f(x) f(- x)

→ P(4) = - f(2) f(- 2) = 11× 9 = 99

→ P(9) = - f(3) f(- 3) = 31 × 29 = 899

UF15.

A, B, C, D

Let k = a + ib and one real root 'c', then (a + ib) c² + c + 1 = 0

Equating real and imaginary parts

$bc^2 = 0 \to c = 0$ (since, $b \neq 0$)

$ac^2 + c + 1 = 0$ when $c = 0 \to 1 = 0$ which is not possible so $k \in \phi$

UF16. 0

As, $3^{\sqrt{\log_3 7}} = 7^{\sqrt{\log_7 3}} \to$ roots are equal hence D = 0

UF17. 4

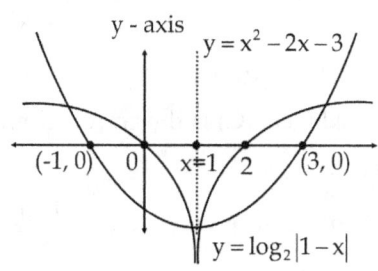

$x^2 - 2x - \log_2|1-x| = 3 \to x^2 - 2x - 3 = \log_2|1-x|$

Now, plot of curves $y = x^2 - 2x - 3$ and $y = \log_2 |1 - x|$

These curves intersect at 4 points as shown in figure.

UF18. 4

$17^2 = (m - n^2)(m + n^2)$

Possible cases: $\begin{matrix} m + n^2 = 17^2 \\ m - n^2 = 1 \end{matrix}$ and $\begin{matrix} m + n^2 = 17 \\ m - n^2 = 17 \end{matrix}$

There are 4 such values of m

UF19. B, C, D

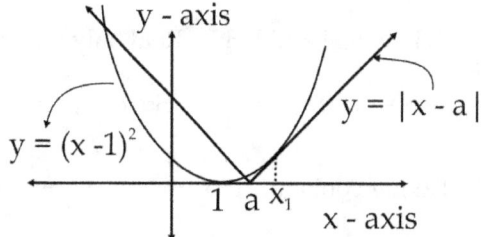

CASE (I) $y = |x - a| = x - a$

I have shown a critical situation

$x = x_1$ is a point where $y = x - a$ and $y = (x - 1)^2$ touching each other.

So, their slope must be same at $x = x_1$

$\to 2(x-1)\big|_{x=x_1} = 1 \to x_1 = \dfrac{3}{2}$

The value of $y = x - a$ and $y = (x - 1)^2$ must be same at $x = x_1$.

So, $3/2 - a = (3/2 - 1)^2 \to a = 5/4$.

CASE (II) $y = |x - a| = -x + a$

For the case, let $x = x_2$ is another contact point so their slope must be same at $x = x_2$

$\to 2(x-1)\big|_{x=x_2} = -1 \to x_2 = \dfrac{1}{2}$

The value of $y = -x + a$ and $y = (x - 1)^2$ must be same at $x = x_2$.

So, $(x_2 - 1)^2 = a - x_2 \to a = 3/4$

CASE (III) $a = 1$

For $a = 1$, we have three roots.

UF20. 1

We determine f(2016) first. From the condition given, we have

$f(x) + 5 \leq f(x + 5) \leq f(x + 4) + 1 \leq f(x + 3) + 2 \leq f(x + 2) + 3 \leq f(x + 1) + 4 \leq f(x) + 5.$

$\to$ Equality holds for all, so we have $f(x+1) = f(x) + 1$.

Hence, from $f(1) = 1$, we get $f(2) = 2$, $f(3) = 3$,......, $f(2016) = 2016$.

Therefore, $g(2016) = f(2016) + 1 - 2016 = 1$.

UF21. C

$$1 + x^2 + x^4 + x^6 + \ldots + x^{2010} = \frac{\left(1 - x^{2012}\right)}{1 - x^2} = \frac{\left(1 - x^{1006}\right)\left(1 + x^{1006}\right)}{(1 - x)(1 + x)}$$

$$= \left(1 + x^{1006}\right)\left(\frac{1 - x^{503}}{1 - x}\right)\frac{\left(1 + x^{503}\right)}{(1 + x)} = (1 + x^{1006})(1 + x + x^2 + x^3 + \ldots + x^{502})(1 - x + x^2 - x^3 + \ldots + x^{502})$$

This is divisible by $1 + x + x^2 + x^3 + \ldots + x^{502} \to n - 1 = 502 \to n = 503$

Hence, (C) is the correct option

UF22. B

$$2x^2 + 2x + 1 = 0 \to x = \frac{-1 + i}{2}, \frac{-1 - i}{2}$$

$$x \text{ satisfies } (x + 1)^n - r = 0 \to \left(\frac{-1 \pm i}{2} + 1\right)^n - r = 0 \to \left(\frac{1 \pm i}{2}\right)^n - r = 0$$

$$\to \left(\frac{1}{\sqrt{2}}\right)^n \left(\frac{1 + i}{\sqrt{2}}\right)^n = r \to \left(\frac{1}{\sqrt{2}}\right)^n \left(e^{\pm \frac{i\pi}{4}}\right)^n = r$$

RHS is real and LHS = Real only when n = multiple of 4

$\to n = 4000 \to r = \dfrac{1}{4^{1000}}$. So, option (B) is correct.

UF23. Taking conjugate of $x + y + z = 1$, we get $\bar{x} + \bar{y} + \bar{z} = 1 \to \dfrac{1}{x} + \dfrac{1}{y} + \dfrac{1}{z} = 1$ as $|x| = 1 \to x \cdot \bar{x} = 1 \to \bar{x} = \dfrac{1}{x}$

Since, $xyz = 1$ so $\dfrac{1}{x} + \dfrac{1}{y} + \dfrac{1}{z} = 1 \to xy + yz + zx = 1$

Now, we have: $x + y + z = 1$, $xy + yz + zx = 1$ and $xyz = 1$

$\to$ x, y and z are roots of equation: $a^3 - a^2 + a - 1 = 0$

$\to a^3 - a^2 + a - 1 = 0 \to (a^2 + 1)(a - 1) = 0 \to a = 1, \pm i$

So, solution set (x, y, z) is the permutation of (1, i, - i)

UF24. A, B, C

$55^{f(x)} + 5^x - 2012 = 0 \to f(x) = \log_{55}(2012 - 5^x)$

for f(x) to be defined $2012 - 5^x > 0 \to 5^x < 2012 \to x < 5$ [since $5^4 < 2012 < 5^5$]

$\therefore$ Integral value or x = -1, 2, 3

UF25. A

$$f(x) = (x - x_1)(x - 1 - x_1)(x - 2 - x_1)(x - \alpha) + 10x$$

$f(12) + f(-8) = (12 - x_1)(11 - x_1)(10 - x_1)(10 - \alpha) + (8 + x_1)(9 + x_1)(10 + x_1)(8 + \alpha) + 40$

If it is independent of 'α' then $x_1 = 1$

$\therefore x_1 + x_2 + x_3 = 6$.

UF26: D

Let $x + \dfrac{y}{8} = a$ & $x - \dfrac{y}{8} = b \rightarrow x = (a + b)/2$ and $y = 4 (a - b)$ so $\rightarrow f(a, b) = \dfrac{(a+b)}{2} \times 4(a - b) = 2(a^2 - b^2)$

Hence, $f(a, b) = 2(a^2 - b^2) \rightarrow f(m, n) + f(n, m) = 2(m^2 - n^2) + 2(n^2 - m^2) = 0$ $\forall$ m and n. So correct option is (D)

UF27. B

$f(x)f(y) = f(xy)$ $\forall$ real x, y $\rightarrow f(1)f(2) = f(2) \rightarrow f(1) = 1$. So $f(1/2)f(2) = f(1) \rightarrow f(1/2) = f(1)/f(2) = 1/4$.

Hence, (B) is correct option.

UF28: A

$f(1) + f(2) + f(3) + \ldots + f(n) = n^2 f(n)$ .... (1)

$f(1) + f(2) + f(3) + \ldots + f(n) + f(n+1) = (n+1)^2 f(n+1)$ ....(2)

From (2) – (1), we get

$f(n + 1) = (n+1)^2 f(n+1) - n^2 f(n) \rightarrow \{(n+1)^2 - 1\}f(n+1) = n^2 f(n)$.

$(n + 2)f(n + 1) = n f(n)$, so $\dfrac{f(n+1)}{f(n)} = \dfrac{n}{n+2}$ $\forall$ n > 1...................(3)

From (3) we get $\dfrac{f(2)}{f(1)} \times \dfrac{f(3)}{f(2)} \times \dfrac{f(4)}{f(3)} \times \ldots \times \dfrac{f(9)}{f(8)} = \dfrac{1}{3} \times \dfrac{2}{4} \times \dfrac{3}{5} \times \dfrac{4}{6} \times \ldots \times \dfrac{8}{10}$.

$\rightarrow \dfrac{f(9)}{f(1)} = \dfrac{1 \times 2}{9 \times 10} \rightarrow f(9) = \dfrac{1 \times 2}{9 \times 10} \times 3600 = 80$. So (A) is correct option.

SAMPLE TEST PAPER FOR IIT JEE MAINS

Mathsarc Education
A learning place to fulfill your dream of success!

Name: ....   M.M. – 120,   Total Time: 1 Hr.

INSTRUCTIONS

❖ The paper is for needy students, who are preparing for IIT JEE Mains / Advanced, MHCET, BITSAT and other State Engineering exams etc.

❖ The paper is matching the frequency of IIT JEE Mains exam so if you solve it then you will be able to crack 40% - 60 % questions of exam easily. There might be chances to get question in exam from this paper. It happens earlier.

❖ The sample paper is self evaluation so be honest while writing this paper. Give exact 60 Minute to the paper.

Mathsarc Education wishes you, best of luck for your future success!! ☺ ☺ ☺

Straight Objective Type (+4, -1, 0)

This paper contains 30 multiple choice questions. Each question has 4 choices (a), (b), (c) and (d) out of which ONLY ONE is correct.

1.   Select the in-correct options

   (A) $\sin(\sin\theta)|_{max} = \sin(1) \ \forall\theta\in R$ (B) $\sin(\sin\theta)|_{min} = -1 \ \forall \ \theta\in R$

   (C) $\sin(\sin\theta)|_{min} = -\sin(1) \ \forall \ \theta\in R$ (D) $\cos(\cos\theta)|_{min} = \cos(1) \ \forall \ \theta\in R$

2.   The value of 'a' so that the volume of parallelepiped formed by $\hat{i}+\hat{j}+\hat{k}$, $\hat{j}+a\hat{k}$ and $a\hat{i}+\hat{k}$ becomes minimum is

   (A) 3   (B) $-\dfrac{1}{3}$ (C) $\dfrac{1}{2}$ (D) –2

3.   A line is perpendicular to $x - 2y - 2z = 0$ and passes through $(1, -1, 1)$. The perpendicular distance of this line from origin is

   (A) $\dfrac{\sqrt{26}}{3}$ (B) $\dfrac{26}{3}$ (C) $\sqrt{26}$ (D) $\sqrt{3}$

4.   The quadratic equations $x^2 - 6x + a = 0$ and $x^2 - cx + 6 = 0$ have one root in common. The other roots of the first and second equations are integers in the ratio 4 : 3. Then the common root is

   (A) 2   (B) 3   (C) 4   (D) 5

5. $\lim\limits_{n \to \infty}\left(\dfrac{1}{1-n^4}+\dfrac{8}{1-n^4}+\dots+\dfrac{n^3}{1-n^4}\right)$ is

(A) $\dfrac{1}{4}$　　　(B) $\dfrac{1}{8}$　　　(C) $\dfrac{1}{2}$　　　(D) none of these

6. The value of $\displaystyle\int_{-\pi}^{\pi}\dfrac{\cos^2 x}{1+a^x}\,dx, a>0,$ is:

(A) 2π　　　(B) π/a　　　(C) $\pi/2$　　　(D) $a\pi$

7. If $f(x)=x^\alpha \log x$ and $f(0) = 0$, then the value of α for which Rolle's theorem can be applied in $x \in [0, 1]$ is

(A) - 2　　　(B) - 1　　　(C) 0　　　(D) $\dfrac{1}{2}$

8. If r is the distance from the point (3, 5) to the line $2x + 3y -14 = 0$ measured parallel to the line $x - 2y = 1$, then the value of $|\sqrt{5}\,r + 1|$ is?

(A) 5　　　(B) 6　　　(C) 7　　　(D) none

9. Let S be the set of all real number. Then, the relation R = {(a, b): 1 + ab > 0} on S is

(A) Reflexive and Symmetric but not Transitive　(B) Reflexive and Transitive but not Symmetric

(C) Reflexive, Transitive and Symmetric　　　　(D) None of these

10. Value of f(0) so that $f(x)=\dfrac{1}{x^2}(1-\cos(\sin x))$ can be made continuous at x = 0, is equal to

(A) $\dfrac{1}{2}$　　　(B) 2　　　(C) $\dfrac{1}{4}$　　　(D) 4

11. The condition $f(x)=x^3+px^2+qx+r\,(x\in R)$ to have no extreme value, is

(A) $p^2<3q$　　　(B) $2p^2<q$　　　(C) $p^2<\dfrac{q}{4}$　　　(D) $p^2>3q$

12. Let $\vec{a},\vec{b}$ and $\vec{c}$ be non-zero vectors such that $(\vec{a}\times\vec{b})\times\vec{c}=-\dfrac{1}{3}|\vec{b}||\vec{c}|\vec{a}$. If θ is the acute angle between the vectors $\vec{b}$ and $\vec{c}$, then sin θ equal to:

(A) 1/3　　　(B) $\sqrt{2}/3$　　　(C) 2/3　　　(D) $2\sqrt{2}/3$

13. The point (4, 1) undergoes the following three transformations successively.

(i) Reflection about line x = y

(ii) Translation through a distance 2 units along the positive direction of x–axis.

(iii) Rotation through an angle of $\dfrac{\pi}{4}$ about the origin in counter-clockwise direction.

Then the final position of point is given by co – ordinates.

(A) $\left(\dfrac{1}{\sqrt{2}}, \dfrac{7}{\sqrt{2}}\right)$
(B) $\left(-\sqrt{2}, 7\sqrt{2}\right)$
(C) $\left(-\dfrac{1}{\sqrt{2}}, \dfrac{7}{\sqrt{2}}\right)$
(D) $\left(\sqrt{2}, 7\sqrt{2}\right)$

14. The value of $\cot\left(\operatorname{cosec}^{-1}\dfrac{5}{3} + \tan^{-1}\dfrac{2}{3}\right)$ is

(A) $\dfrac{5}{17}$
(B) $\dfrac{6}{17}$
(C) $\dfrac{3}{17}$
(D) $\dfrac{4}{17}$

15. The expansion $\left[x + \left(x^3 - 1\right)^{1/2}\right]^5 + \left[x - \left(x^3 - 1\right)^{1/2}\right]^5$ is a polynomial of degree_____?

(A) 4
(B) 6
(C) 7
(D) 8

16. The equation of the ellipse whose distance between the foci is equal to 8 and distance between the directrix is 18, is

(A) $4x^2 + 9y^2 = 180$
(B) $9x^2 + 5y^2 = 180$
(C) $x^2 + 5y^2 = 180$
(D) $5x^2 + 9y^2 = 180$

17. The number of different integral roots of the equation $\left(x^2 - 7x + 11\right)^{\left(x^2 - 11x + 30\right)} = 1$, is/are?

(A) 7
(B) 1
(C) 3
(D) 5

18. Let p: 7 is not greater than 4, q: Paris is in France, be two statements. Then, $\sim\left(p \vee q\right)$ is the statement

(A) 7 is greater than 4 or Paris is not in France

(B) 7 is not greater than 4 and Paris is not in France

(C) 7 is not greater than 4 or Paris is not in France

(D) 7 is greater than 4 and Paris is not in France

19. If n A.M's are inserted between 20 & 80 such that first mean: last mean = 1: 3. Then the common difference of the corresponding AP is.

(A) 4
(B) 5
(C) 6
(D) 8

20. How many four digit numbers can be formed using the digits 1, 2, 3, 4, 5 such that at least one of the digit is repeated?

(A) $4^4 - 5!$
(B) $5^4 - 4^4$.
(C) $5^4 - 4!$
(D) $5^4 - 5!$

21. Range of function $f(x) = \dfrac{x^2 + x + 2}{x^2 + x + 1}, x \in R$ is

   (A) $(1, \infty)$ (B) $\left(1, \dfrac{3}{2}\right)$ (C) $\left(1, \dfrac{7}{3}\right]$ (D) $\left(1, \dfrac{7}{5}\right]$

22. The shortest distance of the line $y = x + 1$ from $y^2 = x$ is

   (A) $\dfrac{3}{8}$ (B) $\dfrac{3\sqrt{2}}{4}$ (C) $\dfrac{3}{4}$ (D) $\dfrac{3\sqrt{2}}{8}$

23. If $\displaystyle\int \dfrac{1 + \cos 4x}{\cot x - \tan x} dx = k \cos 4x + c$, then k equals to

   (A) $\dfrac{-1}{2}$ (B) $\dfrac{-1}{5}$ (C) $\dfrac{-1}{8}$ (D) $-\dfrac{1}{10}$

24. If σ is the standard deviation of a random variable x then the standard deviation of the random variable $ax + b$, where $a, b \in R$ is?

   (A) $a\sigma + b$ (B) $|a|\sigma$ (C) $|a|\sigma + b$ (D) $a^2\sigma$

25. A solution of differential equation $\left(\dfrac{dy}{dx}\right)^2 - x\dfrac{dy}{dx} + y = 0$ is

   (A) $y = 2$ (B) $y = 2x$ (C) $y = 2x - 4$ (D) $y = 2x^2 - 4$

26. A speaks the truth in 60% cases and B in 80% cases. The probability that they will contradict each other in describing a single event is

   (A) $\dfrac{7}{25}$ (B) $\dfrac{12}{25}$ (C) $\dfrac{11}{25}$ (D) $\dfrac{2}{25}$

27. The function $f : R - \{0\} \to R$ given by $f(x) = \dfrac{1}{x} - \dfrac{2}{e^{2x} - 1}$ can be made continuous at $x = 0$ by defining $f(0)$ as

   (A) 2   (B) -1   (C) 0   (D) 1

28. $\displaystyle\lim_{x \to 0} \left\{\dfrac{\sin x}{x}\right\}^{\left\{\frac{1}{\tan x}{x}\right\}}$ is equal to (where $\{x\}$ = fractional part of x)

   (A) 0   (B) $\dfrac{1}{e}$ (C) $\dfrac{1}{\sqrt{e}}$ (D) 1

29. Given $2x - y + 2z = 2$, $x - 2y + 2z = -4$, $x + y + \lambda z = 4$ then the value of λ such that the given system of equation has no solution, is

   (A) 3   (B) 1   (C) 0   (D) -3

30. The circumcentre of the triangle formed by the lines $xy + 2x + 2y + 4 = 0$ and $x + y + 2 = 0$ is

   (A) $(-1, -1)$ (B) $(0, -1)$ (C) $(1, 1)$ (D) $(-1, 0)$

SAMPLE TEST PAPER FOR IIT JEE ADVANCED

Mathsarc Education

A learning place to fulfill your dream of success!

Name: ....

M.M. – 120,   Total Time: 2 Hr.

INSTRUCTIONS

SECTION – A

Straight Objective Type (+3, -1, 0)

This section contains 10 multiple choice questions. Each question has 4 choices (a), (b) , (c) and (d) out of which ONLY ONE is correct.

1. The number of different non-singular matrices of the type $A = \begin{vmatrix} 1 & a & c \\ 1 & 1 & b \\ 0 & -w & w \end{vmatrix}$ where $w = e^{i\theta}$ and

    $a, b, c \in \{z : z^4 - 1 = 0\}$ are
    (A) 44   (B) 48   (C) 56   (D) 55

2. Let $f(x) = \sqrt{|x| - \{x\}}$ (where {.} denotes the fractional part of x) and X, Y be its domain and range respectively. Then

    (A) $X = \left(-\infty, \frac{1}{2}\right]$ and $Y = \left[\frac{1}{2}, \infty\right)$ (B) $X = \left(-\infty, -\frac{1}{2}\right]$ and $Y = \left[\frac{1}{2}, \infty\right)$

    (C) $X = \left(-\infty, -\frac{1}{2}\right] \cup [0, \infty)$ and $Y = [0, \infty)$ (D) $\left(-\infty, -\frac{1}{2}\right]$ and $[0, \infty)$

3. Locus of z given by $|z - 6i| = 6$ is rolled over real axis. If a point P on it have following positions.

At $t = 0$: $P_{initial}$ (12i)   and   At $t = T$: P_{final} $\left((\pi+3)+i(6+3\sqrt{3})\right)$

Then Angle of rotation of point P in span of time T, w.r.t its centre is

(A) $\dfrac{\pi}{12}$ (B) $\dfrac{\pi}{6}$ (C) $\dfrac{\pi}{4}$ (D) $\dfrac{\pi}{3}$

4.   Let $f: R \to R$ be a continuous & differentiable function given by $f(x) = x + \int_0^1 (xy + x^2)f(y)dy$. Then

(A) $\int_0^1 f(x)dx = \dfrac{26}{23}$ (B) $\int_0^1 f(x)dx = \dfrac{25}{13}$ (C) $\int_0^1 xf(x)dx = \dfrac{13}{25}$ (D) $\int_0^1 xf(x)dx = \dfrac{25}{23}$

5.   The probability of non-increasing functions $f: A \to B$, where $A = \{1, 2, 3, 4, 5, 6\}$ and $B = \{1,2,3, \ldots., 20\}$ such that if $f(x_1) = f(x_2)$ for $x_1 \neq x_2$, then it should be true for even number of points in set A, is

(A) $\dfrac{\sum\limits_{r=0}^{3} {}^{20}C_{2r} \cdot {}^{13+3r}C_{2r}}{20^6}$ (B) $\dfrac{\sum\limits_{r=0}^{3} {}^{20}C_{14+2r} \cdot {}^{13+2r}C_{2r}}{20^6}$

(C) $\dfrac{\sum\limits_{r=0}^{3} {}^{20}C_{14+2r} \cdot {}^{13+3r}C_{r}}{20^6}$ (D) none of these

6.   The set of values of x for which the inequality $\sin^4\left(\dfrac{x}{3}\right) + \cos^4\left(\dfrac{x}{3}\right) > \dfrac{1}{2}$ holds, is/are?

(A) R   (B) $\left\{x : x = \dfrac{3n\pi}{2} \pm \dfrac{3\pi}{4}; n \in I\right\}$

(C) $R - \left\{x : x = \dfrac{3n\pi}{2} \pm \dfrac{3\pi}{4}; \ n \in I\right\}$ (D) ϕ

7.   Two equal circles of largest radii having following property
(i)   They intersect each other orthogonally.
(ii)   They touch both the curves $4(y + 2) = x^2$ and $4(2 - y) = x^2$ in the region $x \in [-2\sqrt{2}, 2\sqrt{2}]$.
Then radius of this circle is

(A) $\sqrt{2}$ (B) $\sqrt{3}$ (C) $\dfrac{1}{\sqrt{3}}$ (D) $\dfrac{3}{2}$

8.   Let $P(x,y,1)$ and $Q(x, y, z)$ lies on the curve $\dfrac{x^2}{9} + \dfrac{y^2}{4} = 4$ and $\dfrac{x+2}{1} = \dfrac{\sqrt{3}-y}{\sqrt{3}} = \dfrac{z-1}{2}$ respectively. Then minimum distance between P and Q is

(A) $\sqrt{2}$ (B) $\sqrt{\dfrac{7}{2}}$ (C) 2   (D) none of these

9.   A point $P(\sqrt{2}\sin A, \cot B)$ lies on the curve $(x+y)^2 + 4(y-x)^2 = 8$. If minimum value of

$(2\tan C - \sqrt{2}\sin A)^2 + (2\cot C - \cot B)^2 = a + b\sqrt{2}$ where $C \in R - \left\{\dfrac{n\pi}{2}\right\}, n, a \ \& \ b \in I$.

Then value of a + b is
(A) 0   (B) 2   (C) 4   (D) 12

10. The affixes of a triangle in argand plane are $A(2i\omega+1), B(1-2\omega)$ and $C(1-2i)$ where $\omega = e^{i\frac{2\pi}{3}}$. Then triangle ABC is

(A) Acute, $\angle A = \dfrac{\pi}{6}$

(B) Equilateral

(C) Isosceles, $\angle C = \dfrac{2\pi}{3}$

(D) Obtuse, $\angle B = \dfrac{\pi}{6}$

SECTION – B
(Multiple Correct Answer(s) Type)

This section contains **5 multiple choice questions.** Each question has four choices (A), (B), (C) and (D) out of which **ONE OR MORE** may be correct. Marking (+4, - 1, 0)

11. For all $\theta \in (0, \pi/2)$, select the correct options

(A) $\cos(\sin \theta) > \sin (\cos \theta)$

(B) $\cos(\sin \theta) < \sin (\cos \theta)$

(C) $\sin\theta > \sin(\sin\theta)$

(D) $\cos(\cos \theta) > \sin (\sin\theta)$

12. There exists a triangle ABC satisfying the conditions

(A) $b\sin A > a, A > \dfrac{\pi}{2}$

(B) $b\sin A = a, A < \dfrac{\pi}{2}$

(C) $b\sin A > a, A < \dfrac{\pi}{2}$

(D) $b\sin A < a, A < \dfrac{\pi}{2}, b > a$

13. If $\begin{vmatrix} 1 & e^{|x|} \\ \cos\theta & 0 \end{vmatrix} = \begin{vmatrix} k & \cos x \\ \sin\theta & x(x-1) \end{vmatrix}$ then

(A) $k = 1$

(B) $x = 0$

(C) $x = 1$

(D) $\theta = n\pi + \dfrac{\pi}{4} \forall n \in I$

14. Select the correct statements for the matrix $A = \begin{vmatrix} -1 & 3 & -5 \\ 2 & 1 & 7 \\ 0 & 6 & 1 \end{vmatrix}$

(A) $|\,adj(A)\,| = 625$

(B) $adj\left(\dfrac{1}{5}A\right) = \dfrac{1}{25}adj(A)$

(C) $adj(A^{-1}) = -\dfrac{1}{25}A$

(D) $\begin{bmatrix} -1 & 2 & 0 \\ 3 & 1 & 6 \\ -5 & 7 & 1 \end{bmatrix}^{-1} = \left(A^{-1}\right)^{T}$

15. If $\tan (\pi \cos \theta) = \cot (\pi \sin \theta)$, then the value of $\cos\left(\theta - \dfrac{\pi}{4}\right)$ is equal to

(A) $\dfrac{1}{2\sqrt{2}}$

(B) $-\dfrac{1}{2\sqrt{2}}$

(C) $\dfrac{3}{2\sqrt{2}}$

(D) $-\dfrac{1}{\sqrt{2}}$

SECTION - C
(One Integer Value Correct Type)

This section contains **4 questions. Each question**, when worked out will result in **one integer from 0 to 9.** (*Both inclusive*) Marking (+4, 0)

16. If $\theta \in [0, 5\pi]$ & $r \in R$, then the number of ordered pair (r, θ) such that $2\sin\theta = r^4 - 2r^2 + 3$,

    is / are?

17. The value of $\cot 76° \cot 44° + \cot 16° \cot 44° - \cot 76° \cot 16°$ is?

18. If a, b, c, d are positive real numbers such that $\dfrac{a}{3} = \dfrac{a+b}{4} = \dfrac{a+b+c}{5} = \dfrac{a+b+c+d}{6}$, then value of

    $\dfrac{b+2c+3d}{a}$ is _____?

19. The number of different real roots of the equation $\left(x^2 - 7x + 11\right)^{\left(x^2 - 11x + 30\right)} = 1$?

SECTION - D
(Straight Objective Type only one correct option)(+3, -1, 0)

This part contains 3 comprehension based question. Each group has 2 or 3 multiple choice question based on a paragraph. Each question has 4 choices (A), (B), (C) and (D) for its answer, out of which ONLY ONE is correct.

Paragraph for Question Nos. 20 to 21

Consider a triangle ABC as shown in figure. P is the point of concurrency of line segments AD, BE & CF

Where $AF = \dfrac{4}{\sqrt{3}}$, $FB = \dfrac{5}{\sqrt{3}}$ $BD = 3$,

$CD = \ell$ $AE = \dfrac{6\sqrt{22}}{11}$, $EC = \dfrac{5\sqrt{22}}{11}$

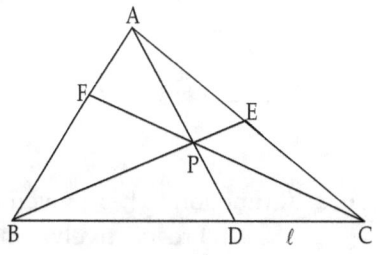

20. The value of $\ell \times$ area ($\triangle PBC$) and point P are respectively

    (A) $5\sqrt{3}$, Circumcentre of $\triangle ABC$

    (B) $5\sqrt{2}$, Circumcentre of $\triangle ABC$

    (C) $5\sqrt{2}$, Orthocenter of $\triangle ABC$

    (D) $5\sqrt{3}$, In-centre of $\triangle ABC$

21. If pair of straight line representing side AB & AC of $\triangle ABC$ is $(y - \sqrt{2}x - 1)(2y + 3\sqrt{2}x + 2 - 15\sqrt{2}) = 0$. Then side BC can be.

    (A) $y = 0$ (B) $y = 1$ (C) $y = 2x$ (D) $y = \sqrt{3}x$

Paragraph for Question Nos. 22 to 23

Consider two points A(1, 2) and B(3, – 1). Let M be a point on the straight line L : x + y = 0.

22. If M be a point on the line L = 0 such that $|AM - BM|$ is maximum, then the distance of M from N(1, 1) is

    (A) $5\sqrt{2}$ (B) 7   (C) $3\sqrt{5}$ (D) 10

23. If M be a point on the line L = 0 such that $|AM - BM|$ is minimum, then the area of DAMB equals

    (A) $\dfrac{13}{4}$ (B) $\dfrac{13}{2}$ (C) $\dfrac{13}{6}$ (D) $\dfrac{13}{8}$

Paragraph for Question Nos. 24 to 26

If $f : X \to Y$ be a function defined by $y = f(x)$ such that f is both one-one and onto then there exists a unique function $g : Y \to X$ such that for each $y \in Y$, $g(y) = x$ iff $y = f(x)$. The function g so defined is called the inverse of f and denoted as $f^{-1}(x) = g(x)$.

24. If $f : R \to R$ be a invertible function such that $f^{-1}(x) = g(x)$ and x_1, x_2 are two distinct roots of the equation $f(x) = g(x)$. Then value of $\dfrac{g(x_2) - g(x_1)}{x_2 - x_1}$

    (A) must be –1   (B) must be 1   (C) may be 3   (D) cannot determine

25. If $f : \left[\dfrac{3\pi}{2}, 2\pi\right] \to [-1, 0]$, where $f(x) = \sin x$ then $f^{-1}(x)$ is

    (A) $\dfrac{3\pi}{2} + \sin^{-1} x$ (B) $2\pi + \sin^{-1} x$ (C) $3\pi - \sin^{-1} x$ (D) $\dfrac{5\pi}{2} - \sin^{-1} x$

26. A function $f : I \to J$ given by $f(i) = j$ where I = {0, 1, 2,....,9} , J = {0, 1, 2, ...,100} and i, j are element of set I, J respectively. Then number of bijective functions of type f: I $\to$ B where B $\subseteq$ J & f(5) = 5 is

    (A) $^{100}C_9 \, 10!$ (B) $^{100}C_9 . 9!$ (C) $^{101}C_{10} \, 10!$ (D) none of these

SECTION E – Matrix Match Type (2×4 = 8, 0)

This section contains 3 questions and each question contains statements given in two columns which have to be matched. Column I are labeled as A, B, C and D. Whereas statements in Column II Labelled as p, q, r and s. Any given statement(s) in column I can have correct matching with ONE OR MORE statement(s) in column II. Only exact match will be awarded 2 marks. Wrong one will get zero.

27. **Match the following and write the correct pairs.**

		Column I		Column II
A		If $\arg\left(\dfrac{z-z_1}{z-z_2}\right)=\dfrac{\pi}{6}$ and $\operatorname{Min}\{|z-z_1|,|z-z_2|\}\le\sqrt{3}|z_1-z_2|$ then value of $\dfrac{3}{\pi|z_1-z_2|}\times$ (length of locus of z) is	(p)	9
B		Point B(3,1) and C(4,6) are two vertices of a triangle ABC having line $y-x+1=0$ as angle bisector of $\angle A$ then twice of area of $\triangle ABC$ is	(q)	4
C		The number of solution of the equation $\cos^{-1}(\cos x)-\sin^{-1}(\sin x)=\dfrac{\pi}{2}$ where $x\in(0,12)$	(r)	2
D		Value of $\displaystyle\int_{-\pi}^{\pi}\left|\sin^{-1}\sin x\right|e^{|x|}dx+4e^{\pi/2}-2e^{\pi}$ is {Where $|.|$ is absolute valued function}	(s)	5

28. **Match the following and write the correct pairs.**
Consider $|.|$ is absolute valued function and $[.]$ = G.I.F.

	Column I		Column II
A	The area of the region $[y]=\big[|2x|\big]$ if $|x|\le 2$	(p)	4
B	If α is a repeated root of $3x^2+bx+c=0$ then $\displaystyle\lim_{x\to\alpha}\dfrac{\tan(3x^2+bx+c)}{(x-\alpha)^2}$ is	(q)	3
C	Let a, b, c, d, e, f $\in$ R such the ad + be + cf = $\sqrt{(a^2+b^2+c^2)(d^2+e^2+f^2)}$ then value of $\dfrac{ae}{bd}$ is	(r)	2
D	Three planes $P_1:tx+(t+1)y+2Z=0$, $P_2:(t+2)x-3y+(t-1)z=0$ $P_3:x+(t-2)y-tz=0$. One of these planes contains the line of intersection of other two planes. The number of possible value of t is	(s)	1

29. **Match the following and write the correct pairs.**

	Column I		Column II
A	The greatest area of the rectangle inscribed in a semicircle of radius 2 unit is	(p)	5
B	If $I_n = \int\limits_{0}^{\pi/2} x^n \cos x\, dx$ and $2^k\left(I_k + 7kI_6\right) = \pi^k$ then k is	(q)	4
C	Find parallel lines of slope 1 is intersected by six parallel lines of slope 3 then 30th part of the number of parallelograms of any size formed by lines is	(r)	2
D	Maximum value of parameter a for which there exist a real number x satisfying $\sqrt{1-x^2} \geq a - \sqrt{3}x$	(s)	8

SECTION – F
(Straight Objective Type only one correct option)(+3, -1)

This section contains 3 assertion Reasoning questions . Each question has 4 choices (a), (b) , (c) and (d) out of which ONLY ONE is correct.

Question No. (30 - 32)

Assertion reasoning and their answer will be based on following condition.

(A)   Statement -1 is true, Statement -2 is true, Statement -2 is a correct explanation for Statement -1.

(B)   Statement -1 is true, Statement -2 is true, Statement -2 is not a correct explanation for Statement -1.

(C)   Statement -1 is true, Statement -2 is false.

(D)   Statement -1 is false, Statement -2 is true.

30. **Statement-1:** Set of values of λ for which $\cos^{-1} x = \cos x + \lambda$ exhibit a solution is [-cos 1, π−cos1]

   Statement-2: For all $\lambda \in R - \left[-\dfrac{\pi}{2}, -1\right]$, equation $\cos^{-1} x = \lambda x + \dfrac{\pi}{2}$ exhibit exactly one solution.

31. Consider a sequence of quadratic expressions
   $Q_r(x) = a_r x^2 + b_r x - 1$, where $a_r = (-1)^r$, $b_r \in (2, \infty)$ $\forall$ $r \in \{0, 1, 2, ...n\}$
   Statement-1: If $b_{r+1}^2 - b_r^2 = 4$ and $b_{r+2} - b_r = k$ where k is positive real constant. Then roots $\in (0, 1)$ of all the quadratic equations $Q_r(x) = 0$, will be in harmonic progression.
   Statement-2: x = 1, lies between and outside the roots of equation $Q_r(x) = 0$ if r is odd and even respectively.

32. **Statement · 1:** $\max\left(\left|e^{|x|} \sin^{-1} x - \tan^{-1} x\right|\right) = \dfrac{(2e-1)\pi}{4}$ and it occurs at two different points.

   Statement · 2: $e^{|x|} \sin^{-1}x = \tan^{-1}x$ has 3 solutions {where $|.|$ represent modulus function}

SAMPLE PAPER KEY – IIT JEE MAINS

1	B	7	D	13	C	19	B	25	C
2	C	8	B	14	B	20	D	26	C
3	A	9	A	15	C	21	C	27	D
4	A	10	A	16	D	22	D	28	C
5	D	11	A	17	D	23	C	29	B
6	C	12	D	18	D	24	B	30	A

ANSWER KEY – IIT JEE ADVANCED

1	D	8	A	15	AB	22	D	26	B
2	C	9	C	16	6	23	A	30	B
3	B	10	D	17	3	24	C	31	B
4	D	11	ACD	18	2	25	B	32	C
5	C	12	BD	19	5	27	$A \to q$ $B \to p$ $C \to q$ $D \to r$		
6	C	13	ABD	20	C	28	$A \to p$ $B \to q$ $C \to s$ $D \to r$		
7	A	14	ABCD	21	B	29	$A \to q$ $B \to s$ $C \to p$ $D \to r$		

NOTES	
Que / Page No.	**NOTES**

POINTS TO REMEMBER:

- ❖ .
- ❖ .
- ❖ .
- ❖ .
- ❖ .
- ❖ .
- ❖ .
- ❖ .
- ❖ .
- ❖ .
- ❖ .
- ❖ .
- ❖ .
- ❖ .

OTHERS: --

After all we are humans and chances of mistakes are always possible
The error may be of any kind like formatting, writing, calculation, grammatical etc.
THE CONCEPT MUST BE DELIVERED
Write to us at pncbyrc@gmail.com

A special thanks to My wife, Daughter and Friends who made this possible to publish this book

If you enjoyed the learning then recommend this book to your Friends, Teachers, and School Libraries etc. for spreading knowledge in effective way.

Mathsarc Education will be thankful to you.

Our Previous released book available at

Mathsarc Education, Notion Press, Flipkart, Amazon and at other online shops

e – book is also available at Kindle, Kobo & Rockstand etc.

Join our Facebook Page & Subscribe YouTube Channel 'Mathsarc Education' to get latest updates & enjoy learning.